THE SECRET ART
OF ALCHEMY

ROBERT M. BLACK

MMXVI

For Beatrice

First published in Great Britain in 2016

Hamilton House Publishing Ltd.

Thanks are due in no small measure to Adam McLean for permission to use his coloured alchemical and hermetic emblems, included in this book

ISBN 978-09928572-8-8

Publishing,
Typesetting & design by
Hamilton House Publishing Ltd.

Rochester Upon Medway,
Kent.

THE SECRET ART OF ALCHEMY

Arma Artis.

**The Arms of the Art showing the The Entrance of the Aspirant
from the *Splendor Solis***

Introduction

The appearance of yet another 'history' of alchemy poses a question which has dominated the study of this subject over the years - what is and was the fascination of alchemy? Veronica Wedgewood, the historian, is said to have held the view that the historian who begins with the question 'Why did this happen?' is starting at the wrong end. It is first necessary to tell the story and delineate the characters who played a part in it. If it can be established what happened and how then the historian will have gone a long way to establish why.In alchemy, however, the establishment of even what happened is often a matter of some difficulty. In their attempts to preserve the secrecy of a process, which they believed if generally disclosed would disrupt the good order of society, alchemists incorporated fact with fiction and fiction with fact. At their clearest they were often most obscure. In the words of Jacques Sadoul, the French historian of alchemy, an alchemist is said to be 'grudging' if he knowingly gives wrong information about his Art, and 'generous' if he reveals the truth. Alas, most alchemists fall into the 'grudging' category!

An important consideration, however, must be the background against which alchemy developed and flourished. Wars and the advancement of science have, in the past, coincided with the emergence of many strange beliefs and have opened up avenues for the resurrection of old concepts and fantasies.

In an exact science such as chemistry what happened is to some extent disclosed in textbooks and in technical papers communicated to learned societies and scientific journals. Practical details of the work carried out are published together with examples and results. Developments in a given field may be the subject of a review in which attention is directed towards the most important aspects from the viewpoint of the reviewer.

In alchemy, however, the motives for publication appear to have been entirely different. There was no attempt to establish priority of any process that was even partially disclosed. On the contrary, an air of verisimilitude was established by reference to the great professors of the Art followed by a coded account of the principles of the work which would appear to have been understood only by those practitioners who had achieved a degree of enlightenment in the mysteries involved.

There remains a problem with all the writings of alchemical practitioners. Why, if it was traditionally forbidden to disclose the nature of the processes involved, did the alchemists not remain silent on the subject? There was apparently no need to set down on paper the details of a secret art, unless, as in the 17th century, such publications were to become best sellers and find a ready market among the intelligentsia.

Historians of alchemy and popular writers on the subject still find an interested readership. Societies such as the Alchemical Society, founded during the early years of the 20th century, and the Society for the Study of Alchemy and Early Chemistry (founded in 1938) were concerned with papers and discussions on alchemical topics.

The present work was started at the request of the Rosicrucian Society of Freemasons in England (Societas Rosicruciana in Anglia), as a general explanatory background to the subject of alchemy and has been extended to include a number of papers read to the Society on individual alchemists and on the development of the subject from the earliest times up to the present day - from Plato to Fulcanelli.

There is no doubt that alchemy possesses a fascination of its own. In a book, *The Lure and Romance of Alchemy*, published in 1932, C.J.S. Thompson wrote:

> 'That alchemy has appealed to the imagination of man for centuries is evident from the prominent part it plays in the legends and romances of the past.'
>
> 'To the artist the alchemist at work in his dim, mysterious laboratory, with its glowing furnaces and fantastic apparatus, formed an attractive subject for his brush, while the poet found in his romantic and picturesque life a fascinating theme for his pen. A great deal of this attraction was doubtless due to the mystery with which the art of alchemy has ever been surrounded.'

Carl Jung, who studied alchemy and the works of the alchemists for many years, was convinced that the illustrations to their works were derived from basic mental processes which he had disclosed in the minds of some of his patients and could be found in the 'collective unconscious'. Be this as it may, the Philosopher's Stone and its mysterious nature akin to the Holy Grail with its purifying effect both on base metals and on the personality of the alchemist, has ensured the fascination of alchemy for the scientist and for the historian of science.

Alchemy, the transmutation of base metals into gold by the methods apparently used by the alchemists, is impossible in terms of current theoretical knowledge of the behaviour of matter. That the nuclear reactions involved could bring about change in the alchemist carrying them out in his laboratory appears absurd! Yet, despite much deception and charlatanism in the past, some accounts have a strange ring of truth which appeals for their further investigation. A fallacy that has influenced many over the centuries, resulted in the publication of thousands of books and manuscripts and intrigued the minds of men is surely worthy of study. Like the metal gold, alchemy has an intrinsic beauty of its own!

The title chosen for this work *The Secret Art of Alchemy* may have been influenced by a delightful publication of rather similar title *Alchemy: The Secret Art* by Prince Stanislas Klossowski de Rola in 1973. Even if the title of this great little work was without influence, and it may well have been, the work itself has been an inspiration to the present author.

Perhaps this introduction might close with the words of the 15th century alchemist, Thomas Norton:

> To the honor of God, One in Persons three,
> This boke is made, that Lay men shulde it see,
> And Clerks alsoe, after my decease,
> Whereby all lay-men which putteth them in prease,
> To fetch by Alkimy great ryches to winn
> May finde good Counsell er they such warke begin;
> And greate deceipts they may hereby eschewe,
> And by this doctrine know fals men from trewe.

Thomas Norton *The Ordinall of Alchimy*: The Proheme lines 1 - 8

Robert M. Black

Contents

Chapter One

An Outline of Alchemy

Since time immemorial Alchemy has formed one of the four Great Pillars of Hermetic Science: Alchemy, Astrology, Magic and the Kabbalah. In its experimental form it gave rise to the science of chemistry, but it was very much more than a predecessor of a modern branch of science – it possessed a spiritual aspect. Just as the work in the laboratory was aimed at bringing base metals to perfection, so that in the oratory was to transmute the soul of the alchemist to a state of grace.

Alchemy is popularly held to be the misguided search for an enigmatic catalyst or reagent, the Philosopher's Stone, by which base metals such as lead and mercury could be transmuted into silver or gold. It is a process with the promise of worldly riches beyond the dreams of avarice! In the following chapters it is intended to give an account of the fundamental aspects of alchemy and to show that it embodied not only a striving for a world view but also a means of self-regeneration for mankind. It is first necessary to set alchemy in an historical context and to trace its development from the aurifiction of the ancient world to the aurifaction of mediaeval times with its occluded concepts of longevity and immortality derived from the East.

Samuel Johnson's 'Dictionary of the English Language' published in 1755 defined ALCHYMY as '... the more sublime and occult part of chymistry, which proposes for its object, the transmutation of metals, and other important operations.' This gives a rather different basis to alchemy from the definition given in the current Oxford English Dictionary as '... the chemistry of the Middle Ages and 16th century; now applied distinctively to the pursuit of the transmutation of the baser metals into gold, which (with the search for the alkahest or universal solvent, and the panacea or universal remedy) constituted the chief practical object of early chemistry.'

A recent *Dictionary of the History of Science* (Bynum, Browne and Porter), however, defines alchemy as 'The art of liberating parts of the cosmos from temporal existence to achieve perfection, which for metals was gold, and for man, longevity, immortality and finally, redemption. Material perfection was sought through the action of a preparation (for example the Philosopher's Stone for metals; Elixir of Life for humans), while spiritual ennoblement could result from the receipt of inner revelation (gnosis or other mystical experience)' This extends Dr. Johnson's definition and is probably more acceptable than that of the Oxford Dictionary.[1]

Origins

It is thought that the term 'alchemy' came from the Arabic definite article 'AL' prefixed to the late Greek word 'KHEMEIA' relating to the Egyptians – CHEMI or CHAM or the Black Land being the ancient name for Egypt, deriving from the dark colour of the soil in the Nile Delta.

It is uncertain precisely where the art of alchemy first began. The earliest practitioners of which there is any detailed knowledge at all derive from the Egypto-Greek world and it would appear that Western alchemy began in Hellenistic Egypt, being centred upon Alexandria and other towns in the fertile Nile Delta. Democritus (c.460 - c.370 B.C.) defined alchemy as '... an art purporting to relate to the transmutation of metals, and described in terminology at once Physical and Mystical'. Reference is made to aurifiction in the *Ebers Papyrus* which was discovered in a tomb at Thebes in 1862 and is said by some to be the oldest book in the world. It was written in Hieratic script and dates from 1550 BC. It is in the form of a roll of papyrus twelve inches wide and some sixty-eight feet long. It contains prescriptions for remedies for human ailments

(811 of them!) and mentions such substances as stibnite, calamine, granite, sulphur, lead, copper, verdigris, lapis lazuli, salt and saltpetre. The Leyden and Stockholm papyri, written in Greek or demotic script, derive from the early centuries of the Christian era. These papyri also contain a collection of recipes for the preparation of metals and alloys which simulate gold and silver and for augmenting these precious metals. Examples of this appear in the papyrus of Leyden:

56. *Asemos (a white alloy resembling silver) one stater (a measure of weight)*
    or copper of Cyprus three staters: four staters of gold; melt them together.
87. *To increase the weight of gold, melt it with a fourth part of cadmia.*
    It will become heavier and harder.

The processes may be referred to as 'aurifiction'. The later concept of the transmutation of base metals into gold may be called 'aurifaction'.[2]

Although the first group of alchemists of which there is any record lived in the Greek speaking parts of the world about 300 BC, the earliest manuscripts are no older than 1000 AD. In the West, the world of learning in which alchemy could flourish was largely centred in the Church and the monasteries. Not only were clerics and monks given to the religious life, an apparent requisite for attempting the search for the Stone, but also they were to some extent protected from the attentions of greedy men. To be an alchemist has always been a precarious occupation. If unsuccessful, as many were, there was the prospect of poverty, and if successful it was necessary carefully to conceal the fact or else such talent as had been developed would be forcibly requisitioned by the local prince or landowner to reimburse his coffers. Even the suspicion of success as an alchemist could result in torture and death or at the best imprisonment until the adept either parted with the Stone in his possession or managed to escape.

The subject of alchemy has suffered by being regarded as a primitive forerunner of chemistry. It is, however, properly a subject in its own right. 'Holy Alkimy', as Thomas Norton called it, seeks to comprehend the whole of creation not just the reproducible part of it that has become the territory of natural science. The alchemist was very well aware of being himself a creature of the Divine Creator. He knew that even to start on the operations it was necessary to obtain divine help and guidance. Without the personal instruction of a 'Father in Alchemy', a sort of chemical guru, and inspiration or enlightenment from above as to the nature of the starting materials, the *prima materia*, the preparation of the Stone would be impossible. A classical example of such a 'guru' is William Backhouse of Swallowfield who is recorded as having initiated Elias Ashmole into some of the secrets of alchemy on April 3 1651. This initiation was thought to link Ashmole with a supposedly long chain of alchemical ancestry who, from Hermes onwards transmitted their secrets by oral tradition to their spiritual sons.

Despite the necessary secrecy which they practiced, alchemists felt it incumbent upon themselves to set out an account of the process for the guidance of those who came after. As has been indicated, the writings of these practitioners are enigmatical in the extreme; not only is the nature of the starting material never disclosed, but also the path to the Stone is heavily veiled in allegory and illustrated on occasion by the most delightful and thought provoking symbols which might stand for one or more things or operations or concepts. It may perhaps be that the obscurity of their writings was in many cases a veil to hide their own ignorance as to the real nature of the work. On the other hand, they would claim that it was imperative at all costs to prevent the knowledge of the devastating forces they invoked from falling into the wrong hands. Besides this secrecy, however, it is thought that the enigmas of the problem provided a challenge, the attempts to solve which by meditation and continual study might trigger off a process of self-enlightenment. Alchemy should therefore be studied with an open mind, free from the confining restraints of a rigorous scientific and materialistic background.

Alchemists of the Egypto—Greek Period

The principal alchemists of this period were Zosimos of Panopolis, Stephanos of Alexandria and Cleopatra (not the Egyptian queen of that name!). Zosimos and his sister Theosebeia in about 300 AD compiled an encyclopaedia (the *Cheirokmeta*) on the subject in twenty-eight books. They were responsible for the famous 'Formula of the Crab' a cryptic inscription said to embody the secret of transmutation. John Read was of the opinion that it was more likely to have been a cipher used by Egyptian craftsmen engaged in aurifictive practices.[3] Stephanos of Alexandria lived during the time of the Byzantine emperor Herakleios I (610 - 641 AD). While he was more of a theoretician than a practical laboratory worker his writings do give a full exposition of the theory of alchemy as it was understood in the seventh century. It was by this time in the West that alchemy had largely become a theme for rhetorical, poetical and religious compositions. The mere physical transmutation of base metals into gold became a symbol of man's regeneration and his transformation to a nobler and more spiritual state. The *Chrysopoeia of Cleopatra* is of particular interest as on a single page of symbolic drawings can be found a diagrammatic representation of a still on a sand bath and the familiar serpent devouring its tail to symbolise eternity. It encloses the Greek words 'EN TO PAN' (One is All).[4]

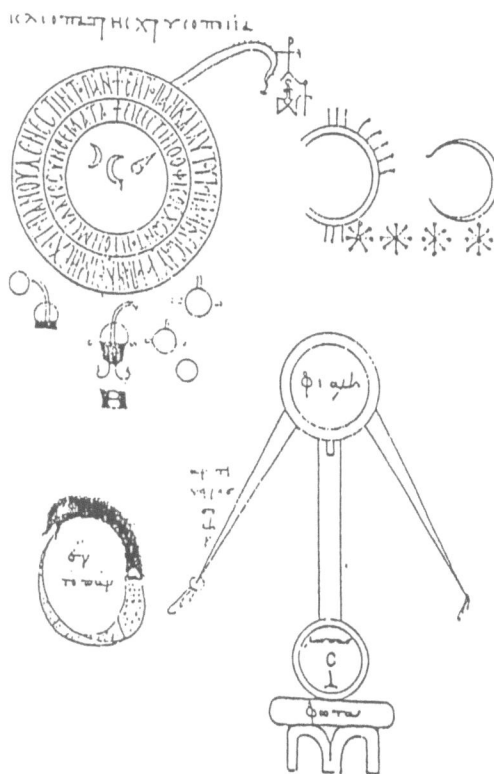

Symbolic drawings from the *Chrysopoeia of Cleopatra*

The Impact of Islam

It was about this time, however, that Western alchemy received stimulation from an unexpected source none other than the foundation of the religion of Islam and the spread of the Moslem empire of the Caliphs, during the seventhth to tenth centuries, to Asia Minor, Syria, Persia, Egypt, North Africa and Spain. While initially hostile to Western learning, after about 750 AD and under the Abbasid Caliphs of Baghdad and such enlightened rulers as Harun Al-Raschid, an apparently insatiable thirst for learning was developed. Many Greek works of science, philosophy and mathematics were translated into Arabic, for the most part by Syriac speaking Nestorians in the newly established University of Jundi-shapur in South West Persia. It was in this way that the Arabs not only preserved valuable manuscripts but also rekindled the ancient lamp of science that had grown dim in Europe. The universities of Islam became great centres of learning, attracting students from all over the known world and later had an important influence on the spread of scientific knowledge.

The most famous alchemists of the Arab world were the Umayyad Prince Khalid ibn Yazid who gave an account of Morienus, a pupil of Stephanus, Jabir ibn Hayyan, the great Geber from whom despite his eminence as the father of Arab Chemistry the word 'gibberish' derives, Rhazes (Abu Bakr Muhammed Ben Zakeriyah er-Razi) a great teacher who emphasised the importance of experimental work in the laboratory rather than the general practice of burning midnight oil in the study! Avicenna (Abu Ali ibn Sina) (980-1037 AD) despite a profound knowledge of the composition of materials expressed complete incredulity on the practicality of the transmutation of base metals into gold.

The Muslim Empire also exerted a broadening effect on Western alchemy for it enabled the traditions of Indian and Chinese alchemy to be incorporated with those of the West. Both these traditions extended far back into the pre-Christian era. In the East, unlike the West, more emphasis was placed on the achievement of longevity or immortality than on aurifiction or aurifaction. It was the search for the Elixir of Life that predominated and this has been linked to the ancient Vedic worship of Soma. The Rig Veda, the oldest of the Vedas, composed between 1700 and 1000 BC, describes Soma Rasa as the Amrita or Ambrosia of the Gods – a potent source of euphoria which when combined with the appropriate mystic rites and incantations could, it was believed, confer immortality, cure all diseases of body and mind, act as an inexhaustible source of strength and vitality, increase sexual energy and stimulate speech!

Alchemy in the Orient

Alchemical knowledge was widely cultivated in ancient India and reached its zenith in the Tantric renaissance period of 700-1300 AD when the SULPHUR – MERCURY conjunction of alchemy was linked to the SIVA – SAKTI psycho-sexual forces of Tantra and their effect on the cosmic consciousness.[5]

There would appear to be very few outstanding names available among the practitioners of alchemy in India. The great alchemical work the RASARATNAKARA originally ascribed to the eminent Buddhist philosopher Nagarjuna is now thought by scholars to be the work of the alchemist Nityanatha. The basic elements of Indian alchemy start at the beginning of all things. It was then that the world existed in an unmanifested state of pure consciousness. From this state the vibrations of the first sound, the almost soundless AUM began to manifest and it was from this vibration that the AKASA, the element space or ether, gave rise to air (VAYU) and produced friction from which light or fire (TEJAS) was formed. The fire dissolved or liquefied parts of the ether giving rise to water (JALA), which then solidified to form the element earth (PRTHIVI). These five elements could be perceived by the five senses of hearing, touch, sight, taste and smell. The five subtle elements of the PANCHA MAHABHUTAS formed the basis of Indian alchemy and also played an important part in the healing art of AYURVEDA.

In China the theory of the Five Elements was first formulated by Tsou Yen (350-270 BC) and comprised water, fire, wood, metal and earth. The theory of the Two Fundamental Forces dates from the beginning of the fourth century BC and were represented by the characters YIN and YANG which were concerned with darkness and light respectively. In alchemy they represent the SULPHUR – MERCURY binary, the former being the active essence or spirit and the latter the receptive and passive role of the soul itself. This latter theory is the Eastern equivalent of the Western *Mysterium Conionctionis*. The legendary Fu Hi is said to have stated this philosophy as follows: 'The Illimitable produced the Great Extreme; the Great Extreme produced the Two Principals; the Two Principals produced the Four Figures and fromn the Four Figures the eight Trigrams if the *Book of Changes* (*I Ching*) were produced.' The Great Extreme was depicted by the circle of the YANG and YIN.

A number of Chinese alchemists are worthy of note. These include: Wei Po-Yang, who lived during the second century of our era and produced the earliest book on alchemical theory the *Tshan Thung Chhi* (The Kinship of the Three) and Ko Hung, a celebrated Taoist philosopher and alchemist and probably the greatest of the Chinese alchemical authors. In his great work the *Pao Phu Tzu* (Book of the Preservation of Solidarity Master) there are numerous recipes for the preparation of the various forms of Elixir and even for the transmutation of base metals into gold. The alchemist of the 'Golden Age' of alchemy in China was Thao Hung-Ching (456-636 AD). He followed in the tradition of Ko Hung and is said to have prepared several successful Elixirs. He appears to have been more concerned with aurifiction than transmutation. The physician and alchemist Sun Ssu-Mo was born in 581 AD and was noted for his work *Essentials of the Elixir Manuals for Oral Transmission*. This gives lucid descriptions of preparations he had found efficacious and details of the essential alchemical apparatus. With the end of Taoism in China in the 13th century when the empire fell into the hands of the Mongols who replaced Taoism with Buddhism, alchemy went to ground, being no longer encouraged by the emperors as heretofore.

The Chinese also had an enigmatic summary of the Great Art not unlike the 'Emerald Tablet' of Hermes Trismegistus. About 1111 AD a stele was discovered in a Taoist temple somewhere in the Szechhuan Province. It had the following inscription and has been attributed to the philosopher Yin Chhang-Sheng the teacher of Wei Po-Yang:

> 'There is a thing, which contains another thing,
> It can be augmented, it can be prolonged,
> It must be plucked before it can be gnawed by silk-worms
> And used after being transformed by fire.
> Thang the completer showers down from above,
> Khua-Fu being empty can receive his fill.
> The chhi responds to the light of the morning.
> The process accords with the night clepsydra.
> White flowers accumulate, putting the snow to shame,
> Yellow flakes solidify, surpassing gold in beauty.
> The cyclical process continuously goes on,
> Now there is rapid steaming and gassing,
> Now there is drastic solidifying shrinkage.
> What is it that appears, gold or jade?
> It brings longevity, eternal as the heavens.
> All this must never be recorded in writing,
> Only oral instruction can transmit it.'

Over the years the concepts of Indian and Chinese alchemy travelled the trade routes of the East and enriched the science of Islam. The full extent of this science has yet to be realised as many of the relevant manuscripts in Arabic, which lie in the libraries of the world, have yet to be discovered and translated.

The Literature of Alchemy

The literature of alchemy is complex, repetitive and to modern readers often difficult to understand and elucidate. Often it is full of quotations from previous exponents of the Art and many tracts are attributed to well-known and eminent professors who may not, in fact, have been responsible for their composition. Other works are attributed to characters, such as Basil Valentine or Abbot Cremer of Westminster, who may have adopted pseudonyms to conceal their identity or may never even have existed! Rich symbolism is employed and it may well be that the literature of alchemy seeks to communicate with more than one level of the mind.

The Tabula Smaragdina

Perhaps the most famous passage in the alchemical canon is the Emerald Tablet attributed to Hermes Trismegistus and which was included among the works of Jabir ibn Hayyan dating from the 8th century of our era.

'True, without error, certain and most true; that which is above is as that which is below, and that which is below is as that which is above, for performing the miracles of the One Thing.
And as all things were from one, by the mediation of one, so all things arose from this one thing by adaption.
The father of it is the Sun, the mother of it is the Moon;
The wind carries it in its belly; the nurse thereof is the earth.
This is the father of all perfection, or consummation of the whole world.
The power of it is integral.
If it be turned into earth. Thou shalt separate the earth from the fire
the subtle from the gross, gently with much sagacity;
It ascends from earth to heaven, and again descends to earth;
And receives the strength of the Superiors and of the Inferiors.
So thou hast the glory of the whole world therefore let all obscurity
flee before thee.
This is the strong fortitude of all fortitudes, overcoming every
subtle and penetrating every solid thing.
So the world was created.
Hence were all wonderful adaptions of which this is the manner.
Therefore am I called Thrice Great Hermes, having Three Parts of the philosophy of
the whole world.
That which I have written is consummated concerning the operation of the Sun.'

This translation by Mrs. Atwood,[6] appears to be all that remains to us from Egypt of their Sacred Art. It is believed to conceal the secrets of alchemy and it has thus merited the rapt attention of prospective alchemists throughout the ages. It contains the well-known aphorism 'as above so below' and has been the subject of several commentaries. The reference to the 'planets' Sun, Moon and Earth implies the importance of astrology in the regimens for the production of the Philosopher's Stone and texts such as the *Splendor Solis* of Solomon Trismosin[7] elaborate on which stage should be carried out at which period of the year. Operations commence with the Sun in Aries when the process of separation of the *materia prima* into its components and recombination by conjunction of the King and Queen should take place.

Western Alchemy in Mediaeval Times

The various strands of alchemical tradition, gathered from Egypt, Persia, India and China, were preserved and added to in the great Moslem empire, from where they were recovered by the efforts of such translators from the Arabic as Robert of Chester who completed one such work, the *Book of the Composition of Alchemy* on February 11 1144. This particular work tells the story of Khalid ibn Yazid and the alchemist Morienus. In this way the earliest figure of Muslim alchemy became also the earliest in European alchemy. Other translators were Adelard of Bath[8] and perhaps the greatest of them all, Gerard of Cremona who worked in Toledo in Spain and translated works by Jabir and Avicenna.   Translations such as these stimulated scholars like Albertus Magnus, his pupil Thomas Aquinas and Roger Bacon, all of whom added insatiably to the new learning. Arnold of Villanova, a Catalan born near Valentia in 1235, produced, it is thought, the famous *Rosary of the Philosophers* and accepted the sulphur-mercury theory of metallic constitution. Arnold was associated with Ramon Lully, although it is uncertain whether the latter ever contributed to the alchemical scene as he did not believe in the possibility of transmutation of the elements. There is, however, a legend that Lully assisted in the production of Rose Nobles from alchemical gold produced in the Tower of London!

The literature of alchemy is full of the most interesting and fascinating characters such as Theophrastus Bombast Philippus Aureolus von Hohenheim, known to us as Paracelsus, who travelled the world with a sample of Azoth, the Philosopher's Stone concealed in the pommel of his sword.   The English alchemists included Sir George Ripley, Thomas Norton of Bristol, Thomas Charnock from Faversham in Kent, the remarkable partnership of John Dee and Edward Kelly, and the part-time alchemist Sir Kenelm Digby. Scotland had its Michael Scot and Alexander Seton, while in France Nicolas Flamel was able to endow churches with the proceeds of his experimentation. Denis Zachaire also is said to have achieved the Stone after many mishaps described in his amusing autobiography.

During the last thousand years an immense amount has been written and published on the subject.   Scientists as eminent as Robert Boyle and Isaac Newton have laboured valiantly to establish the truth behind the apparently successful reports of transmutation carried out during the seventeenth century by mysterious personalities such as one referred to as 'Elias of the Goldmakers'. These adepts travelled about Europe seeking to convince eminent scientists of the truth of transmutation. Even during the present century claims for aurifaction have been made. In 1922 Eugene Canseliet claimed to have transmuted 100 grammes of lead into an equal quantity of gold using a minute quantity of the Philosopher's Stone given to him by the contemporary alchemist known as Fulcanelli in the laboratory of a gasworks at Sarcelles near Paris.

Archibald Cockren, a physiotherapist living in London in the 1930s, had, in the Holborn area, a well-appointed alchemical laboratory, which was destroyed during the blitz, and in which he prepared efficacious Elixirs derived from the techniques described in Basil Valentine's *Triumphal Chariot of Antimony*.   In order to produce 'oil of gold' he achieved the mercury of the philosophers and filled his laboratory with a potent and subtle odour described by a friend as '... resembling the dewy earth on a June morning, with the hint of growing flowers in the air, the breath of wind over heather and hill and the sweet smell of rain on the parched earth.'

In recent years specialised courses on alchemy have been held and seminars on the subject held in England under the auspices of the late Frater Albertus (Albert Richard Riedel) of Salt Lake City.   Despite all this and the apparently reliable accounts of transmutations carried out either by or in the presence of eminent scientists such as Helvetius (Johann Friedrich Schweitzer – 1625-1709) and the Hon. Robert Boyle there is still much doubt as to whether aurifaction was ever achieved. Production of gold on a minute scale is, of course, possible using the techniques of nuclear physics and bombarding a platinum target with neutrons, but like the recent attempts at 'cold fusion' there would appear to be no way in which the process described by the alchemists can be achieved under modern laboratory conditions.

Alchemy may well be classed as a paranormal phenomenon that has yet to be understood!

Chapter 2

The Literature of Alchemy

Over the years the literature of alchemy has grown. It is written in many languages and in every case the subject is treated with respect and with the object of communicating not with any particular group of readers necessarily, but with other adepts to whom the basic secrets of the art were known. In no case is the information imparted in cipher but rather by allegory and association. Use is made, however, in some cases of the kabbalistic language called the 'Language of the Birds' or the 'Green Language'. This is a form of argot or slang intended to be understood only by informed adepts.[1]

The literature of alchemy might be described as 'literature in depth' as it seeks to communicate on more than one level of the mind. To do this use was made of both prose and verse, together with enigmatic illustrations full of fascinating association and allegory. In spite of the more direct approach of the present day, words have particular associations whether derived from their context in the *Authorised Version* of the Holy Bible, the poems of John Donne or those of T.S. Eliot. It is by these associations and the mental concepts to which they give rise that some degree of communication with the sub-conscious levels of the mind is achieved. Even music has been composed by the alchemists for this purpose and Michael Maier in his *Atalanta Fugens* has left some fifty canons set to the alchemical texts, but whether these were intended for performance during the course of the laboratory work is not clear. Robert Fludd also included a number of musical allusions in the course of his writings. It is thought that constant involvement with and meditation upon the enigmas of such alchemical symbolism tended to induce a process of enlightenment or self-revelation.

The *Emerald Tablet* of Hermes Trismegistus is considered to be the basic document of Western alchemical literature and Mrs. Atwood's version is reproduced in Chapter One. It is generally held to contain the necessary secrets of the Art and in consequence has been the subject of a number of commentaries. The text has always been highly considered by alchemists of all centuries. Not only does it contain the aphorism 'As Above So Below' but it also confirms the analogies between the macrocosm, represented by the circle 'O', and the microcosm represented by the axial point '.', to give a point within a circle without which the infinite would remain uncreated, incomplete and without a centre. A commentary on the *Tablet*, written by the fourteenth century adept Hortulanus, is worth quoting as an example of the care required for the elucidation and interpretation of alchemical manuscripts:

> '*I. The philosopher says:* **It is true**, *that is that the art of alchemy was given to us.* **Without lie**, *he says this to confound those who say that the science is a lie or false. Certain, that is to say experienced for any thing experienced is most certain.* **And most veritable**. *For the most veritable Sun is procreated by the art. He says* **most veritable**, *in a superlative degree, for the Sun begotten through this art exceeds all natural Sun in all medicinal and other properties.*'

> '*II. Consequently he touches upon the operation of the Stone saying* **that which is below is like that which is above**. *He says this because the Stone is divided into two principal parts by the Magisterium, into the superior part which rises above, and the inferior part that remains below, fixed and clear. And however these two parts are concordant in virtue. And for this he says* **that which is above is like that which is below**. *This division is certainly necessary.* **To perpetuate the miracles of one thing**, *that is to say the Stone. For the inferior part is the Earth which is called the nurse and ferment; and the superior part is the soul, which vivifies and resuscitates the*

whole Stone. And for this the separation is made, the conjugation celebrated, and many miracles come to be perpetrated and done within the secret work of Nature.'

'III. **And as all things have been, and come from One by the mediation of One**. *He gives here an example saying:* **As all have been, and come from One**, *that is to say, from a chaotic globe, or a chaotic mass.* **By the mediation**, *that is to say by the cogitation and creation of One, that is to say Almighty God.* **Thus all things have been born**. *That is to say have sprung.* **From this single thing**, *that is to say from a confused mass.* **By adaption**, *that is to say by the sole commandment and miracle of God. Thus our Stone is born and sprung from a confused mass, containing within itself all the elements, which has been created by God, and by his sole miracle is our Stone sprung and born.'*[2]

Another enigmatic paradox of alchemy is contained in the Enigma of Bologna, an epitaph said to have been found in that city, and known as the 'Aelia-Laelia-Crispis Inscription'. It was appropriated by the alchemists, who claimed, in the words of Michael Maier, that '... it was set up by an artificer of old to the honour of God and in praise of the Chymic Art.'

AELIA LAELIA CRISPIS

Nor male, nor female, nor hermaphrodite,
Nor virgin, woman, young or old,
Nor chaste, nor harlot, modest hight,
    But all of them you're told -
Not killed by poison, famine, sword,
But each one had its share,
Not in heaven, earth or water broad
    It lies, but everywhere!

LUCIUS AGATHO PRISCUS

No husband, lover, kinsman, friend,
Rejoicing, sorrowing at life's end,
Knows or knows not, for whom is placed
This - what? This pyramid so raised and graced,
This grave, this sepulchre? 'Tis neither,
'Tis neither - but 'tis all and each together.
    Without a body I aver,
    This is in truth a sepulchre;
    But notwithstanding, I proclaim
    Both corpse and sepulchre the same![3]

While Carl Jung was of the opinion that this epitaph together with 'Lucius Agatho Priscus' was sheer nonsense, may this not have been used by the alchemists as a contemplative means for triggering the inner understanding of the whole alchymic process in the manner similar to the use of koans in Zen Buddhism?

Alchemy was full of paradoxes if only to emphasise the difference between it and materialistic and rational science that is based upon a series of cause and effect. By questioning the causation hypothesis alchemy opened itself to a mystical world which is ultimately more akin to reality, able as it was to encompass the depths of the human mind and such phenomena as can be found there.

Michael Maier attempted an interpretation of the enigma by maintaining that Aelia and Laelia represent two different persons who are united in a single subject, namely Crispis.[4] Nicholas Barnard, who lived in the second half of the sixteenth century, also gave an alchemical interpretation of the inscription. He suggested that Aelia represented Sol and Laelia Luna and their combination 'our materia' .[5] These two persons, says Maier, are neither man nor woman, but they once were; similarly, the subject was in the beginning an hermaphrodite but no longer is so, because though the arcane substance is composed of *sponsus* and *sponsa*, and thus it was bisexual. As a third thing, it is new and unique: neither is the subject a maid or virgin, because she would be 'intact'. In the Great Work the virgin is called a mother although she has remained a virgin. Nor is the subject a boy, because the consummation of the *conjunctio* contradicts this, nor a crone, because it still retains its full strength, nor a whore, because it has nothing to do with money, nor is it virtuous, because the virgin has cohabited with a man The subject he says is a man and a woman, because they have completed the conjugal act, and an hermaphrodite because two bodies are united in one. It is a girl because it is not yet old, and a youth because it is in full possession of its powers. It is an old woman because it outlasts all time. It is a whore because Beya prostituted herself to Gabritius before marriage. It is virtuous because the subsequent marriage gave it absolution. Maier concludes that the Aelia-Laelia section refers to the Philosopher's Stone, while a similar analysis of the Agatho-Priscius section concludes that it signified the chief requisite necessary for the fulfilment of the Art (the *materia prima*).

Having given examples from the alchemical literature and of their interpretation, it still remains to define what alchemy was, or at least what it was believed to be. As has been mentioned, alchemy was one of the four great branches of Hermeticism: Alchemy, Astrology, the Kabbalah and Magic. It is of great antiquity for it came into being before the birth of Jesus Christ and there are still practitioners of alchemy to be found throughout the world. It is of a two-fold nature, having an outward and exoteric form and an inward or esoteric one. Exoteric alchemy is concerned with the preparation of the mysterious additive or catalyst, the Philosopher's Stone, by which base metals may be transmuted into one of the precious metals, gold or silver. Esoteric alchemy, deriving from the belief that the Stone could only be prepared by the intervention of divine grace and favour, was concerned with a devotional system in which the mundane transmutation of metals became merely symbolic of the transformation of sinful man into the perfect being that God intended at the creation of the world. These two facets of alchemy are superimposed upon each other, the material progress in the laboratory being indicative of that in the oratory - *laborare est orare*![6]

The literature of alchemy is immense, ranging from translations of Greek and Arabic texts (and many such Arabic texts have yet to be studied and translated) to modern commentaries and histories of science. It may be broken down into two main categories: firstly, writings by the alchemists themselves, or by others in their names, on the subject of the Philosopher's Stone. Accounts are given of its preparation, which vary from the apparently lucid and often deliberately misleading to the obscure, and, secondly, the more academic studies of the subject by historians of science.

Some of the works by the alchemists such as the *Mutus Liber* are almost entirely pictorial. Some such as the *Splendor Solis* are of extreme beauty, while yet others exhibit a compelling, if rather amateur, artistic presentation. The poetic nature of the subject, the need to express concepts appreciable only by the deeper levels of the mind or even with the sub-conscious by word association, is shown by the collection of alchemical poems by the seventeenth century antiquary and astrologer Elias Ashmole (1617-1692) in his *Theatrum Chemicum Britannicum* in which the well-known *Ordinall* of Thomas Norton appears together with the *Compound of Alchymie* of Sir George Ripley.

There are also a number of poems and plays written by laymen knowledgeable on the subject, such as 'The Canon Yeoman's Tale' in the *Canterbury Tales* of Geoffrey Chaucer (*c*.1340 -1400) and Ben Johnson's wonderful play *The Alchemist* which was performed in London at the time of the Centenary Celebrations of the Chemical Society, now the Royal Society of Chemistry, and was supported by that august body one delightful evening in 1947. Paracelsus was the subject of a long narrative poem by Robert Browning (1812-1889).

Next are the historical works on alchemy, from the viewpoint of the forerunners of chemistry, the so-called father of which, the Hon. Robert Boyle (1627-1691), although credited with dealing a death blow to the old beliefs in his book *The Sceptical Chymist*, was himself sufficiently interested in alchemy to petition for the repeal of the laws against gold makers. His experiments with his 'red earth' were of considerable interest to his contemporaries Isaac Newton (1642-1727) and Henry Oldenburg (1615?-1677), both members of the Royal Society. Newton himself practiced as an alchemist and a considerable volume of his work on this subject has survived although overshadowed by that on optics, gravity and the calculus, or the method of fluxions as he called it. It is strange that King Charles II, who kept an alchemical laboratory in a room beneath his bedchamber, should appoint Newton, an alchemist, as Master of the Royal Mint. Perhaps not!

Foremost among the histories is the monumental work on the *History of Magic and Experimental Science* by Lynn Thorndyke of Columbia University, while as a sympathetic study of the alchemist in life, literature and art, the books of the late Professor John Read (1884- 1963) of St. Andrew's University such as *Prelude to Chemistry* can be recommended. The most important psychological study of alchemy is that by Carl Jung (1875-1961), whose three volumes on the subject: *Psychology and Alchemy, Alchemical Studies* and *Mysterium Conjunctionis* are of exceeding importance. On the more mystical side, the recent books by Prince Stanislas Klossowski de Rola and Titus Burkhardt are well worth reading, particularly those of the former for his concise statement of the Art and for its collection of alchemical illustrations and engravings.

The literature of alchemy poses one additional enigma. Why, if the secrets of alchemy were so dangerous should they fall into the wrong hands, did the workers in the field feel it as their duty to go into print, or manuscript, on the subject? Perhaps it was some instinct causing them to perpetuate a secret tradition – a tradition that was sought without apparent success by Arthur Edward Waite (1857-1942). Like the legend of the Tarot where the eternal symbolism is said to be perpetuated in the harmless form of a pack of playing cards, so will the intriguing enigmas of alchemy retain its secrets.

According to Fulcanelli some alchemists were adept in using the 'Language of the Birds' or the 'Green language', sometimes referred to as 'the Gothic Art'. This used words, the explanation of which derived from their cabalistic origin rather than from their literal root. For the modern alchemist such as Fulcanelli, for example, 'gothic art' (*art gothique*) was a corruption of the word *argotique* (cant) which sounded exactly the same. It is an example of that which he refers to as phonetic law, and which governs the traditional Kabbalah in every language and does not pay any attention to spelling. It is this language that teaches the mystery of things and unveils the most hidden truths. 'Argot' is defined in dictionaries as '... a language peculiar to all individuals who wish to communicate their thoughts without being understood by outsiders'. *The Concise Oxford Dictionary* defines it as '... the jargon of a group or class, formerly especially of criminals'.

An excellent dissertation on the Green language or Language of the Birds used by Nostradamus has been included by David Ovason as an appendix in his book *The Secrets of Nostradamus*. Its use in alchemy is discussed by Fulcanelli in his work *Les Demeures Philosophales* and in *Le Mystère des Cathédrales*. The basics of this 'language' derives from the Hebrew *Kabbalah* and its three basic methods of interpretation: *Gematria, Notaricon* and *Temurah. Gematria* is based on the fact that each letter of the Hebrew alphabet has a numerical equivalent so that words that add up to the same sum may be substituted for each other. In *Notaricon* the letters of a word may be expanded to represent a whole sentence and vice versa. For example the alchemical phrase 'VISITA INTERIORA TERRAE RECTIFICANDO INVENIES OCCULTUM LAPIDEM' can be reduced to 'VITRIOL' which in the *Geheime Figuren der Rosenkreuzer* has been used in an illustration to the *Tabula Smaragdina* of Hermes Trismegistus. *Temurah* is a permutation in which one letter is substituted for another preceding or following it in the alphabet. In the method known as *Albath* the Hebrew alphabet is bent in half in the middle and one half is put over the other:

11	10	9	8	7	6	5	4	3	2	1
K	I	T	Ch	Z	V	H	D	G	B	A
M	N	S	O	P	Tz	Q	R	Sh	Th	L

in this way the word 'RUACH' (RVCh) for example becomes 'DETZAU' (DTzO), (S.L. MacGregor Mathers *The Kabbalah Unveiled*, p. 9).

Other techniques used in the 'Language of the Birds' include the *anagram* in which the letters of a word may be re-arranged to form another word, as for example, 'deal' = 'lead' or 'Crymeru' = 'Mercury'. In *Anastrophe* – in which a word is reversed, either by sound or letter – 'Hiram' represents 'Maria'. In *Aphesis* a letter or syllable is omitted from the beginning of a word so that 'lution' may stand for 'solution' while in its complement *Apocope* a letter or syllable is omitted from the end so that 'Sol' might refer to solution rather than to gold! Letters may be added or subtracted from words to disguise their import. A popular device, particularly in France, during the sixteentth century was the *Rebus*, a riddle in which pictures, letters or sentences were read in terms of sound values. The word *Rebus* itself is derived from the Latin '*Rebis*', 'the thing twice', a thing seen from two different aspects. Thus the 'Red' and the 'White' may refer to the King and the Queen, the imperfect and perfect Philosopher's Stone or even to the constitution of an egg, full of symbolism to the alchemist. The names of alchemical authors were often concealed in *anagrams* as for example in *The New Chemical Light* of Michael Sendivogius (or Alexander Seton!) in which the anagram was '*Divi Leschi genus amo*' and in its sequel '*Angelus Doce Mihi Jus*' stood for the name of the writer. Fulcanelli, writing in French, declared that language came directly from archaic Greek and not from Latin, as was generally held to be the case. Consequently all the words chosen to define certain secrets may have their phonetic Greek equivalents.

It follows from all this that the reading of an alchemical text is not the same as studying a chemical textbook. The secret of the Art cannot be obtained by analysis and deduction even if a large number of texts are correlated and studied. Not only that but the materials used were different from those of the same name obtainable from modern chemical suppliers. The substances, even if correctly identified, originally contained levels of impurities that may have altered the final products of any chemical reaction. A contemporary investigator, Professor Lawrence Principe, of the Johns Hopkins University in the United States, has, however, attempted with some success to reproduce these reactions under conditions likely to have been experienced by the alchemists. He has demonstrated fairly conclusively the experimental work of the alchemists can be translated into modern chemical terminology if due account is taken of the condition of the materials that they used.[10]

Chapter 3

The Theory of Transmutation

The world of the alchemist was very different from that of a nuclear physicist in the 20th or 21st century. To begin with, his understanding of the nature of matter was based on the views of Aristotle (384-322 BC) and those before him. Empedocles (*c*.495-435 BC) taught that there were four elements – Earth, Air, Water and Fire – while Aristotle added a fifth, the Ether. These elements were regarded not as different kinds of matter but rather as different forms of the one original matter, by which was manifested its different properties.

These properties were based on the dry, the moist, the warm and the cold which gave rise to the cold and dry EARTH, the hot and dry FIRE, the hot and moist AIR and the moist and cold WATER. These four elements were again not quite as we understand them. The term WATER, for example, covered liquids in general which possessed the properties of moistness and coldness. All liquids were known as 'waters'. Because the four elements were not regarded as different kinds of matter, transmutation was thought to be possible, one substance being converted into another by an appropriate adjustment of the quantities of the elements present.

The theory of transmutation is, in fact, very simple and is based on a statement of Arnold of Villanova:

> 'That there abides in Nature a certain pure matter, which being discovered and brought by Art to perfection converts to itself proportionally all imperfect bodies that it touches.'

This 'pure matter' was the Philosopher's Stone.[1]

It has been said that natural species cannot be transmutable because species are indestructible. But the alchemists never professed to transmute species, nor taught in their theories that, for example, lead as lead or mercury as mercury can be changed into gold. It was the subject matter of these metals, the radical moisture of which they were uniformly composed, that they thought could be withdrawn by Art and transformed from inferior forms, being set free by the force of a superior ferment or attraction.

There were several phenomena, well known at that time, which would encourage the belief that transmutation was possible. If, for example, an iron rod is immersed in a solution of copper sulphate a deposit of copper on the rod is produced. This might be considered on superficial examination as a case of transmutation of the iron into copper. The fact that some of the iron would have entered the solution as ferrous sulphate would not be apparent without refined analytical techniques which were beyond those available to the alchemist.

The preparation of yellow coloured alloys by melting copper with, for instance, arsenic might also suggest that the copper had been transmuted into gold.

As mentioned above alchemists did not think of matter as composed of atoms and molecules but rather of the four 'elements' or qualities. Transmutation appeared possible only if the proportions of these 'elements' in the material or base metal could be adjusted to those proportions present in silver or gold.

To some extent this is born out in nuclear physics in which the properties of a chemical element depend upon the extra-nuclear electrons and on the charge on the nucleus. If the nuclear structure is modified by proton or neutron bombardment of a target of the element different elements can be produced. For example, an isotope of gold is formed in minute quantities when a target of platinum is bombarded with fast neutrons.

The views of the alchemists on the composition of metals in the earlier days of alchemy gave rise to what is known as the 'Sulphur-Mercury Theory'. Needless to say, neither the 'Sulphur' nor the 'Mercury' was the common material designated by those names! The 'metals' were more examples of properties rather than substances, as were the elements of Aristotle. Alchemists were, perhaps deliberately, rather imprecise in writing about these substances and gave the impression they were by no means certain what they did mean. Frequently in the literature there is mention of different sorts of 'sulphur' and 'mercury'. The most acceptable understanding of the terms is achieved by regarding 'sulphur' as the principle of combustion and colour and 'mercury' as the metallic principle to which was attributed such properties as fusibility, malleability and lustre. Even metals such as gold when subjected to alchemical processes became known as 'our gold' so throughout the literature the reader can never be certain precisely what is intended!

Writing of gold the alchemist Eirenaeus Philalethes (George Starkey) says:

> 'But our gold is twofold; one kind is mature and fixed, the yellow latten, and its heart or centre is pure fire, whereby it is kept from destruction, and only purged in the fire. This gold is our male, and it is sexually joined to a more crude white gold - the female seed: the two together being indissolubly united, constitute our fruitful Hermaphrodite. We are told by the sages that corporeal gold is dead, until it be conjoined with its bride, with whom coagulating sulphur, which in gold is outwards, must be turned inwards. Hence it follows that the substance which we require is mercury …. our mercury is spiritual, female, living and life-giving.'[2]

The Sulphur-Mercury dichotomy represents the polarities of the Lunar Queen and the Solar King, the bride and the bridegroom. Their union in the Chymical Marriage, the supreme union of hostile opposites, as Jung describes it, is the central symbol of alchemy.[3] According to Titus Burckhardt, it is only on the basis of the interpretation of this symbol that a distinction can be made , on the one hand, between alchemy and mysticism and, on the other, between alchemy and psychology.[4] From the point of view of mysticism, the soul has become alienated from God, while from that of alchemy, man as the result of the fall from his original state of grace, is divided within himself. It is the reconciliation of these opposites that is the concern of spiritual alchemy. The King represents gold or 'sulphur' and the Queen 'mercury'.

Perhaps the best symbol for this sulphur-mercury couple is the Chinese device of Yin-Yang, with the black pole in the white vortex and the white pole in the black vortex, as an indication that the passive is present in the active and the active in the passive. Just as man contains in part the nature of woman, so woman contains in part that of man. This symbolism also bears a direct relation with the Indian esoteric tradition of Tantra. The sulphur of gold represents the PINGALI NADI (Sun breath) and the mercury of silver the IDA NADI (Moon breath).

The addition of 'salt' to the duad of 'sulphur-mercury' resulted in the *Tria Prima* of the basic components of matter: salt, mercury and sulphur. This 'salt' or 'Secret Fire' can also be related to the Tantric concept of KUNDALINI, energy flowing in the SUSHUMNA, or central channel, through the seven CHAKRAS, the energy centres of the body and the points of contact between the psychic and physical bodies.

The Salt or Secret Fire is often referred to as the 'Chalybs of the Sages' and is:

> '... the true key of our art, without which the torch could in no wise be kindled, and as the true magi have delivered many things concerning it, so among vulgar alchemists there is great contention as to its nature.'

Again there is some confusion. The secret fire might be the name given to the mercury when prepared, or might be a kind of water which acts as a catalyst. These two concepts are always mixed up deliberately so that one is often named for the other and mistakes may be made, but really they are two different things. The secret fire, which might be termed the fiery water, dissolves the metals, whilst the latter is a natural product found everywhere and in everything – as is the Philosopher's Stone!

Eirenaeus Philalethes included many useful comments on the art of alchemy in his various works which were highly esteemed by such 17th century alchemical practitioners as Isaac Newton and Robert Boyle. The following five precepts, taken from the work *Ripley Reviv'd* are of great interest.[5]

Cbe Precepts of Eirenaeus Philalethes

1.   As all things are multiplied in their kind, so may be metals, which have in themselves a capacity of being transmuted from the imperfect into the perfect.

2.   The main ground for the possibility of transmutation is the possibility of the reduction of all metals, and such minerals are of metallic principles into their first mercurial water.

3.   Among so many metalline and mineral Sulphurs and so many Mercuries there are but two Sulphurs that are related to our work, which Sulphurs have their Mercuries essentially united to them.

4.   He who understands the two Sulphurs and Mercuries aright, shall find that one is the most pure red Sulphur of Gold, which is *Sulphur in manifesto* and *Mercurius in occulto* and that other is the most pure white Mercury, which is indeed *Quicksilver in manifesto*, and *Sulphur in occulto*, these are our two principles.

5.   If a man's principles be true, and his operations regular, his Event will be certain, which Event is no other than the true Mystery.

Commentary on Philalethe's Five Principles

1.   The statement that 'all things are multiplied in their kind' relates to the belief held by the alchemists that as animals and plants propagate themselves by the processes of reproduction so also do minerals. Man is a composite being having in his nature the three basic components of Body, Soul and Spirit.   This is symbolically illustrated in the *Book of Lamspring* by two fishes swimming in a sea, the two fishes being Soul and Spirit and the Sea the Body. The Body is the outward manifestation and form, the Soul is the inward individual spirit and the Spirit the universal Soul in all men. In metals these components exercise a similar role. The Body is the outward form and properties of the metal, the Soul the metalline Soul or Spirit of the individual metal and the Spirit the all-pervading essence of all metals.

Throughout alchemy the importance of following Nature is stressed. As Sendivogius remarks:

> 'As Nature has her being in the Will of God, so her will, or seed, is in the Elements. She is one, and produces different things, but only through the mediate instrumentality of seed.' (*The New Chemical Light*: Second Treatise)[6]

2.   The concept of the 'First Mercurial Matter' follows on from the all-pervading essence of all metals and their individual souls or spirits. It was the subject matter of these metals, the radical moisture of which they are composed both on the material and on the spiritual plane. It has been said that *Species non transmutantor sed subjecta specierum optime et propriissime* (It is not species that are transmuted but their underlying essence). The first stage in the operations of alchemy is to reduce the body to water, that is to its Mercury and this is known as Solution, the foundation of the whole art. This is part of the well-known alchemical saying *Solve et Coagula*.

3.   The two sulphurs referred to by Philalethes are those of the metals Gold and Silver, for these two metals were, in accordance with the Sulphur-Mercury theory, composed of 'sulphur' and 'mercury' in different proportions and degrees of purity. In addition, the two metals Gold and Silver were known as the Solar King and the Lunar Queen and, paradoxically, referred to as Sulphur and Mercury respectively. This pair exhibited the perverse power of the Duad. The King and Queen form the Bridegroom and the Bride, and their union in the Chymical marriage is, as has been mentioned, the supreme union of hostile opposites. This Marriage, the central symbol of Alchemy, it is the *Mysterium Coniunctionis* – copulation followed by death and re-birth, giving rise to the phenomenon of the alchemical Nigredo and the formation of the Hermetic Androgyne.

On the interpretation of this symbol depends the relationship between Alchemy and Mysticism on the one hand and between Alchemy and Psychology on the other. From the point of view of Mysticism, the soul has become alienated from God and turned towards materialism and the things of Mammon, while Alchemy would insist that Man, as the result of the Fall is divided within himself. It is imperative that in this life man should strive to regain his original state of grace and Alchemy was a key process to attain that end. Integral with this is the need to know oneself – the ἔῃῐἔ ῐἐὰᾳῆῳῒ of the Oracle at Delphi. In Mysticism the soul is reunited with God by discovering itself in this manner, while in Alchemy the union is achieved by regaining the integral nature of man. This reconstitution is pictorially expressed by the masculine-feminine androgyne, illustrating the Sulphur-Mercury conjunction.

4.   The understanding of these two principles is rightly stressed by Philalethes as being essential to the alchemical process. It introduces the concept of the alchemical paradox. Not only is the literature of the subject deliberately obscure and confusing, but so also are the interpretations of alchemy by its commentators. Jung would have us believe that Alchemy had a close relationship with the psychic life of man and not with material reality in the laboratory. Others have stressed the spiritual and mystical components of the subject. Alchemy was, however, both spiritual and material and, in fact, its operations could only be carried out successfully on two planes at the same time.

5.   The final comment of Philalethes could be interpreted as relating to the qualities needed for a successful alchemist. A rational approach to the subject was clearly a disadvantage! In the 17th century two of the most brilliant natural philosophers, Robert Boyle and Isaac Newton, approached the subject but as far as is known without achieving its ultimate goal. Those alchemists who would appear to have been more successful lacked the mental calibre of Boyle or Newton, but this did not appear to have been any disadvantage. In the 20th century Archibald Cockren, the author of *Alchemy Rediscovered and Restored*, had only the elementary knowledge of chemistry necessary for his vocation as a physiotherapist yet he managed to progress far in the field of alchemy.

The amount of energy involved in the action of transmutation must have been considerable and by means of Einstein's equation:

$$E = mc^2$$

where c is the velocity of light, m is the mass involved and E the energy produced, it can be calculated that the conversion of even a few grams of matter into energy would give rise to the output of a large power station. If alchemy involved some latent psychokinetic process, inherent in the human mind, to bring about transmutation, the fears of the alchemist that his knowledge might pass into the wrong hands is understandable. An equal prudence on the part of present day nuclear physicists would have been desirable. Nuclear science has given the physicist certain power but, unlike the alchemist, lacks the wisdom or reverence for life to ensure that the knowledge would be used solely to the glory of God and the benefit of mankind.

Chapter 4

The Philosopher's Stone

The Philosopher's Stone was that catalyst or reagent by which imperfect matter could be rendered perfect, base metals transmuted into gold and health, and longevity imparted to a recipient. It was sought for by many and supposedly found by only a very few. In the alchemical literature it is called by many names, though never confused with the 'Stone of the Philosophers' that was, in fact, the *prima materia* or mysterious starting material or ore from which the Philosopher's Stone was derived. Amongst other synonyms it was called the Essence, the Stone of the Wise, the Magisterium, the Magnum Opus, the Quintessence and the Universal Essence. As an alchemist commented:

> 'It is called a stone not because it is like a stone, but only because of its fixed nature and that it resists the action of fire as successfully as any stone. Its appearance is that of a very fine powder, impalpable to the touch, sweet to the taste, fragrant to the smell, in potency a most penetrative spirit, apparently dry yet unctuous and easily capable of tingeing a plate of metal. If we say its nature is spiritual it would be no more than the truth; if we describe it as corporal the expression would be equally correct.'[1]

This is not unlike the passage in Sendivogius, the Cosmopolite's *The New Chemical Light*:

> 'It is and is not a stone. It is called a stone because it looks like one; first, because it really is stone when it is extracted from the depths of the earth – it is hard and dry and can be broken up and pulverised like stone. Secondly, because, after the stinking sulphur and other impurities in which it is embedded have been eliminated, it can be reduced to its essence by slow degrees, as by a natural process, making it an incombustible stone, resistant to fire, and as easy to melt as wax: which implies that it has regained its universality.'[2]

Raymund Lully described the Stone as *Carbunculus*, Paracelsus said it was solid like a ruby, transparent and flexible, Beregard that it had the colour of a wild poppy with the smell of heated sea salt, Helvetius describes it as yellow, the colour of saffron and in the form of a heavy powder, while Van Helmont says it is pale yellow like sulphur. Others maintain it is of great weight and is small or it is 'a certain, heavenly, spiritual penetrative and fixed substance which brings all metals to the perfection of gold or silver and that by natural methods' or, again, its colour is sable with intermixed argent which 'marks the sable fields with veins of glittering argent'.

This Philosopher's Stone mentioned above was, however, only the Mineral Stone, which had the power of transmuting any imperfect earthly matter into its utmost degree of perfection, for example base metals into gold or silver, and flints into all manner of precious stones such as rubies, sapphires, emeralds and diamonds. There were, in addition, three other stones, the Vegetable, the Magical, and the Angelic Stones, each of them differing in operation and nature. By the Vegetable Stone could be known the nature of all animals and plants and how to make them grow, flourish and bear fruit. By the Magical or Prospective Stone it was possible to discover any person wherever they were in the world and to understand the language of birds and animals and to obtain a knowledge of the future. By the Angelic Stone which could, apparently, only manifest itself by its taste, the power of conversing with angels could be obtained. Such was the power of the Angelic Stone that no evil spirit could approach the place where it was because it was a quintessence wherein there is no corruptible thing. To the alchemist the attainment of these Stones was secondary to the changes that the search brought in itself. As Elias Ashmole put it:

'Certainly he to whom the whole course of nature lies open, rejoiceth not so much that he can make gold or silver or the devils to become subject to him, as that he sees the Heavens open, the Angels of God ascending and descending and that his own name is fairly written in the Book of Life.'[3]

It is clear from the writings of the alchemists it was the search for the Stone that was important and the operations of 'Holi Alchymy' were never intended to be carried out only in the laboratory. The operations of the work were not only physical on the material plane but also spiritual on a celestial plane linked by the Hermetic maxim 'as above so below'. Writers in the past have tended to concentrate more on one than on the other of these two aspects, but it cannot be emphasised too strongly that alchemists apparently operated on these two planes at the same time. The full maxim in the Emerald Tablet reads:

'True without error, certain and most true; that which is above is as that which is below, and that which is below is as that which is above, for performing the miracles of the One Thing.'

A typical example of the use of the Philosopher's Stone to bring about transmutation is given in the account of Jean Baptiste van Helmont in his *Oriatrike or physick refined* of 1662:

'For truly I have divers times seen it and handled it with my hands, but it was of colour, such as is saffron in its powder, yet weighty and shining like unto powdered glass. There was once given unto me one fourth part of a grain but I call a grain the six hundredth part of one ounce: this quarter of one grain therefore, being rolled up in paper, I projected upon eight ounces of quicksilver made hot in a crucible; and straightway all the quicksilver, with a certain degree of noise, stood still from flowing, and being congealed, settled like into a yellow lump: but after pouring it out, the bellows blowing, there were found eight ounces and a little less than eleven grains of the purest gold. Therefore only one gram of that powder had transchanged (transmuted) 19,186 parts of quicksilver, equal to itself, into the best gold.'[4]

The Green Lion from *Atalanta Fugiens*

Three things suffice for the work: a White Smoke, which is Water; a Green Lion, which is the Ore of Hermes, and a Fetid Water. (*Tria sufficiente ad Magisterium Fumus Albus, hoc est Aqua, Leo Viridis, id est Aes Hermetis et Aqua Fœtida*).

Chapter 5

Alchemical Operations and Apparatus

There are in alchemy a number of technical terms for the various operations employed. The twelve principal operations are described in a poem by Sir George Ripley *The Compound of Alchymie.* [1] They are as follows:

1. **Calcination** – the reduction by heat but not by burning.
2. **Dissolution** – the slow separation of a body into its components in a liquid.
3. **Separation** – the breaking up into light and heavy parts.
4. **Conjunction** – the amalgamation of several elements.
5. **Putrefaction** – the first change to be seen in the alchemical process, the appearance of blackness or the *Nigredo*.
6. **Congelation** – the solidification of the liquid in the reaction vessel.
7. **Cibation** – the wetting of the dried matter.
8. **Sublimation** – extraction by volatilisation or distillation.
9. **Fermentation** – achieved by adding the desired metal as a 'yeast' to the Philosopher's Stone so enabling it to transmute base metals into silver or gold..
10. **Exaltation** – the raising of the power or virtue of the Philosopher's Stone to enable it to transmute.
11. **Multiplication** – the process of increasing the quality and quantity of the Philosopher's Stone.
12. **Projection** – the final work of transmutation of a base metal into either silver or gold.

Dom Antoine-Joseph Pernety in his *Dictionaire Mytho-Hermetique* gives the following correspondence between the twelve steps or operations in the Great Work and the signs of the Zodiac:

1. Calcination   Aries, the Ram
2. Congelation   Taurus, the Bull
3. Fixation   Gemini, the Twins
4. Solution   Cancer, the Crab
5. Digestion   Leo, the Lion
6. Distillation   Virgo, the Virgin
7. Sublimation   Libra, the Scales
8. Separation   Scorpio, the Scorpion
9. Ceration   Sagittarius, the Archer
10. Fermentation   Capricorn, the Goat
11. Multiplication   Aquarius, the Water Carrier
12. Projection   Pisces, the Fishes

The work is also held to proceed under a series of regimens, each controlled by one of the planets and its associated sign(s). Solomon Trismosin, who wrote in the latter half of the 15th century, in his *Splendor Solis* places the regimens in the following order:

1. **Regimen of Saturn** (Aquarius and Capricorn) solution of the *materia prima*.
2. **Regimen of Jupiter** (Pisces and Sagittarius) separation of the Three Principles.
3. **Regimen of Mars** (Aries and Scorpio) the unification of the Three Principles.
4. **Regimen of Sol** (Taurus and Libra) the solidification of the Three Principles.
5. **Regimen of Venus** (Taurus and Libra) the resolution and integration of the Three Principles.
6. **Regimen of Mercury** (Virgo and Gemini) the preparation of the White Stone (Philosophical Salt).
7. **Regimen of the Moon** (Cancer) the preparation of the Red Stone (Philosophical Sulphur).

Eirenius Philalethes in his *Open Entrance* placed the regimens in a different order – Mercury, Saturn, Jupiter, the Moon, Venus, Mars and the Sun. He remarked, however, 'Let me assure you that in our whole work there is nothing hidden but the regimen, of which it was truly said by the Sage that whoever knows it perfectly will be honoured by princes and potentates.' The order of the regimens is therefore concealed, certainly by Philalethes, although Trismosin placed his in the order of the planets in a geocentric universe. This could be significant.

These operations were carried out on the physical plane in the laboratory of the type made familiar to us by such Dutch and other painters of the 17th century as David Teniers, Witt, Van der Veldt, David Ryckhaert, Hendrick Heerschop and Jan Steen. To quote C.J.S. Thompson again:

'They delighted to portray the bearded and venerable alchemist, in cap and gown, intent upon some operation in his quest for the Stone, or seated in a chair peering at some manuscript or ancient tome that might perhaps give him the key to the mystery. From the black and smoke begrimed beams of the roof hangs the crocodile, alligator, or strange shaped fishes, and from the corner glares a great owl, symbolic of wisdom. On the shelves that line the walls stand quaint jars, curiously shaped flasks and bottles, or the skulls and bones of beasts, birds and fishes. On one side of the adept stands the great still, and on the other his aludel or sublimatory, while on a bench or table close by are his globe and hourglass. The floor is littered with jars, pots, dishes, crucibles, mortars and pestles, funnels, tongs and other implements of his art. Away in the background the assistant is often seen working the great bellows, feeding a furnace, or luting some apparatus and making ready for the next operation. The twisted condenser, like a monstrous snake, and the big globular recipients or long-necked matrasses are tinged with gleams of light from a burning brazier, while alembics glow like fireflies in a dim corner.'[2]

This delightful description, which covers fairly accurately the majority of alchemical interiors, mentions the type of apparatus used by the alchemist. The great still emphasises the importance of distillation that was a much repeated operation of mystical significance. It not only purified the liquid in the still but also induced in the alchemist himself a state of tranquillity as anyone who has carried out such operations well knows.

The following is an alphabetical list of many of the different types of apparatus likely to have been found in typical alchemical laboratories:

Alembic the upper part or head of a still; also known as a **Limbec** or **Helm** by being of a similar shape to a helmet. The word Alembic is often incorrectly used to denote an entire still. Dr. Johnson defines it as:

> '... a vessel used in distilling, consisting of a vessel placed over a fire in which is contained the substance to be distilled, and a concave closely fitted on, into which the fumes arise by the heat; this cover has a beak or spout into which the vapours rise, and by which they pass into a serpentine pipe, which is kept cool by making many convolutions in a tub of water; here the vapours are condensed, and what entered the pipe in fumes comes out in drops.'[3]

Aludel this is a sublimating pot used in chemistry. Aludels are without bottoms and fit into one another, as many as there is occasion for, without luting. At the bottom of the furnace is a pot that holds the matter to be sublimed, and at the top is a head, to retain the flowers that rise up.

Athanor - a digesting furnace, sometimes regarded as an incubator, as in reference to 'gentle heat' or 'the House of the Chick', to keep heat for some time; so that it may be augmented or diminished at pleasure, by opening or shutting some apertures made on purpose with sliders over them, called registers. The same name is applied analogically to the secret furnace of the philosophers, wherein a fire is always maintained at the same grade. It is unlike the vulgar athanor; it is actually the matter itself animated by a sophic fire which is innate therein, and is developed by art.

Balneo - a bath of sand or ashes which can be heated.

Balneum Mariae - the furnace of the sages, the secret furnace, to be distinguished from that of ordinary chemistry. Sometimes the name is applied to philosophical mercury. The term bath is also given to a matter which is reduced into the form of a liquor. For example, when it is desired to make projection upon a metal, it is said it must be in the bath, that is, in a state of fusion. The term is also given to a bath of warm water, and the name is still applied to a large cooking pan.

Bolt-Head - a round-bottomed, long-necked flask, also called a matrass.

Cineritum - is an ash pan often mentioned by Paracelsus, but can also be an equivalent of Regale, an amalgam of gold and silver.

Coppel - a dish, made of ashes, well washed to cleanse them from all their salt; or of bones thoroughly calcined. Its use is to try and purify gold and silver, which is done by mingling lead with the metal, and exposing it in the coppel to a violent fire for a long while. The impurities of the metal will then be carried off in a dross, which is called the lithage of gold and silver.

Croslet - a crucible.

Crucible - a clay melting vessel, capable of withstanding a severe degree of heat. It had a narrow base and widened out into a round or triangular body. The cupel or coppel was a species of crucible.

Curcurbite - the lower section of a still, made of earthenware or glass; sometimes known as a gourd.

Egg - usually the philosophic egg or glass vessel containing the philosophic substance for the final coction.

Helm - an alembic, the upper part of a retort.

Lute - a cement form of a composition like clay and used for sealing apertures and joints of laboratory equipment.

Matrass - a glass vessel made for digestion and distillation, being sometimes bellied and sometimes rising gradually by taper into a conical figure.

Mortar - a strong vessel in which materials are broken by being pounded with a pestle.

Pelican - a pot-bellied, two-armed circulatory vessel for use in continuous distillation.

A Pelican

Retort - a glass vessel with a bent neck to which the receiver is fitted and which is used for distillation.

Sepulchre - the glass vessel which contains the matter of the philosophical work; the dissolver of the sages; the black stage or *nigredo* of the matter.

Spatula - a broad and flat instrument like a slice, it is also used as a term for the registers of certain furnaces for regulating the heat.

Vessel of the Philosophers - Geber describes this instrument as a round glass vessel with a flat round bottom. This and all simple explanations concerning it are supposed to be wholly deceptive, and it is regarded as one of the most profound mysteries of the alchemical art.

A treatise attributed to St. Thomas Aquinas (*c.*1225-1274) says:

'... there is but one vase, one substance, and one way and only one operation'.

Before a start could be made on the processes of the Great Work, it was necessary to find the *materia prima* or subject, sometimes confusingly called the Stone of the Philosophers. This had the outward characteristics that are common to many ores. It would also appear necessary to journey to the mine, where this material was to be found, and personally take possession of the raw material. This must be done at the most favourable time as calculated by casting a horoscope. It is uncertain whether this refers to the technique of mundane astrology or to the natal horoscope of the alchemist himself (and of his *Soror Mystica*). It should be remembered that in earlier centuries people were very conscious of the influence of the heavenly bodies and

The *Mutus Liber* frontispiece

for any major event in their lives would consult the nearest astrologer for a horoscope or advice on the appropriate talisman to wear. The **'Diary of Simon Forman'**, recently published by the historian A.L. Rowse, is of particular interest in this respect.[4]

Once the ore had been obtained, it was to be purified by conventional metallurgical techniques before starting upon the preparation of the Stone. The First Work was completed when the starting material had been sufficiently purified by repeated distillation and solidification, and reduced thereby to a pure 'mercurial' substance. The second degree of perfection was obtained when the substance had been cooked, digested and fixed into an incombustible 'sulphur', while the third stage appeared when the substance had been fermented, multiplied and brought to the ultimate perfection. This took the form of a fixed permanent tingent tincture – the Philosopher's Stone.

The progress of the Work was marked by colour changes within the reaction vessel. The first stage ended in blackness, the *nigredo*; the second in the appearance of whiteness, the *albedo*; and the third in redness, the *rubedo*.

The various alchemical operations leading to the completion of the *nigredo* were: calcinations to reduce the purified *material prima* to the consistency of a fine powder or ash; solution in which the calcined powder was dissolved in a mineral water '... that shall not wet the hands', separation into its basic components of mercury and sulphur. Here the alchemists emphasised that these were not the common elements of that name; conjunction in which the 'mercury' and 'sulphur' were married and conjoined together, followed by putrefaction, in which the reconstituted subject in its conjoined state is killed by the continuous application of moist heat when it decays to a black featureless mass.

The 'mineral water that does not wet the hands' is also known as the 'secret fire', although it has been suggested this may be produced from cream of tartar and spring dew that has been collected very much in the manner depicted in the *Mutus Liber*, by pegging out linen cloths in a field

Detail from the *Mutus Liber*

overnight during the months of March and April. The alchemist collects the deposited dew by wringing out the cloths before the warmth of the sun has had time to dry them again.[5]

After the *materia prima* at the *nigredo* stage has been finely powdered, mixed with the secret fire and moistened with dew, it is hermetically sealed into an egg shaped reaction vessel (the Philosopher's Egg) and placed in the athenor or furnace which is capable of continuous and regulated heat from about blood heat to very much higher temperatures, probably above that of molten lead or tin. Nowadays, electric thermostatic heating has greatly simplified the operation and solved the difficulties experienced in the past by alchemists who found it imperative to maintain a constant temperature over prolonged periods using only fuels such as coal or charcoal.

The basic reaction in the vessel was between the hot male sulphur and the cold female mercury, sometimes called the King and Queen respectively. This reaction resulted in death followed by decay, which lasted until '... all is putrified and the opposites dissolve in the liquid *nigredo*. This darkness, darker than darkness, is the first sure sign that the alchemist is on the right path. Hence was applied the aphorism 'no generation without corruption'. The *nigredo* phase ended with the appearance on the surface of the reactants of a 'starry aspect', the so-called metallic volatile humidity, the Mercury of the Wise.

When Archibald Cockren carried out this work he wrote:

> 'The first intimation I had of this triumph was a violent hissing, jets of vapour pouring from the retort and into the receiver like sharp bursts from a machine gun, and then a violent explosion, while a very potent and subtle odour filled the laboratory and its surroundings. A friend has described this odour as resembling the dewy earth on a June morning, with the hint of growing flowers in the air, the breath of the wind over heather and hill, and the sweet smell of the rain on the parched earth.'[6]

As the heating of the vessel continued, the volatile principle or spirit that had left the *nigredo* rises and falls again as the new earth. The spirit apparently recombines gradually with the killed subject, and as the outer fire in the athanor is slowly intensified the moist *nigredo* gives rise to a dry continent within the vessel. This is accompanied by the appearance of a great number of beautiful colours, culminating in the white of the *albedo*. This stage has also been called 'The Peacock's Tail'.

The third stage of the work is to some extent a recapitulation of the earlier stages with such added ingredients as silver, mercury, arsenic, antimony or tin. Despite the maxim *nihil extraneum* (nothing from outside), the matter in the vessel is fed with these additives or with the secret fire, the King being reunited with his blessed Queen in its presence and from this fresh union the ultimate perfection is brought about and the Philosopher's Stone is born.

The process leading to the final *rubedo* involved cibation in which the vessel was fed with the secret fire and other ingredients, sublimation in which the solid matter was heated until it vaporized and the vapour on cooling returned to the solid state; and fermentation in which the material turned yellow. At this stage gold was sometimes added to hasten the development of the Stone. Finally came the exaltation resulting the colour change to the glorious red of the *rubedo*. The material in the vessel had at last become the Red Tincture, or Red Elixir, the ONE. There only remains the stages of multiplication, in which the quality and quantity of the Stone is raised and, ultimately, projection, the actual process of adding a portion of the Stone, wrapped in wax or paper, to the molten base metal to be transmuted into gold.

A fairly typical example of an alchemical operation is given in a clearly phrased book spuriously ascribed to Raymund Lully (1225-1315) and dated 1330:

> 'The 33rd Experiment of Sol - Take aqua fortis (nitric acid) with his form, as I have taught you before, and in it dissolve three ounces of Lune (silver); then purify it twenty days, then take three ounces of Sol (gold) and dissolve it in eighteen ounces of the same aqua fortis with his form in yet four ounces of the fixed salt of Urine (common salt or sodium chloride)

ought to be first dissolved, as you have it in his experiment. Then putrefy these two bodies by themselves severally for twenty natural days. (This results in solutions of silver nitrate and gold chloride.) Then enanimate them both severally by themselves as well Lune as Sol, (Both solutions are distilled to dryness) even to the rule delivered you before. Now when everyone shall be exanimate by themselves and their quickened waters (fairly strong nitric acid and nitric acid containing some chlorine) shall be severally kept by themselves, and also the earth shall yield no more smoke (the dry salts are heated until they give no more fumes), then shall there be a sign that the Sol and Lune do suffer eclipse. Beat then the earth of either of them, and likewise mingle them in a little glass ball strongly luted (they are then ground, mixed and heated in a glass vessel). Put them to a fire of reverberation twenty-four hours'.

'Then take it out and give it first the water of Lune quickened and rectified first seven times by ashes. (The nitric acid is distilled seven times on the gentle heat of ashes.) And when it shall have drunk up all his water by little and little, in the same order as you had it before in the other experiments, then give it the Water of Sol, without any rectification, by little and little after the order which you kept in imbibing that earth with Water of Lune. (Add the chlorine-containing gold distillate.) Then you shall ferment it in this manner. Take one part of Sol and three parts of Mercury and one part of the Medicine, that is to say as much as there was gold. Put it all together in a glass vessel upon warm ashes, and in a short time it shall be turned into powder (gold amalgam with a mixture of gold and silver salts). Then you shall incere it with the third Oil of Sol (a solution of gold chloride in a mixture of hydrochloric and nitric acids). Now when it is all very well cerated and brought into the form of an oyle, project one part thereof upon one hundred of mercury: and it shall all be turned into Medicine. Of which take again one part and project upon five hundred parts. It shall turn the mercury itself into Sol better and purer than mineral gold.[7]

Chapter 6

An Interpretation of Alchemy

The Occult Theory of Alchemy

One of the most important contributions to the understanding of the alchemical work was the publication in 1850 by Mary Anne South, *later* Mrs. Atwood, of a massive prose work entitled *A Suggestive Enquiry into the Hermetic Mystery with a Dissertation on the More Celebrated Alchemical Philosophers*. Mary Anne, the daughter of Thomas South of Bury House, Gosport in Hampshire, was born in 1817 and, at the time of writing this her second book, was a comparatively young woman. Thomas South has been described as 'a gentleman of leisure and certain means, a scholar and somewhat of a recluse, and the possessor of an exceptionally fine specialised library of classical, philosophical and metaphysical works'. He was a member of a circle that performed experiments in hypnosis and spirit communication and had become convinced that modern students of the paranormal were simply re-discovering secrets known to the ancients. He was in an ideal position to investigate this hypothesis and in these studies was actively assisted by his daughter Mary Anne.

In summarising their work in this field the father embarked on the writing of a long epic poem about the ancient wisdom while the daughter communicated her findings in prose. Her book was eventually finished and published and about one hundred copies were sent out to libraries and reviewers. At this stage, Thomas South, who had apparently not read the work, studied it and believing that it contained undue disclosure of the ancient secrets, immediately went to the enormous trouble of calling in all the copies that had been sent out and burning them in a bonfire on the lawn of Bury House. A few copies managed to survive this holocaust and with the approval of the author an amended copy of the book was finally reprinted in 1918 with an introduction by W.L. Wilmhurst.[1]

This re-issue had a mixed reception. Arthur Edward Waite (1857-1942) in his book *The Secret Tradition in Alchemy* treats the work unduly critically and condenses the detailed analysis in Wilmhurst's introduction, based on the conclusions of the work, as follows:

1. Hermeticism, 'or its synonym Alchemy', is primarily the science of the soul's regeneration.
2. In a secondary sense it is a method of raising metals and other sub-human things to a nobler form.
3. It postulates an event corresponding to the theological fall of man.
4. A supernatural principle submerged in man can be awakened and a way is thus provided for escaping the consequences of this event.
5. The supernatural principle cannot be awakened, however, by evolutionary processes alone.
6. There must be assistance from outside Nature, and this is the office of religion.
7. The proper function of religion is to bring about a second birth of man.
8. A 'definite and exact science' of rebirth and its process has been always in the world.
9. It was 'possessed and administered' by the Mystery Schools of Antiquity.
10. There was never a time when it could be taught except in secret.
11. Were it known generally now it could not be put in general practice.

12.   The reason is 'moral unpreparedness'.
13.   Personal and general perils are alike involved, 'apart from
      the privacy' which attaches to something called 'sacrosanctities'.
14.   The process is affirmed to lay open the 'most secret recesses
      and properties of the human organism'.
15.   The candidates for the second birth are said to have been 'prepared'.
16.   They were prepared in order that they might be identified with
      'the universal substrate of life', which is the Light termed 'Life
      of Men' in the fourth Gospel, the Azoth, Magnesia and First Matter
      of the Alchemists.
17.   It is the 'primal fire' and 'free ether'.
18.   Alchemy is the science of this ether, which man also has
      within himself because he is 'the measure and image of the
      universe'.
19.   The antique sanctuaries purified the natural man and set
      free the 'divine ether' within him.
20.   By the concentration of his spiritual energies its light
      became polarized within him, and was thus consolidated into
      a Philosophical Stone, otherwise a vehicle of consciousness
      and a body of regeneration.
21.   The process for discovering the Stone 'was accomplished in a
      condition of magnetic trance hypnotically induced by some wise
      and skilled operator.
22.   The consciousness of the subject was awakened upon the plane
      of the polarised ether, which was elicited into objective form.
23.   Thus awakened, it beheld 'the interiors and causes of things'
      and could proclaim divine truths.
24.   The enlightened masters of the past were experienced in the
      conditions of experiment and the limits to which it should be
      carried in different cases.
25.   It follows that Hermeticism was a 'science of applied magnetism'.[2]

A study of *The Suggestive Enquiry ...*, however, produces a rather different impression. Mary Anne South not only gave examples of the physical production of gold from base metals by the alchemists but offered the view that the basic aim of the alchemist was to discover the *material prima* from which the Stone might be prepared.

Modern Work on the Transmutation of the Elements

On 28 November, 1936 Lord Rutherford (1871-1937) delivered the Henry Sidgwick Memorial lecture at Newnham College, Cambridge and later expanded the subject matter of the lecture into a book entitled *The Newer Alchemy*.[3] In this he concluded '... that there is good evidence that an isotope of gold can be formed by bombardment of platinum with fast neutrons, but it is uncertain which of the isotopes of platinum is involved'. Since that time nuclear physics has developed considerably and the structure of the atomic nucleus has gradually been elucidated.

In chemistry, an atom is defined as the smallest part of an element that can take part in a chemical reaction. Atoms of the same element usually combine to form stable molecules, the smallest part of a chemical substance that can exist independently. Rutherford's (1911) model of an atom comprised a massive nucleus no larger than 10^{-12} cm in diameter and possessing a positive charge. About this central nucleus was a cluster of negative electrons and outside this was an outer group of a few electrons which were much less rigidly attached to the atom. These were the valency electrons. The old concept that these electrons rotated about the nucleus in a similar manner to a miniature solar system of a sun with its encircling planets has now largely been replaced by the electrons occupying not orbits but probability functions which depend upon the energetic state of the atom. As the radius of an atom is about 10^{-8} cm, it is obvious that an atom must have a very 'empty' structure. It has been calculated that the total volume of the nucleus and electrons in a single atom is only about 10^{-36} cc, compared with the total effective volume of 10^{-24} cc for the atom as a whole.

If an atom is regarded as a dense central nucleus surrounded by a number of orbital electrons, the structure of the nucleus is important. Originally, it was thought to consist solely of protons as these were the lightest positively charged particles known prior to 1932. The mass of a proton is approximately unity on the ordinary atomic weight scale and it carries a single positive charge. A nucleus could not consist entirely of protons, however, as this would result in the positive charge on the nucleus being the same as the atomic weight, whereas, in fact, it is rather less, about half of the atomic weight. After the discovery of the neutron, also of unit mass, in 1932, it was proposed by Werner Heisenberg (1901-1976) that the nucleus contained only protons and neutrons. He showed by means of wave mechanics that the attractive forces existing between these elementary particles would be sufficient to account for the existence of stable nuclei.

Rutherford also showed that by bombarding a target of lithium, for example, with protons accelerated to high energy in a cyclotron a lithium nucleus could be transformed into two helium nuclei. The use of other high energy particles resulted in the building up of elements of higher atomic number. He was of the opinion that in the interior of a hot star like the Sun, where the temperature is very high, the protons, neutrons and other light particles present must have thermal velocities sufficiently large to produce transformations in the material of the Sun. There must be a continuous process of building up of new atoms and of disintegrating others to give rise to the range of elements found on earth at the time, 3,000 million years ago, when the earth separated from the sun.

If lead is to be transmuted into gold, for example, some nuclear process must take place which will perhaps result in the formation of lithium as a side product. The nuclear reaction might be expressed:

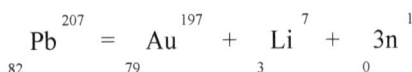

$$\mathrm{Pb}^{207}_{82} = \mathrm{Au}^{197}_{79} + \mathrm{Li}^{7}_{3} + 3\mathrm{n}^{1}_{0}$$

the exact nature of this process is, of course, uncertain. It may be purely nuclear, in which the lithium may subsequently transform into two helium atoms or alpha particles that, having a limited range, would be absorbed in the walls of the reaction vessel or crucible. Again, the process may be psycho-nuclear in which some concentrated force from the mind of the alchemist embodies itself in the material of the Philosopher's Stone and can thus bring about an otherwise unlikely transformation. We just don't know!

Alchemical Symbolism

Throughout its history alchemy has employed symbolism of a highly pictorial kind to express ideas that it wished to keep undisclosed to the uninitiated or popular world. In the earliest manuscripts, such as those attributed to Cleopatra, the figure of a serpent or dragon is used to represent matter in its imperfect or unregenerate state. This serpent or dragon had to be slain, implying that the metals which are the subject of the Art must be reduced to a non-metallic state and rendered susceptible to receiving a new spirit. The dragon has a sister Mercury and as he is the matter, metal and body so his sister is spirit, metallic mercury or soul.

The second great symbol of alchemy is the *mysterium coniunctionis*, the 'Chymical

Marriage' or combination of Sol and Luna, 'our gold' and 'our silver'. This appears with a frankness of sexual symbolism`. Sol is to impregnate Luna in order to generate the Stone.

In the Middle Ages the concept of impregnation and generation was very different from that of today. It was symbolised as death followed by resurrection because at the back of the alchemist's mind was the teaching of St. Paul in the New Testament:

> 'Whatever you sow in the ground has to die before it is given new life and the thing that you sow is not what is going to come.' (1 Corinthians, ch. 15, v. 36)

The product of the Chymical Marriage is the hermaphrodite or 'rebis', which containing elements of both partners is symbolised as a dead body, a hermaphroditic corpse in a sepulchre, becoming black and putrefied. Yet the grave is a place of renewal and not destruction and the soul arises into heaven and the celestial power descends as dew upon the earth.

Another symbol is that of birds, eagles, who fly up to heaven and descend again. These are an obvious symbol for sublimation, distillation and for any process whereby a 'spirit' is raised from a body. If a symbol, such as a hermaphrodite, possesses wings this always implies volatility or a spiritual body in which the spirit has mastery over all the elements.

The tree is an important symbol of the Work, growing out of the earth, which is mineral, and bearing fruit, which is spiritual. There are in the literature pictures of trees bearing flasks or birds, and also fruit representing the Sun and the Moon, symbols both of gold and silver and of the Red and White Stones. The alchemical reaction vessel is often depicted as an egg which, after incubation, hatches into the glorious child. In the treatise *The Splendor Solis* of Salomon Trismosin the egg is used as a symbol of the four elements and of the quintessence. The shell represents the earth, the space between the shell and the inner membrane the air, the liquid 'white', water and the yolk fire. The quintessence was located in the middle of the yolk.

The 'Green Lion' represented the devouring corrosive acid, no doubt because of the colour imparted to it either by copper compounds, always present as an impurity in gold and silver, or by solution of the oxides of nitrogen in the nitric acid.

The most important symbolism used in alchemy, however, was that based on religious concepts. Devotion to God was never far from the mind of the medieval alchemist. Life and death were very real to him and there was no blasphemy in applying religious symbolism to everyday affairs. Thus the death of Our Lord Jesus Christ and his Resurrection in a glorified body could be, and was, compared with the death of base metals and their resurrection in a transformed state either as gold or, better still, as the Philosopher's Stone itself. The Stone was often represented as the Risen Christ.

Alchemical engravings and pictures were not intended to be interpreted in words in the same way that modern chemical symbols are. They were meant for looking at, for meditation upon and for deriving from them a degree of 'innerstanding'. The concepts of cause and effect and the classification of chemical changes had yet to be achieved. The world view of the alchemist was very much more primitive. He knew about life and death and the duties owed to God and his neighbour. It was only when expressed in these terms that alchemical processes and operations became intelligible to him. Carl Jung (1875-1961) has related the individuation processes of modern psychology to the symbolism of hermetic philosophy and alchemy. As Jolan Jacobi comments:

> 'For probably alchemy was not at all a matter of chemical experiments but, in all likelihood like psychological processes expressed in pseudo-psychological language. And the gold sought was not the ordinary aurum vulgari but rather the philosophic gold or even the 'marvellous stone', the lapis invisibilitatis, the 'alexipharmakon', the 'red tincture', the 'elixir of life'.[4]

It is true that there are similarities between the pictures produced by psychotic patients and those of the alchemists, but the former have neither the depth nor the implicit symbolism of the latter which are often works of art designed like icons as gateways to the inner world. This is not to say that Jung was misguided in his approach to alchemy. In the absence of a scientific dogma to guide him the alchemist must have depended upon his own mental processes for guidance. These processes resulted in the type of symbolism that he produced and the continuous meditation upon these symbols resulted in a feed-back which enhanced the whole process and resulted in the fascinating series of engravings and drawings which make alchemy a bridge between the celestial and terrestrial worlds.

The Chymical Marriage

Throughout the alchemical literature there are references to the alchemist being assisted in his operations by his wife or female companion, his *soror mystica*. The classical example is that of Nicholas Flamel and his wife Perrenelle who transmuted half a pound of mercury into gold on 17 January 1392 - 'Perrenelle only being present'. The *Mutus Liber* includes the alchemist and his wife in many of its plates and it would appear that her participation was very much a part of the process. Thomas Vaughan (1622 - 1666), the alchemist and poet, mentions in his *Aqua Vitae* that:

> 'On the same day my deare wife sickened, being a Friday, and at the same time of the day, namely in the evening: my gracious God did put into my heart the Secret of extracting the Oyle of Halcali, which I had once accidentally found att the Pinner of Wakefield, in the Days of my most deare Wife.'[5]

Waite was of the opinion that this 'Oyle' was a reference to the *prima materia*.

A more modern case is that of Armand Barbault who, in his book *Gold of a Thousand Mornings* states:

> 'On Sunday, August 3rd, 1947 I went for the first time with my female collaborator to the site in order to inspect the ground and fix the exact point at which, from that day on, intensification of certain currents should take place. These currents were destined to preside over the most daring attempt we could ever make: to seize the First matter and capture the etheric forces, which would then gradually increase in intensity during irradiation of the Matter'.

The symbolism of the *Mysterium Coniunctionis* has led some commentators to ascribe the alchemical secret as being essentially sexual in nature, the various vessels employed being symbolic of the male and female genitalia. While it is not intended to develop this approach to any great extent it should be mentioned that the tantric forces of a psycho-sexual nature are of such intensity that they may, perhaps, be able to trigger some psycho-nuclear processes.

Magic and Alchemy

There are close relationships between Magic and Alchemy for the latter involves the induction of chemical change by means of operations upon the spiritual plane while the former involves change brought about by means of the will of the magician. The four major divisions of the Hermetic Art are, of course, Alchemy, Astrology, the Kabbalah and Magic. These four divisions are not, however, self-contained and each is involved to some extent with the other. An alchemist might be expected to have a good practical knowledge of all four branches of Hermeticism.

In the Golden Dawn manuscripts an apparently elaborate process is described for carrying out an alchemical operation. As far as is known this aspect of the Golden Dawn's work was never carried very far during their early years and 'Aleister' Crowley (1875-1947) ventured the opinion that the process as set out was complete nonsense. It is doubtful whether he employed so mild a description! Since that time, societies deriving from the Golden Dawn have experimented in this field, although with little comment upon any results obtained. Details are given of one such attempt by the writer Francis King in a chapter of his book *Ritual Magic in England*.[6] These details relate to the preparation of an elixir of life type of product rather than of the Philosopher's Stone, although similar steps might well be expected to be used but with an initial invocation of Solar force instead of the planet Jupiter.

The Golden Dawn operation is carried out in a temple arranged as for the grade referred to as 0 = 0. After the usual banishing rituals are performed, including the invocation of the spirit pentagram, the first matter (of undisclosed nature) is placed in a flask and the planetary force 'Sol' is invoked. The flask is then sealed and heated gently on a water bath for three days. The flask is then unsealed and a distillation head fitted. The contents are then transferred by distillation into another flask. The residue, the *caput mortem*, is removed from the still, ground to a fine powder and returned to the flask with the distillate. This flask is sealed and placed on the flashing tablet of the planet Sol while the general forces of the associated kabbalistic sephira are invoked. In the ritual so far there are a number of magical terms which require some explanation:

invocation - to call down a force from the macrocosm
evocation - to call up a spirit or being, and
flashing tablet - a tablet of wood or wax some 4.5 inches in diameter which has equal areas of two contrasting colours associated with the planet or element to which they are consecrated and the sigil of the planet involved. The tablet has been charged with the force from the planet in a manner similar to that adopted in the preparations of talismans.

As the operation continues, the flask is allowed to stand on a water bath until the contents turn black (the *nigredo*), when it is placed on the North side of the altar and the planet Saturn evoked. The distillation, grinding of the *caput mortem* and recombination is then repeated. The flask is transferred to the west of the altar and the *Cauda Draconis* invoked. The flask is then exposed to moonlight for nine nights while placed on the flashing tablet of Luna, the first night being that of the full moon. The distillation, grinding and recombination is again repeated. The flask is next placed on the east of the altar and the waxing moon of *Caput Draconis* invoked. The flask is then replaced on the flashing tablet of Luna and exposed to moonlight for a further nine nights, ending at the full moon. The contents are again redistilled and recombined.

The flask is then moved to the south of the altar at a time when the Sun is in the constellation Leo and the forces of Tiphareth and Sol are invoked. The flask is placed on the flashing tablet of Sol and exposed to sunlight from 08.30 to 20.30 hours for six days. The elemental from the material in the flask is next invoked to check if the correct state has been reached and, if not, the lunar and solar workings are repeated. Mars is then evoked and the flask placed between the black and white pillars of the temple while standing on the associated flashing tablet. The contents of the flask are again distilled and the distillate kept separate from the residue.

After invocation of the forces of Mercury the liquid is exposed to the Sun for a further eight days while the residue is ground up and, after invoking the forces of Jupiter, left in the dark for four days on the flashing tablet.

The next stage of the operation takes place in the presence of the flashing tablet of the elements and of the elementary weapons; the dagger, the cup, the wand and the pentacle which represent the four elements – fire, water, air and earth respectively and, incidentally, also the Grail Hallows and the Tarot suits. At this stage the Greater Ritual of the Pentagram is performed and fire invoked to work upon the solid residue while water is invoked to work on the distillate.

The flashing tablets of the elements are then replaced by the white/gold tablet of Kether, while the alchemist identifies himself with his Holy Guardian Angel and invokes the sephira. Liquid and solid are recombined and exposed to the rays of the Sun for a further ten days. The flask is finally placed on the flashing tablet of Venus and the forces of the planet invoked. After a further seven days heating on a water bath the product is distilled, hopefully, to give the Elixir.[7]

Astrology and Alchemy

The use of astrology in alchemy has been amply illustrated in the work of the contemporary alchemist Armand Barbault who used astrological charts to arrive at the time when carrying out the operation would be most successful. In the *Splendor Solis* of Solomon Trismosin the various regimens are associated with corresponding astrological signs and planets and it is generally held that the operations of the Secret Art should be commenced in the Spring with the Sun in the sign of Aries the Ram, and dew plentiful upon the ground.

The correspondence between the planets and the seven metals of antiquity has been pointed out in an earlier chapter.   It was there emphasised that the operations of alchemy occur on two planes and the importance of the Hermetic doctrine 'as above so below'.

The Kabbalah and Alchemy

A.E. Waite, that profound scholar of the Secret Tradition in all branches of Hermeticism, gives an appendix on Kabbalistic Alchemy in his work *The Secret Tradition in Alchemy*. He mentions that the connection between these two pillars of Hermeticism is a traditional notion which is taken for granted and referred to frequently in the literature concerning transmutation. There is, however, only one tract relating to this subject the *Aesh Mezareph* or 'Purifying Fire' which was probably compiled in the sixteenth century and which ascribes the metals to the Kabbalistic Tree of Life. This ascription is as follows:

KETHER (The Crown) = Metallic Root

CHOKMAH (Wisdom) = Lead	BINAH (Understanding) = Tin,
CHESED (Mercy) = Silver	GEBURAH (Severity) = Gold
TIPHERETH (Beauty) = Iron	NETZACH (Victory) = Brass
HOD (Glory) = Brass	YESOD (Foundation) = Mercury

MALKUTH (The Kingdom) = Medicine of Metals

A second scheme of allocations derived from the same source has:

KETHER = Philosophical Mercury,

CHOKMAH = Philosophical Salt	BINAH = Philosophical Sulphur
CHESED = Silver	GEBURAH = Gold
TIPHERETH = Iron	NETZACH = Tin
HOD = Copper	YESOD = Lead

MALKUTH = the Metallic Woman (Moon Lady)

Waite was of the opinion that in so far as the *Aesh Mezareph* was a Kabbalistic light on Alchemy it was such only on the basis of being an analogy between things above and things below.

Chapter 7

Alchemy – Hoax or Hyperchemistry?

It would appear from the consensus of modern scientific opinion the transmutation of molten base metals into silver or gold by the addition of any type of substance or catalyst such as that called the Philosopher's Stone is impossible. Although, as has been noted, Lord Rutherford presented evidence as long ago as 1936 for the formation of an isotope of gold by the bombardment of a platinum target with fast neutrons, the production of massive quantities of gold by the techniques said to have been employed by the alchemists has yet to be achieved under strict laboratory conditions![1] In the present chapter the claims of the alchemists will be examined, as will the methods by which some unscrupulous deceivers sought to falsify and hoax unsuspecting and gullible patrons.

The first alchemists of which there is any detailed knowledge at all derive from the Egypto-Greek world and centred upon the city of Alexandria which was founded in 332 BC. Their writings, contained in the great papyri, are largely concerned with aurifiction, the fabrication of metals resembling gold. Aurifaction, the process of transmutation of base metals into gold in accordance with Empedocles' (c.495-435 BC) concept of matter based on the four elements (Earth, Air, Fire and Water) came later, as did the concept of the Philosopher's Stone as a transmuting agent with its implications as an elixir for longevity. The physical transmutation of base metals into gold became the symbol of man's regeneration and transformation to a nobler and more spiritual state and a link was then formed with the Christian religion: the Lapis-Christi parallel with Christ as the Philosopher's Stone. Thereafter alchemy developed a spiritual and mystical side in which the work in the laboratory was carried out in conjunction with that in the oratory – *laborare est orare*!

It was in the twelfth century of the present era that Western alchemy, which had almost lapsed following the destruction of the extensive library at Alexandria in the 4th century AD, received stimulation from an unexpected source. The enthusiasm of the Abbasid caliphs of Baghdad following the formation of the new religion of Islam in the fifth century AD had resulted in the translation into Arabic of many of the alchemical manuscripts of the ancient world. A study of these manuscripts by Western scholars and their subsequent translation into Latin was to stimulate afresh the study of alchemy and, at the same time, to introduce concepts derived from the East where alchemy was pursued not only for wealth in the form of gold but also for longevity!

Alchemy appears always to have been attractive to the human mind. It was a subject that gave rise to a number of fascinating phenomena associated with the Great Work, some of which were concerned with the basic psychology of man himself. The search for the Philosopher's Stone was long and complex and while frustrating at times must have given the alchemist considerable personal satisfaction. Its secret nature must have had a fascination of its own. In a world ravaged by plague, wars and brutality, and possessing a more clearly defined social structure than our own, the prospect of an increase in status by attracting the interest of one's peers or even of the princes of the courts which ruled the land, could well have been a motive which would attract not only genuine aspirants but also the less honest members of society. There is little doubt that there were a number of genuinely honest alchemists throughout history who laboured to disclose the wonders of God's creation, but the literature of the past also contains a number of accounts of clever rogues who adopted a pretence of the art to deceive others to their own advantage!

It is from the world of Islam that one of the earliest accounts of attempts at alchemical trickery derives, and is delightfully described by E.J. Holmyard (1891-1959) in his book *Alchemy* published in 1957.[2]

'An ingenious pseudo-alchemist arrived in Damascus with a quantity of gold filings which he mixed into a paste with charcoal, various drugs, flour and fish glue. He then rolled the paste into small pellets that he allowed to dry. Clothing himself as a Dervish he took the pellets to a druggist and sold them for a few pence under the name of 'Tarbamaq of Khorassan'. Next he assumed a rich cloak, engaged a servant, and went to the mosque, where he scraped acquaintance with several notable persons. He told them that he was an expert alchemist and could make untold wealth in a single day, a boast that soon came to the ears of the Vizier, who ordered his presence at the court. The Sultan expressed his desire to witness a transmutation, which the charlatan readily agreed to demonstrate if he could be provided with the requisite chemicals. The recipe he produced included a certain amount of Tarbamaq of Khorassan, and while all the rest of the drugs were easily obtained no trace of Tarbamaq could at first be discovered. The charlatan insisted, however, that Tarbamaq was essential, and when the druggists' shops had been well searched the discovery was at length made – of course in the shop of the druggist to whom the Tarbamaq had been sold earlier, and who said that he had obtained it from a Dervish.'

'The pellets were bought, and the charlatan ordered the ingredients to be placed in a crucible and strongly heated. When all was sufficiently hot, "Take out the crucible", he said. It was taken out, cooled and turned upside down, when a fine lump of gold rolled out.'

'The Sultan deeply impressed by the success of the experiment, ordered the self-styled alchemist to be rewarded, and the next step was now to find a further supply of Tarbamaq. Search failed to reveal even a single further pellet in Damascus, but the alchemist said he knew of a cavern in Khorassan where large supplies were to be found. He suggested to the Sultan that an expedition should be sent to bring back a goodly quantity of the rare substance. But the Sultan – no doubt reacting just as the alchemist had foreseen – was unwilling that the provenance of Tarbamaq should become known to everyone else, and commanded the alchemist to go alone. After a show of reluctance the man accepted, and was furnished by the Sultan with everything needed for the journey: a tent, a travelling kitchen, sugar, carpets, stuffs and silks, manufactured objects from Alexandria, and, in addition, a large sum of money. Thus equipped he set out – and that was the last that was seen of him!'

A tale worthy of the *Arabian Nights*!

The literature of alchemy contains a number of descriptions of the methods adopted for transmutation of base metals into gold. Eirenaeus Philalethes writing on 'projection' in his work *An Open Entrance to the Closed Palace of the King* gives the following account:

'Take four parts of your perfect stone, either red or white (of both for the Medicine); melt them in a clean crucible. Take one part of this pulverizable mixture to ten parts of purified mercury till it begins to crackle, then throw in your mixture, which will pierce it in the twinkling of an eye; increase your fire till it be melted, and you will have a Medicine of an inferior order. Take one part of this, and add it to a large quantity of well purged and melted metal, which will thereby be transmuted into the purest silver or gold (according as you have taken white or red sulphur). Note that it is better to use a gradual projection for otherwise there may be a notable loss of the Medicine. The better the metals are purged and refined, the quicker and more complete will the transmutation be.'[3]

Nicholas Flamel (*c.*1330 - 1418) in his *Exposition of the Hieroglyphical Figures* tells of projection in the following words:

'The first time I made projection, was upon Mercury, whereof I turned half a pound, or thereabouts, into pure silver, better than that of the Mine, as I myself assayed, and made others assay many times. This was upon Monday, the 17th January, about noon, in my house, Perrenella only being present, in the year of the restoring of mankind, 1382.' (4)

These published accounts of transmutations using the Philosopher's Stone sound as precise as the account of any chemical experiment but there is no means currently available for establishing the truth of the observation. There was and still is considerable interest in the alchemical literature, particularly in the 17th century when the printing presses of the time were in peak production to supply the demand of the literary public. It may well have been a reaction against the new enlightenment which had displaced alchemy and magic from its part in the understanding of God's creation. In Victorian times occultism surfaced as a reaction against the impact of science on faith, particularly Darwinism. Just as the non-fiction publications reproduced the alchemy of the past so the fiction literature of the time also makes reference to alchemical transmutation or the search for the Stone.

Geoffrey Chaucer (*c*.1433-1400), the author of the *Canterbury Tales*, had a good knowledge of the science of his day as shown in his *Treatise on the Astrolabe*. In the 'Canon's Yeoman's Tale' he gave an account of the contemporary practice of alchemy and of the wiles of its practitioners. The Canon in the story is the alchemist and the Yeoman his laboratory assistant. As Chaucer describes it, the alchemical scene in the fourteenth century contained both 'puffers' and charlatans. The former as bellows blowers were earnest and honest seekers after the Philosopher's Stone and as Professor John Read has said:

'Their unending labours, although arbitrary and uninformed, and usually activated by sordid motives, were sustained by a fervent faith in the existence of a potent transmuting agent.'

The Yeoman possessed a discoloured face due to his continual labours in the laboratory, where he was exposed to the heat and smoke from the furnace. He goes on to recount how they borrowed gold from their patrons, convincing them that 'of a single pound we can make tweye' and even the Yeoman himself contributed 'Al that I hadde, I have y-lost therby'. Employed by the Canon for some seven years, apparently without wages, he goes on to describe his work in the laboratory: 'I blowe the fyr til that myn herte feynte.' It is this sentence that epitomises work in practical alchemy – the laboratory resembled a blacksmith's forge with hearth, bellows and even an anvil and a selection of his tools.

The principal apparatus. beside the Athanor or furnace, was the crosslet or crucible with urinals (small flasks), alembics (still-heads) and receivers of various kinds. Heating by fire was continuous, requiring constant vigilance. The temperature had to be strictly controlled to achieve success and that without such modern aids as electrical heating elements and thermostats!

In the second part of the tale the Yeoman gives an account of the activities of a dishonest Canon who set out deliberately to deceive, unlike the Yeoman's master who was a genuine puffer labouring hopefully, although beggaring himself as well as his clients. This false Canon (the foule feend him fetche!) set out to gain the confidence of a priest by means of a series of rigged experiments. In the first of these he showed the priest a crosslet and invited him to put some mercury in it and he would add a small quantity of the Stone ('I have a powder here that cost me deere'), but thought by the Yeoman to be either powdered chalk or glass. While the priest was setting the crucible on the fire the Canon produced a piece of charcoal in which some silver had been concealed and the hole stopped with wax. Distracting his attention the Canon placed the false charcoal in the fire above the crucible and plying the bellows with vigour caused the molten silver to replace the mercury that had by then volatilised from the crucible. This left the impression that the mercury had been transmuted into an equal amount of silver! Further experiments involved the concealing of the silver in the Canon's sleeve or in a hollow stirring rod stopped with wax.

The degeneration of alchemy into pure confidence trickery is well displayed in that wonderful play by Ben Jonson (1572-1637) *The Alchemist* in which the character Subtle and his companion Dol Common join Face, the housekeeper, in taking over the residence of his master, one Lovewit, who had left London for fear of the plague. The audacious trio set out to obtain money from a range of people using the vocabulary of alchemy to 'lend an air of verisimilitude to an otherwise bald and unconvincing narrative'!

However, there are several, presumably non-fictional, narratives of transmutations carried out by those not professing to have prepared the Philosopher's Stone themselves. One of the most interesting of these is that due to Johann Friedrich Schweitzer, known as Helvetius, (1625 -1709). He gives an account of a meeting he had with a stranger who came to his house on 27 December 1666. The stranger's object was to enlighten Helvetius who had, in his writings, expressed the view that the Grand Arcanum of the Sages might be only a gigantic hoax. Helvetius, demonstrating a degree of gullibility, identified the stranger as the Artist Elias whose coming was, apparently, prophesied by Paracelsus. The stranger demanded a gold coin and while Helvetius was looking for one, produced a green silk handkerchief in which were folded up five medals 'the gold of which was infinitely superior to that of my gold piece'. These medals had inscriptions glorifying God and implying that they had been made of alchemical gold on 26 August 1666. The stranger admitted that his knowledge of these matters had, in fact, been bestowed upon him by a friend, his Master, who had demonstrated to him the transmutation of lead into gold. A length of lead water pipe had been melted in a pot

> '... whereupon the Artist had taken some sulphurous powder out of a little box on the point of a knife, and cast it into the melted lead, and that after exposing the compound for a short time to a fierce fire, he had poured forth a great mass of molten gold upon the brick floor of the kitchen.'.[5]

The stranger departed and returned after three weeks when he gave Helvetius a small piece of the Stone 'no larger than a grain of rape seed' which when Helvetius expressed the doubt that this would be sufficient 'to tinge more than four grains of lead' demanded it back and divided it into two with his thumb-nail, throwing away one half and returning the other saying 'Even now it is sufficient for you.' He instructed Helvetius to melt half an ounce of lead in a crucible and to throw in the Stone, protected against the fumes of lead with yellow wax. Helvetius apparently had confessed to having extracted a few crumbs of the Stone during Elias' previous visit but having had no success with an attempted transmutation. The Stranger departed promising to return on the following day at the ninth hour when he would demonstrate the manner of projection. He did not return, and Helvetius began to doubt 'of the whole matter'. His wife, who he tells us 'was a most curious student and enquirer after the art, whereof that worthy man had discourse', was of sterner stuff and suggested that at least he try and make the projection for himself. Helvetius ordered a fire to be prepared while his wife

> '... wrapped the said matter in wax, and I cut half an ounce or six drams of old lead, and put it in a crucible in the fire, which being melted, my wife put in the said medicine made up into a small pill or button, which presently made such a hissing and bubbling in its perfect operation, that within a quarter of an hour all the mass of lead was totally transmuted into the best and finest gold, which made us all amazed as planet-struck.'

Helvetius ran immediately to the nearest goldsmith who judged it as excellent gold, while next day the general Assay Master (Examiner of Coins in this Province of Holland), a Mr. Porelius, suggested that the gold be subjected to the usual trials. This they did and found that when a part of the gold was mixed with three or four parts of silver, laminated, filed and granulated and dissolved in nitric acid, the precipitated gold was not only of high purity but had increased in weight due to a scruple of the silver having been transmuted by the remaining amount of the 'medicine'.   The gold was then mingled with seven parts of antimony, which was melted and poured into a cone, and the regulus blown off in a test,

> '... where we missed eight grains of our gold, but after we blowed away the rest of the Antimony, or superfluous scoria, we found nine grains of gold more for our eight grains missing, yet this was somewhat pale and silver-like, which easily recovered its full colour afterwards.'

They concluded that in total 'the said medicine (or elixir) had transmuted six drams and two scruples of the lead and silver into most pure gold'.

This account, taken from that of the bookseller William Cooper's *The philosophical epitaph of W.C(ooper) [...] also, a brief of the Golden Calf...* by J.F. Helvetius, London (1674-5), was reproduced in publications by E.J. Holmyard[6] and F. Sherwood Taylor[7] whilst slightly different versions are given in the *Hermetic Museum* and in Manget's *Bibliotheca Chemica Curiosa.*[8] Sherwood Taylor concludes from this extract that '... there seems in this account to be no room for any mistake or illusion: Helvetius either transmuted lead to gold or has lied prodigiously.' Holmyard commented 'It is not surprising that a circumstantial narrative of this kind, related by a man of high standing, brought conviction to the minds of many doubters.'

Another scientist of eminence in the 17th century who claimed to have seen and himself performed transmutation of base metals into gold was Johann Baptista van Helmont (1579 - 1644). In his treatise *De Vita Aeterna* from his *Ortus Medicinae* he makes the following statement

> 'I have seen and I have touched the Philosopher's Stone more than once; the colour of it was like saffron in powder, but heavy and shining like powdered glass.I had once given me (by a stranger whom he saw only once) the fourth part of a grain - I call a grain that which takes six hundred to make an ounce. I made projection therewith, wrapped in paper, upon eight ounces of quicksilver, heated in a crucible, and immediately all the quicksilver, having made a little noise, stood still from flowing and congealed into a yellow mass. Having melted it in a strong fire, I found within eleven grains of eight ounces of most pure gold, so that a grain of this powder would have transmuted into very good gold, nineteen thousand one hundred and fifty-six grains of quicksilver.'[9]

Professor J.R. Partington (1886-1935) was of the opinion that van Helmont was not free from the superstition and credulity of his time any more than Boyle for his later time.[10] Sherwood Taylor expressed the view that van Helmont's account has internal evidence against it. Since the density of gold is 19.3 and that of mercury 13.6, the sudden conversion of the latter into the former should be accompanied by a shrinkage of about one third in volume, which would be very noticeable, although it is not mentioned by van Helmont.

In Robert Boyle's *Dialogue on the Transmutation and Melioration of Metals* as reconstructed recently by Professor Lawrence Principe[12] there is a lengthy narrative, spoken by the character Philoponus, which appears to be Boyle's own eyewitness description of a transmutation of lead into gold. As summarised by Principe, Philoponus, while visiting a 'forraigne doctor of Phisick' was introduced to a certain foreigner who attempts to show him a method whereby lead can be turned into running mercury. The experiment miscarries when the crucible is upset and a servant is then sent to obtain further lead and crucibles. When the fresh sample of lead was melted, the foreigner, after offering to allow Philoponus to perform the operation himself, cast a small quantity of red powder onto the molten metal (no wax or paper involved!) and covered the crucible. After heating strongly for fifteen minutes, the crucible was allowed to cool and on knocking out the cooled metal Philoponus was amazed to find a ponderous yellow metal which he later found to be pure gold!

It was perhaps this incident that finally convinced Boyle of the reality of the claims of the alchemists – but was this not the whole object of the exercise? It is known that the first part of the experiment doubtless involved a sample of lead, provided by Boyle but this was lost in the accidental upsetting of the crucible. Was it established that the second sample of lead was really only lead – might it not have been a gold-rich lead mixture? If the demonstration took place away from Boyle's usual laboratory, as the text suggests that it did, from where did the servant obtain the supplies? The history of deception has shown that even scientists can readily be deceived by those skilled in the art!

Throughout the centuries there have been those who claimed the transmutation of base metals into gold was possible and that they had themselves achieved it following successful preparation of the Philosopher's Stone. Others, more sceptical, maintained that transmutation was impossible and that the whole subject was based upon a confidence trick to obtain money from rich patrons. If, as we now know, a modern stage magician could substitute a crucible of gold-rich lead for the base metal and deceive his audience into believing that transmutation had taken place, then all such transmutations could have been the subject of such trickery!

Some of the principal works on alchemy,[13] notably the *Summa perfectionis* of the Latin Geber, begin with a discourse on the arguments of the sophists denying the truth of the art. The present state of chemical knowledge while defining the atomic constitution of the elements and their compounds pays little attention to the building up of complex atoms from the basic elements of hydrogen and helium at the high temperatures pertaining in the Sun and over prolonged periods of time from which the elements of the periodic table are derived. Minerals were found in the rocks and strata of the earth and both philosophers and the alchemists tried to understand how this diversity of substances was produced. As animals and vegetables exhibited a diversity of forms so also did minerals. The metals known to the alchemists: gold, silver, copper, iron, tin, mercury and lead (perhaps not forgetting antimony) are those most readily available in the ancient world and associated with such heavenly bodies as the Sun, Moon, Venus, Mars, Jupiter, Mercury and Saturn respectively, Gold, probably the most beautiful and chemically resistant metal, occurred naturally in its metallic form, sometimes alloyed with small amounts of silver or copper, in alluvial deposits and rocks from which it could be extracted by crushing, washing and refining. The fact that these minerals and ores were dug out from the earth and that gold was the perfect metal would imply that Nature produces metals over many thousands of years while the alchemist sought to bring about this same perfection in his laboratory in a matter of weeks or even days. In Avicenna's *Congelatione et Conglutina Lapidum* (in the *Kitab al-Shifa),*[14] he divides mineral bodies into four groups – stones, fusible substances, sulphurs and salts - and describes the geological processes in the depths of the earth that result in the formation of the then known metals – gold, silver, copper, tin, lead and iron. Surprisingly, however, Avicenna denies the possibility of transmutation in the laboratory! He argues that artificial and natural products are intrinsically different and that Art is inferior to Nature and while the alchemist may produce passable imitations of the precious metals these are not the same as the real thing. He says:

'As to the claims of the alchemists, it must be clearly understood that it is not in their power to bring about any true change of species. They can, however, produce excellent imitations, dyeing the red white so that it closely resembles silver, or dyeing it yellow so that it closely resembles gold. They can too, dye the white with any colour they desire, until it bears a close resemblance to gold or copper; and they can free the leads from most of their defects and impurities. Yet in these the essential nature remains unchanged; they are merely so dominated by the induced qualities that errors may be made concerning them, just as it happens that men are deceived by salt, qalqand, sal-ammoniac etc.'

The alchemists themselves were aware of Avicenna's views and put forward counter-arguments which asserted that man's ability to transform the natural world was virtually unbounded.[15] The arguments against transmutation or aurifaction can be briefly summarised – the incredibility of the change, the stubborn nature of the metals, the small proportion of the Philosopher's Stone employed and the celerity of the action.

Principe is of the opinion that:

> '... the fact that some of the phenomena described in alchemical texts - even those dealing with the arcana maiora - can be successfully reproduced in the modern chemical laboratory clearly refutes the claim of those who assert that alchemy had no real experimental component. Had the alchemists not actually performed and described real laboratory operations, or had they been mere victims of self-induced delusional states, or described the 'irruption' or 'projection' of the unconscious on matter (as many followers of Carl G. Jung claim), they obviously could not have accurately described actual, reproducible chemical phenomena.'

Despite the obscurity of the writings of many of the alchemists, it has been found possible to follow their procedures in a modern laboratory 'transforming from the obscure and foreign language of alchemy into a form comprehensible and meaningful in a chemical sense'. Principe has drawn attention to the fact that the chemical substances used by the alchemists were not of the same degree of purity as those used today and that these impurities, unknown to the alchemists, contributed to the results obtained. He attempted to follow a recipe of Basil Valentine given in his *Triumph-Wagen antimonii* for producing glass of antimony, extracting this glass with 'well-rectified, distilled vinegar' and from this forming 'tincture of antimony'.

He found that in order to produce a transparent yellow glass it was necessary to add silica to the antimony ash formed by roasting antimony trisulphide slowly in air. In following Valentine's next step of extracting the glass in a curcurbit with distilled vinegar, he found it necessary to add ammonium acetate to the vinegar prior to distillation. Basil Valentine was aware, for example, that *aqua fortis* could be 'strengthened' to dissolve gold by the addition of the substance he referred to as 'Salmiac', composed mainly of ammonium chloride and ammonium carbonate. Principe found that when the distillate prepared in this way was added to the powdered glass of antimony the extraction readily occurred giving a product which was, in the words of Valentine, 'coloured a beautiful deep-yellow tending to redness like a lovely, clear, transparent gold.' The final stage in the operation was to isolate this coloured material and dissolve it in water to form a 'tincture'. Using a water bath, as prescribed by Valentine, a reddish powder was obtained which could be dissolved in absolute alcohol giving a red solution and leaving some residue behind. This solution on analysis, however, was found to contain no significant quantity of antimony, the major component being ferric acetate! Principe concluded that the iron came from the impurities in the native stibnite and in the work of Valentine from the possibility he had used an iron rod to stir the ore during calcining. This work has shown that the operations of the alchemists can be translated into modern chemical terminology if due account is taken of the condition of the materials they used.[16]

It seems possible that a number of the operations carried out by the alchemists can be duplicated in a modern chemical laboratory if only the composition of the reagents used originally and the coded symbolism of their texts are interpreted correctly. There remains, however, a side to the chemical reactions involved that has little in common with conventional chemistry. This side might be referred to as 'hyperchemistry'. Examples of this 'so-called' hyperchemistry may well include the recent phenomena of anomalous water, cold fusion and the extreme dilutions of the medicines of homeopathy, in addition to the seventeenth century claimed transmutations of base metals into silver or gold. Attempts to formulate the transmutation of mercury into gold results in a hypothetical equation such as:

$$^{200}_{80}\text{Hg} \rightarrow \, ^{197}_{79}\text{Au} + \, ^{1}_{1}\text{p} + 2\, ^{1}_{0}\text{n}$$

Some radiation in the form of protons and neutrons might be produced along with the gold, which would not be noticed. Dependant upon the rate of the transmutation reaction, the contents of the crucible would be subject to a considerable increase in temperature during the emission of the high energy radiation, although nothing can be predicted concerning the energy produced in the reaction.

The alchemists always held that Art copies Nature and that which was achieved in the bowels of the earth in thousands of years might be achieved in the laboratory in a very much shorter time. But how were the chemical elements formed in the first place? Even modern geochemistry has little to say on the subject. It is now suspected that the elements were built up by a series of thermonuclear reactions at very high temperatures when protons or hydrogen nuclei could interact to form deuterium and then helium. In fact, from a study of the spectra of the stars and nebulae it has been concluded that hydrogen is the most abundant element in the universe and may comprise some three quarters of its mass. Helium atoms account for most of the remainder and all other elements contribute only slightly more than one per cent of the mass. Studies of solar spectra have led to the identification of nearly seventy of the known elements. If the planet Earth was formed from matter removed from the Sun, where the necessary high temperatures pertain, then the elements were distributed in its crust as the liquid mass solidified from the outside. Studies by F.W. Clarke and others have established that oxygen accounts for about 47% of the earth's crust by weight, while silicon comprises about 28% and aluminium about 8%. These elements plus iron, calcium, sodium, potassium and magnesium account for about 98% of the earth's crust.

Lord Ernest Rutherford, writing in 1936 expressed the opinion:

'In the interior of a hot star like our Sun, where the temperature is very high, it is clear that the protons, neutrons and other light particles must have thermal velocities sufficiently high to produce transformation in the material of the sun. Under this unceasing bombardment there must be a continuous process of building up new atoms and disintegrating others, and a stage at any rate of temporary equilibrium would soon be reached. From a knowledge of the abundance of the elements in our earth, we are able to form a good idea of the average constitution of the sun at the time, 3,000 million years ago, when the earth separated from the sun'.[18]

As the legendary alchemist of modern times, Fulcanelli, has remarked:

'Chemistry, incontestably, is the science of facts, just as alchemy is that of causes. The first, confined to the material domain, is supported by experiment. The second preferably takes its directives from philosophy. While the object of the first is the study of natural bodies, the other tries to penetrate the mysterious dynamics which preside over their transformations. Therein lies their essential difference, enabling us to say that alchemy, compared with our positive science, the only one permitted and taught today, is a spiritualistic chemistry, for it allows us to catch a glimpse of God through the darkness of substance.'[19]

The alchemists sought to understand how the various metals were formed in the earth and to copy this process in their laboratories.

The philosophers of ancient Greece sought to explain the diversity of material things in Nature in simple terms. To them the 'world of appearances' was brought about by one or more 'roots' or simple substances termed 'elements' by Plato. hales (624-526 BC) identified water as the unique element from which everything was formed, other philosophers based their reasoning on different candidates. By the fifth century BC Empedocles (*c*.492-432 BC) had developed a comprehensive cosmology based upon the four elements – Earth, Air, Fire and Water – whose commingling in different proportions, separation and motions under the power of 'love' (attraction), 'hate' (repulsion) explained all material substances! However, Aristotle (384-322 BC) imparted a degree of mutability to the four element theory by associating each element with its tactile quality. Thus fire was hot and dry, air moist and hot, water cold and moist and earth cold and dry. Each element could be transformed into any other by the exchange of one or more of its qualities. This theory of the elements, which dominated European matter theory until the 16th century, was to become the theoretical basis of alchemy.

The alchemists of Islam, such as Avicenna and Jabir ibn Hayyan, believed that all minerals were formed not in the sun but in the bowels of the earth by the co-mingling of dry and moist exhalations. They identified these exhalations with the 'principles' of Sulphur and Mercury. All metals were supposed to be formed from, and hence composed of, these two principles. Paracelsus in the sixteenth century added a third principle that of solidity which he called 'Salt' to give the *Tria Prima*. There were thus a number of theories involving two, three, four or even five principles. The latter was formed by the addition of the Quintessence to the four elements. These were later rejected on biblical grounds by van Helmont who pointed out that God made the Universe from Air and Water.

In 1661 Robert Boyle argued in *The Sceptical Chymist* that lacking general agreement on the components of matter it would be better to abandon the concept altogether for a corpuscular theory of homogeneous matter possessing only geometrical and mechanical properties. The corpuscular theory that was fundamentally consistent with laboratory experience, was given a rigorous mathematical framework by Newton in his *Principia*. Both Newton and Boyle were, however, strong believers in the claims of the alchemists – that matter could be transmuted and base metals achieve perfection as gold.

Gold has always been a special metal with almost religious rites adopted during its refining.[20] All other metals easily become tarnished with an oxide layer but gold remains untarnished, beautiful and unique. It has been calculated that all the gold in the world, so far refined, could be placed in a single cube of approximate dimensions 50ft. x 50ft. x 50ft. With a specific gravity of 19.32 this corresponds approximately to 68.4 thousand tonnes.

In recent years much effort has been spent in the study of nuclear fusion particularly at the Fusion Research laboratory at Culham in Oxfordshire. Here the Zeta machine was constructed to contain plasma at extremely high temperatures by means of magnetic fields. Under these conditions it was hoped that deuterium atoms could be induced to fuse together to produce helium, emit a neutron and large amounts of heat energy.

In 1989 two scientists, Martin Fleischmann and Stan Pons, of Southampton and Utah Universities respectively, claimed to have achieved fusion in the laboratory at room temperature! They used a cell containing lithium, heavy water and electrodes of platinum and palladium. It appeared to them that under the influence of powerful electric currents the deuterium produced by electrolysis of the heavy water was absorbed by the palladium and a fusion type of reaction occurred leading to the release of large amounts of heat (and neutrons). This claim was contested by some of the leading fusion research centres and several attempts were made to reproduce the findings at Utah. One, at the Lawrence Livermore National Laboratory in California, lead to an explosion in the apparatus used! Cold fusion was never confirmed and became something of a 'nine days wonder'. It was, however, noted in *The Times* that the price of palladium on the London Metal Exchange leapt after 23 March of that year from $145 to $179 an ounce after years of stagnation! Nothing further has been heard of this cold fusion process, despite experiments supervised by Professor Martin Fleischmann at the UK Atomic Research Authority's Harwell laboratory with a team led by Professor David Williams and massive screening comprising paraffin wax and concrete surrounding the electro-chemical reaction vessel - about the size of a milk bottle.

Another interesting 'red herring' of recent years was the discovery of 'anomalous water' that it was claimed had properties differing considerably from those of ordinary water. Produced in capillary tubes it demonstrated higher surface tension and the ability to dissolve silica from the tube. Scientists generally were naturally sceptical about these claims, attributing them to impurities introduced into the water. The discovery of anomalous water is, however, of relevance as the alchemists have always attached importance to the solvent properties of dew. Prince

Stanislas Klossowski de Rola describes alchemical dew as the 'Vitriol of the Wise' and as 'The Philosopher's Green Lion'. Morning dew is a component of the Secret Fire. The collection of dew by the technique so beautifully illustrated in the *Mutus Liber* was vital to the processes of the alchemists and was adopted quite recently by Armand Barbault, as described in his book *Gold of a Thousand Mornings*. The difference between dew and rain water is immediately apparent to those taking dogs with furry feet, such as Cavalier King Charles Spaniels, for their walks along grass edged pavements! Feet wet with dew become far dirtier than those moist with rain!

Barbault collected his dew by sliding lengths of fine but porous canvas up and down rows of 'green yet fairly stiff corn' when the canvas became saturated with the dew drops which formed on the tips of the plants. Every twenty metres or so they would stop and squeeze the canvas out over a glass vessel. While the method depicted in the *Mutus Liber* – lengths of canvas pegged out on the grass – resulted in dew of a purer, more etheric composition, the dragging method gave higher yields. The dew was used to act as a basis for fermented vegetable matter and to nurture his *prima materia* (the alchemist's first matter). It would appear that this *prima materia* was a sample of clean healthy soil taken from a special place at a time predicted by astrology and by the alchemist and his wife (*soror mystica*) in a special manner. This material was nourished with dew and with the sap of growing plants and used to produce a 'gold elixir' with health giving properties. According to Barbault :

> 'All the evidence leads us to believe that the action of the vegetabilised gold takes place at the level of the etheric forces, that is to say at a higher level than that reached by normal medicine'.[21]

Chemical analysis of the product failed to disclose any quantity of gold in the elixir although considerable quantities of the metal were used in its preparation. There is an analogy here with homeopathic preparations in which the effect of the remedy is inversely proportional to its substance - the smaller the amount of material that is present in the dose, the more potent is its effect! Homeopathy is another subject that proves anathema to the scientific community as in some cases the dilutions used in treatment are such that no molecules of the 'active product' are present!

The question has yet to be answered: did any alchemist actually prepare the Philosopher's Stone and were base metals ever transmuted into silver or gold? It is known when such transmutations were said to have taken place the gold produced was even purer than that used for coinage and it was usually subjected to rigorous analysis and assay. On a number of such occasions coins and special medals were struck to commemorate the occasion. Clearly, if the gold used for these artefacts was of such high purity then chemical analysis of such coins and medals still preserved in museum collections would indicate not only if this were so, but might even identify the source of the gold used. Following a survey of alchemical numismatics carried out by E.J. Holmyard last century, a study of coins and medals made of alchemical metal has been made by V. Karpenko of the Charles University in Prague.[22] Although it was not found practical to make analyses of the gold involved, a large silver medallion, partially transmuted into gold, was of sufficient size that samples between 15mg. and 30mg. could be taken from different parts of the medallion and analyses performed. The results were disappointing. Analysis of a typical sample gave – silver 42.04%, gold 49.04%, copper 7.21%, the remainder being small quantities of zinc, iron and tin. If the original medallion had been composed of pure silver this is very encouraging. Despite the large amounts of coinage said to be produced from alchemical gold and medals struck to mark demonstrations of transmutation it would seem that in only one case has the density been measured. A gold ducat made of gold produced by an Augustinian monk, one Wenzel Seyler, in 1675 and preserved in the Imperial cabinet of Coins in Vienna, was examined in 1888 and found to have a specific gravity of only 12.7. The specific gravity of gold is 19.3.

Chapter 8

Aspects of Greek Alchemy

It would seem that the first alchemists of which we have any detailed knowledge at all derive from the Egypto-Hellenist world. Western alchemy began in Hellenistic Egypt and was centred upon Alexandria and other towns in the Nile Delta. Alexandria, founded in 332 B.C., developed into the most important city of the ancient world and, under its rulers, the Ptolemies, gathered together a great library that attracted scholars from all parts of the Greek world. It was about this time the concepts of aurifiction and aurifaction were arising in India and the Far East. These concepts were, however, of less importance than longevity to the oriental mind, as was the achievement of immortality and the search for the Elixir of Life. Some knowledge of this quest may have travelled with the merchants along the trade routes of the ancient world to contribute to the alchemical goals of the West.

The background to the Greek alchemists were the teachings of their earlier philosophers such as Orpheus and Pythagoras (6th century BC), Plato (*c*.428-347 BC) and Aristotle (384-322 BC). The teachings of these early philosophers are interwoven into the fabric of alchemy.

Orpheus the Thracian[1]

Orphicism, as the teachings of Orpheus were later to be called, may be summarised in three basic precepts:

1. The soul is of celestial origin: a fallen God.
2. The body is the sepulchre of the soul and the soul only reveals its true nature when out of the body in a state of ecstasy (stepping out).
3. The soul can be restored to its former state by ascetic life, sacraments and purification which release it from the 'wheel of birth' which restrained the original soul to a succession of reincarnations in human, animal and even vegetable forms.

It is thought that the underlying concept of these doctrines is Indian in origin and may have reached Greece by way of Persia.[2]

Pythagoras of Samos

Pythagoras taught a rather similar doctrine, a way of life in which release from transmigration, the cycle of reincarnations, was again to be obtained by asceticism, purification, ritual and the study of philosophy, particularly mathematics. These principles are closely related to Orphicism and his followers, the Pythagoreans, have also been called reformed Orphics. Pythagoras was not without influence on Plato and his school.

Plato the Athenian

Plato was born about 428 BC of a distinguished Grecian family and in due time became a pupil of the great philosopher Socrates. Indeed, much of the detail of the life and teachings of Socrates is owed to Plato who, in the dialogue entitled *Phaedo*, has given us a sublime account of the last hours and death of Socrates. In 387 BC Plato visited Sicily and Southern Italy and on his return to Athens founded the Academy in about 385 BC. This was to continue as a famous seat of learning of the ancient world until its dissolution by Justinian in 529 AD.[3]

Teaching in the Academy was by lecture and discussion, and included the subjects of philosophy, mathematics and biology. Like Pythagoras before him, Plato had practical ideas on the formation of an ideal state and in 367 BC he journeyed to Syracuse with the object of founding a Philosopher's State under Dionysius. Unfortunately, the whole project was a failure and by 360 BC Plato had returned to Athens and resumed his teaching at the Academy.

It is said that Plato made contact with the secret tradition of the ancient world when he visited Egypt as a young man about 398 to 395 BC, travelling to Heliopolis and conversing with its priests. While the fact of this visit is still controversial, Plato frequently mentions Egypt in his discourses. For example in the *Timaeus* he mentions that the priests claimed to have written records over 8,000 years old, and in the *Critias* in his account of the inundation of the City of Atlantis he refers to the Egyptian records transcribed by Solon:

> '... and if I can recollect and recite enough of what was said by the priests and brought
> hither by Solon, I doubt not that I shall satisfy the requirements of this theatre'.[4]

The philosophy of Plato is a tremendous subject which has stimulated and influenced many generations of students since his time up to the present day. He was the first author to give a synthesis of earlier Greek philosophy combined with his own, and for this he drew upon four main sources – the Orphic and Pythagorean, that of Parmenides (*fl. c.*480 BC) (reality is eternal and change illusory), that of Herakleitos of Ephesus (*fl. c.*500 BC) (nothing in the sensible world is permanent) and that of his teacher and friend Socrates, (the world was created so as best to serve the purpose of man; the real is the good).

Plato's concept of the origin of the universe and of the elements, not forgetting the five regular Platonic bodies, contributed to the theory of transmutation in alchemy. In the dialogue called *Timaeus'* Plato gives an account of the creation of the world and of all that therein is. The dialogue is one of extreme verbal complexity, being partly mythical, or figurative, and partly literal. It searches into the concepts which are upon the frontiers of human intelligence.

Before embarking on the description of the creation of the universe, Plato asks and answers four questions, here reproduced at length as they convey the atmosphere of philosophical discussion so favoured in an earlier age when time's winged chariot was not so insistent!

> 'Firstly, is the world created or uncreated? It is created, being visible and tangible, and having a body and therefore sensible; and if sensible then created; and if created made by a cause, and the cause is the ineffable father of all things who had before him an eternal archetype. For to imagine that the archetype was created would be blasphemy, seeing that the world is the noblest of creations and God is the best of causes.'

> 'Secondly, why did the Creator make the world? He was good and therefore not jealous, and being free from jealousy he desired that all things should be like himself. Wherefore he set in order the visible world which he had found is disorder.'

> 'Thirdly, in the likeness of what animal was the world made? The form of the perfect animal was a whole and contained all intelligible beings, and the visible animal made after the pattern of this, included all visible creatures.'

> 'Fourthly, are there many worlds or one only? In order that the world might be solitary, like the perfect animal, the creator made not two worlds or an infinite number of them, but there is and ever will be one only-begotten and created heaven.'

In Plato's account of creation, which follows, it is made clear, or as clear as Plato can make it, that God first created the world soul, which he then diffused throughout and over the surface of the body of the universe, the perfect animal. So that the macrocosm, although not called that, was also a body-soul relationship similar to that of the microcosm, thus implying the grand hermetic doctrine - as above so below.

The Body, being visible and tangible, was composed of the elements FIRE and EARTH, but as these two elements could not rightly be put together without at least a third to act as a cement, God placed the other two elements AIR and WATER between them and arranged them in a continuous proportion mathematically:

FIRE : AIR : AIR : WATER : and
AIR : WATER : WATER : EARTH

These elements were taken into the universe body whole and entire to ensure the perfection of the animal by leaving no remnants from which another such animal could be created. The animal or body of the universe had no limbs or organs as these functions were not required of the animal, there being no use for them.

The creation of the Soul is more difficult to comprehend. It was formed by the mixing of the UNCHANGEABLE and INDIVISIBLE with the DIVISIBLE and CORPOREAL, and with a MEAN NATURE derived from them. This was then operated upon mathematically in a complex way, which involved dividing the mass into portions related to each other in the ratios of 1 : 2 : 3 : 4 : 9 : 8 and 27. This series is derived from the two Pythagorean progressions: 1 : 2 : 4 : 8 and 1 : 3 : 9 : 27 of which the number 1 represents a point, 2 and 3 lines and 4 and 9 and 8 and 27 the squares and cubes respectively of 2 and 3. This series is thought to represent the diatonic scale in music, the order and distances of the heavenly bodies, principally the sun, moon and the planets, and to make an allusion to the music of the spheres.

After this mathematical operation performed upon the primary soul compound, it was divided lengthwise into two parts that intersected at a point, but formed two great circles or spheres. The OUTER CIRCLE or SPHERE OF THE SAME, and the INNER CIRCLE or SPHERE OF THE OTHER. The former rotated horizontally to the right and formed a belt wherein the fixed stars were located, a forerunner of the astrological zodiac, while the latter rotated diagonally to the left and was sub-divided into seven orbits round the earth containing the Moon, Sun, Mercury, Venus, Saturn, Mars and Jupiter. The inner motion he divided in six places and made seven unequal circles having their intervals in ratios of 2 and 3, three of each, and made orbits proceed in a direction opposite to one another; and three he made move with equal swiftness to the three and to one another but in due proportion.

The body, soul and the universe having been created, God made them eternal by introducing the concept of time. The planets performed uniform motion according to number and this in different ways, indicating the passage of months (the moon), days (the sun), and years (the progress of fixed stars in the zodiac). Time was also divided into past, present and future.

Plato observed that these divisions all applied to becoming in time and had no meaning in relation to the eternal nature which ever is, and never was or will be, for the unchangeable is never older or younger.

The universal animal was so far made in the DIVINE image, but other animals were to be included: Gods, Birds, Fishes and Beasts. The Gods were formed as fixed stars in the zodiac. From the remains of the soul compound left over from the mathematical process of creation, God made souls in number equal to the number of the stars, each soul being assigned to a star (foreshadowing Aleister Crowley's *Book of the Law*!). These souls were to be implanted in bodies and permitted a range of sensations that included Love, a combination of pleasure and pain, Fear and Anger, and their opposites (not specified). The reactions of the souls to these sensations during life, whether they overcame them or submitted to them, would determine whether after life they would return to their native star, or be reincarnated into some lesser form, specifically stated as a woman or animal, etc., until such time as the soul achieved mastery over the sensations or achieved a righteous life.

The different elements in the universe – fire, air, water and earth – are based, according to Plato, on geometrical combinations of two basic types of triangles, one derived from a Square and the other from the most pleasing form of Isosceles Triangle. This concept was based on the supposition that as the elements are bodies and bodies are solids and solids are formed by the intersection of planes and plane rectilinear figures are made up of right-angled triangles of the two sorts. By combinations of these triangles regular geometrical figures were formed; the equilateral triangle, the square and the pentagon: from these figures the five regular solid bodies can be constructed. Those formed from triangles are the Tetrahedron (fire), the Octahedron (air) and the Icosahedron (water); from the square a Cube (earth) and from the pentagon the Dodecahedron (the sphere of the universe or soul).

Plato further subdivided the elements. Fire was divided into flame, the light that burns not, perhaps 'phosphorescence' observed in rotten wood and in some gem stones, and red heat of the embers of a fire. Air was divided into pure ether, opaque mist and various nameless forms (perhaps gases), water in liquid was formed from small unequal particles and fusile water formed from large uniform particles. Examples of fusile water were gold, adamant (lodestone or diamond) and copper. Juices such as wine, oil or pitch and honey, and vegetable acids that dissolve flesh, were also regarded as examples of such water. Earth comprised materials such as stone and rock, pottery, soda and salt.

Because the elements were composed of these minute basic triangles in geometrical combinations there was apparently no difficulty inherent in separating them from their existing combinations and building them up into new forms. In this way one element could pass into another. From this followed quite naturally the idea of the transmutation of the elements. Alchemy and the basic principle behind it can thus be traced back to Plato, and so also can the close correlation between physical operations upon the body and their corresponding effects upon the soul, both of the world and of the individual.

Plato had much to say about the nature of the soul. In some ways it was to him an abstract entity, the impersonation of ideas, and as ideas are immortal so also must the soul be. The wicked is not released from his evil by death, that would be too easy, but everyone carries with him into the world below that which he is or has become, and that only. After death, the soul is carried away to judgement and when she (the soul) has received her punishment she returns to the earth in the course of ages. Associated with this punishment is the concept of the underworld. This is described in the **Phaedo** as formed from a huge cavern in the surface of the earth, which is pictured as a globe placed in the centre of the heavens. This chasm or opening is called Tartarus into which streams of fire, water and liquid mud are always flowing. There is a perpetual inhalation and exhalation of the air, rising and falling as the waters pass into the depths of the earth and return again in their course forming lakes and rivers, but never descending below the centre of the earth for on either side the rivers flowing either way are stopped by a precipice.

There were four principal rivers – the Oceanus which circles the earth, the Acheron which takes the opposite direction and after flowing under the earth, reaches the Acherusian lake. This is the river at which the souls of the dead await their return to earth. The other two rivers are the Pyriphlegethon and the Cocytus. The former is a stream of fire and the latter falls into a savage region called the Stygian and passes into and forms the lake Styx from the waters of which it gains new and strange powers. These rivers also fall into Tartarus.

The dead are judged according to their deeds during life. Those who are incurable are thrust into darkness. Those who have committed only venial sins are first purified and then rewarded for the good that they have done. Those who have committed crimes, which may not be great but not unpardonable, are cast into Tartarus but carried forth by way of the rivers Pyriphlegethon or Cocytus at the end of a year. These rivers carry them as far as the Acherusian lake, where they call upon the victims of their crimes to pardon them and let them into the lake where their sufferings may cease. If they fail to obtain pardon, they are borne unceasingly into Tartarus and back again until they obtain mercy. The pure souls also receive their reward. They have their abode in the upper earth, while a select few dwell in still fairer mansions.

Plato's concept of the soul emphasises both its aspiration after another state of being and of the necessity of retribution for those souls that fail.   As he writes in a significant passage in the *Phaedo*:

'All virtues, including wisdom, are regarded by the philosopher only as purification of the soul.   Temperance and justice and courage are in reality a purging away of all these things, and wisdom herself may be a kind of baptism into that purity. The founders of the mysteries would appear to have had a real meaning and were not devoid of sense when they intimated in a figure long ago that he who passes unsanctified and uninitiated into the world below will lie in a slough, but he who arrives there after initiation and purification will dwell with the Gods. For many, as they say in the mysteries, are the thyrsus bearers, but few are the mystics, meaning, as I interpret the words, the true philosophers.'

Aristotle (384-322 BC)

Aristotle was born at Stageiros in Chalkidke, northern Greece, an Ionian settlement in Macedonia, at that time under the rule of King Amyntas II, to whom Aristotle's father was physician.   Both his parents died while he was young and at the age of sixteen Aristotle journeyed to Athens with the intention of studying under Plato at the Academy. Plato, however, was at this time conducting his Philosopher's State experiment in Syracuse, so Aristotle spent the next three years in intensive private study and in laying the foundations of what was to become a very fine and famous library. On Plato's return, Aristotle was able to fulfil his original purpose and he spent the next twenty years with Plato, at first as a pupil and then as a friend.[5]

About 343 B.C. Aristotle entered the service of Philip, King of Macedon, as tutor to Alexander the Great, who was at that time only thirteen years old. He remained as royal tutor for the next three years until Alexander became Regent, when he returned to Athens about 335 BC and obtained permission to teach at the Lykeion, a gymnasium attached to the Temple of Appolo – Lykeios. This school was later called the Lyceum and became the site of Aristotle's great library and museum.

On the death of Alexander in 323 BC, Aristotle left the Lyceum in charge of Theophrastos and retired to Chalkis in Euboia were he died a year later of a gastric complaint. The description of Aristotle that has come down to us describes him as thin-legged, bald in later life, speaking with a lisp, noticeably well-dressed and with a mordant wit.

While Plato's teaching seems to have been rather informal, it is known that Aristotle delivered carefully prepared lectures and set problems to his students. He was very much more scientific than his Master and laid more stress on empiricism, the use of observation and experiment rather than on theory. His interests centred more in chemistry and biology and less in mathematics than those of his teacher.   Aristotle in his teaching emphasised that Nature does nothing in vain and that a problem must be studied from all sides and not from one side alone. Experience gives an accumulation of material evidence, which may then be converted into general precepts. Scientific consideration must be limited to what is valid universally.

The following is a synopsis of Aristotle's views on the universe and on the formation of the elements, and how the concept of coming-to-be and passing-away relates to the transmutation of the elements. Some remarks on Aristotle's ideas on the nature of the soul will be included.

Aristotle held that everything comes into being from its opposite, or from some intermediate stage that is derived from that opposite. Pairs of opposites were fairly well established as, for example, full and empty, hot and cold, dry and moist, dense and rare, love and hate, being and not being. All things come from something else and Aristotle distinguished five ways in which things could come into being: by transformation – as a statue is cast from bronze, by accretion – as plants and animals grow, by reduction - as a block of marble is carved into a statue by cutting away the excess, by combination – as blocks of stone are brought together in building a house and by qualitative alteration – as one substance is changed into another by heating or some such process.

There has, however, to be a starting point for this process and for this purpose he defined matter as something from which something comes into being, and which is a component of the thing produced. Matter is not just a purely physical entity, but with other individual things it is striving to become immortal as far as it can. In this it largely fails due to its own nature.

As coming-into-being is a material change, Aristotle proceeded to analyse the process of the change itself. Change pre-supposes a substrate which can take on the change. Everything that is comprises the substance and the form in which the substance clothes itself. A thing cannot come into being because it already exists, nor can a thing come into being from non-being because nothing can come from nothing. Things, therefore, come into being from something intermediate between being and non-being. This something has the potentiality of becoming something else, but is not that thing actually. Matter is imperishable and uncreated; it neither comes into being nor ceases to be.

Aristotle supposed that there are as many distinct species as there are bodies. The substance of a body is made up of matter and form. An element is one of those bodies into which other bodies can be decomposed. It is contained in those bodies, actually or potentially, and is itself incapable of being divided further. It is thus the first inherent component out of which a thing is constructed. Aristotle uses the terms analysis and synthesis much as is currently done for separation of a body into and formation from its elements. Elements are simple bodies, and distinguished from the so-called elements, namely earth, air, fire and water.

Every real thing is made up of the so-called elements in varying proportions. They all contain earth, as this element predominates in this sub-lunary region in which they are found. They also all contain water, since compounds have a definite outline and of the elements water alone has the property of being readily adaptable in shape. Also earth cannot cohere without moisture. Because compounds are made out of contraries, or opposites, they must also contain air and fire, the opposites of earth and water.

All material bodies are cognisable by touch, at least in principle, and all the properties so felt, except weight and levity, which cannot apparently be included since they are neither active nor passive, are – in the end – reducible to four basic properties, all of equal rank. These comprise two active properties, heat and cold, and two passive ones, dryness and moistness. The moist (liquid) has no determinate form of itself, but readily adapts its shape to that of its container, whilst the dry (solid) has a determined form within its proper limits which can be changed only with difficulty. Other secondary properties, such as fine, coarse, brittle, hard, soft, viscous and so on, are all derivable from the basic properties, dry and moist. The four basic properties could give rise to six possible pairs in combination, but since opposites cannot be coupled together as, for example, heat with cold, or moistness with dryness, there remain four pairs:- heat and dryness, corresponding to fire, heat and moistness, to air (steam); cold and dryness, to earth and cold and moistness to water. Each so-called element therefore contains one active and one passive quality. This may be represented on the well-known diagram that has found its way into many works on alchemy (**figure 8.2**).

The heaviness or lightness of a body depends upon its make-up and the elements it contains. Aristotle postulated that the universe had a definite up and down direction, with the element fire rising and that of earth sinking. The other elements behave as might be expected, air rising in water, but sinking below fire and water rising in earth but sinking below the other two (air and fire). This has given rise to their specific dispositions in the universe that, while not developed at any length by Aristotle, approximates to the geocentric world model accepted by Plato. The main difference lies in the concept that the celestial or heavenly sphere above the moon contains the fifth element called 'ether'. This is perhaps Aristotle's major contribution to ancient cosmology.

Below the ether is the terrestrial region, made up of the other four elements in order of their density: fire, air, water and earth, earth being at the centre. The fire and air regions in the universe overlap, there being a greater proportion of fire in the upper and more air in the lower regions of the atmosphere. According to Aristotle, there is only one heaven and nothing, not even a vacuum, outside it. The heaven is ungenerated and indestructible, as is the whole cosmos or universe. The natural places of the elements offer proof that there is only one universe, with the earth at its centre. Because the universe is complete on all sides, it is spherical in shape, but not necessarily infinite in size. The earth at its centre is at rest, and the spherical heaven rotates around it, moved by the unmoved mover, in contact with its exterior limit.

The fifth element, the ether, is a divine substance of which the heavens and the stars are made. Because the ether is devoid of opposites, some doubt has been raised as to whether Aristotle regarded it as an element in the same sense as the other four.

As the ether has a divine nature, however, the question probably does not arise. The sun is regarded as the sole source of light and warmth on earth, its annual movement in the ecliptic or zodiacal circle is the efficient cause of the ceaseless alternation of the transformation phenomena, coming to be and passing away.

There are two aspects of a single transformation which are as follows. Consider two substances A and B. A is not B and B is not A, but B can turn into A and A into B, A, in fact, is potentially B. When B comes-to-be from A, what was potentially B is now actually B. The passing-away of one substance involves the coming-to-be of another.

In the case of the four so-called elements, transformations of this kind are possible and can occur. The transformation of fire into air, air into water, water into earth and back is rapid and easy, but transformations which involve a change of two qualities as in fire into water, and air into earth, are slow and difficult. Apparently Aristotle had not considered consecutive reactions!

The generation of metals and minerals in the earth takes place by the interaction of two exhalations that are given off by the earth in the heat of the sun. The first exhalation is known as the vaporous exhalation, and this reacts with the earth to produce metals (iron, copper and gold), while the second, or smoky exhalation, produces minerals (realgar, ochre, ruddle and sulphur).

By the operation of heat and cold, the two active qualities of matter, on substances changes are produced. The operation of heat is called concoction, while that of cold inconcoction. The former has three degrees: the mildest is termed ripening, the next boiling and the most severe roasting. In concoction the water element in the substance becomes denser and hotter. The corresponding degrees of inconcoction are rawness, scalding and scorching. These operations foreshadow the importance of the degrees of heat in the alchemical processes.

Aristotle held the soul to be the capability of a body to produce changes in itself. An animate being is the union between a body and a soul. When the being is asleep, the soul predominates as the first reality. When awake, however, the body takes precedence. Aristotle recognised three types of soul, the vegetative, the animal, and the intellectual. Furthermore, he divided the faculties of the soul into five major categories:

1. Nutritive, or generative, as in all plants.
2. Sentient, that is, having the power of sense perception, as in all animals.
3. Appetitive, having the desire to satisfy natural instincts, as in some animals.
4. Locomotive, the power of movement as in some animals and
5. Rational, the power of thought in man only.

Man possesses all these faculties and therefore can be considered a microcosm. The human soul is in an ultimate sense that principle by which we live, feel and think. Aristotle sometimes compares the soul with heat or breath and with the ether, the element of the stars.

The union between body and soul, the living being, is not just a compound of body plus soul, but something different. The soul is the moving force of the body that, in turn, is the form of the soul. The body is the instrument of the soul and in life they are inseparable, each body having its own particular soul. There is no transmigration of souls in Aristotle's philosophy. Death and old age he ascribes to the gradual loss of innate heat that, not being completely replenished by natural processes, finally becomes a flickering flame that the slightest puff will extinguish.

There remain two other components of man's being. Between the body and the soul is that which Aristotle calls the innate spirit. This is situated in the heart. It is associated with the mind and with a motive power and is the faculty of thinking, the rational faculty of the soul. It is still uncertain whether he regarded the mind as immortal, when he wrote:

> '... respecting mind and the theoretic faculty, nothing as yet is evident, but it seems to be another kind of soul and it alone is capable of separation as the everlasting from the perishable.'

Plato and Aristotle, though both influenced by the teaching of Socrates and the Orphic and Pythagorean concepts of the nature of reality, were very different in their approach to these subjects. Plato, it seems, experimented with ideas, while Aristotle adopted a more scientific approach. This lead the former to the grand description of the macrocosm and the latter to that of the microcosm.

The Egyptian Source of Alchemical Concepts

The authors of the Greek alchemical texts had no hesitation in attributing the source of their art to ancient Egypt. If alchemy is defined in the words of Democritos as:

> '... an art, purporting to relate to the transmutation of metals, and described in terminology at once Physical and Mystical',

then Egyptian alchemy might be described as mainly aurifiction, the imitation of gold, and not aurifaction – the making of it! Indeed, the step from one to the other, from believing that something that looked like gold was gold, must have been all too easy for the alchemist who lacked the technical training of the artisan goldsmith. Workers in metals had their own rituals and beliefs and were acquainted with assaying techniques such as cupellation which were ignored by the alchemists who occupied a different class in the structure of Greek society. It is known that there were expert goldsmiths in Egypt as early as 3000 BC. and in the Euphrates valley metal workers practiced their craft some 500 years earlier still.

The Ebers Papyrus

This papyrus has been called 'the oldest book in the world' and was found in a tomb at Thebes about 1862. It is now preserved in the University of Liepzig. Written in Hieratic script it dates from about 1550 BC and consists of a roll of papyrus 12 in. wide and 68 ft. long. It contains some 811 prescriptions for remedies for various human ailments and mentions such mineral ingredients as stibnite (antimony sulphide), calamine (zinc carbonate), granite, sulphur, lead, copper, verdigris (copper acetate), lapis lazuli, salt and saltpetre.

The Leyden and Stockholm Papyri

These manuscripts, written in Greek and Demotic script, derive from the early centuries of the Christian era. The Leyden Papyrus is a collection of a large number of recipes for the preparation of metals and alloys simulating gold and silver, together with processes for augmenting these precious metals. The chemical parts of the Leyden Papyrus (Papyrus V) have been translated by Marcellin Berthelot in his *Introduction a l'étude de la chemie des anciens et du moyen age* published in Paris in 1889.

The Stockholm Papyrus was published in 1913 in Upsala as *Papyrus Graecus Holmiensis*, edited by O. Lagercrantz. Like the Leyden Payprus it contains recipes for the preparation or imitation of gold, silver and asemos (a white, silver-like alloy). Metallic gold was not only debased by alloying with other metals but also subjected to surface treatments that enriched the surface layers of a gold-base metal alloy to give the appearance of pure gold and to resist any chemical tests which might prove to the contrary. The following recipe is typical of those given in the Papyrus:

'38.   To give objects of copper the appearance of gold so that neither the feel nor rubbing on the touchstone will discover it; particularly useful for making a fine-looking ring. This is the method. Grind gold and lead to dust fine as flour: two parts of lead for one of gold, then mix them and incorporate them with gum, coat the ring with this mixture and heat. This is repeated several times until the object has taken the colour. It is difficult to discover because the rubbing gives the mark of an object of gold and the heat consumes the lead and not the gold.' [Bertholet, op cit. pp. 37-38][5]

Some records, such as this, survived but many were destroyed by the Romans following a decree of the Emperor Diocletian in 292 AD abolishing the practice of so-called 'alchemy' in Egypt. They are not accounts of the work of alchemists but rather of experienced practitioners in metal working. This, of course, may be the reason of the survival of the Papyri following the destruction of all texts relating to alchemy.

Alchemy as generally known appears to have made its appearance in the Greek-speaking and Latin world only at a comparatively late date. No allusion to it can be traced prior to the Christian era.   An early reference, due to Pliny, relates to the making of gold from Orpiment (arsenic trisulphide) by the Emperor Caligula. Manuscripts referring to Greek alchemy, of which the oldest is of the 11th century in the library of St. Mark's in Venice, includes works dating from the 3rd century AD.[2]   Some 52 alchemical texts are listed and these include works by Stephanos of Alexandria, Zosimos, Cleopatra, Heraclius, Democritus and Archelius.

Stephanos of Alexandria

After Zosimos, probably one of the better known of the Greek alchemists, was Stephanos of Alexandria.   Stephanos lived during the time of the Byzantine Emperor Herakleios I (610-641), known mainly for his defeat of the Persians at the battle of Nineveh in 627 AD and for the restoration of the 'True Cross' to Jerusalem. Herakleios had a considerable interest in alchemy and besides making his own experiments in the art encouraged Stephanos. Stephanos was an authority on the philosophies of Plato and Aristotle and gave public lectures on geometry, arithmetic, astronomy, and music. He wrote several books, including two on alchemy. He was believed to have been the teacher of Morienus Romanus, a traditional figure in Islamic alchemy. The principal work *Stephanos of Alexandria the Universal Philosopher and Teacher of this Great and Sacred Art of the Making of Gold* is divided into nine 'Lectures' all given 'with the help of God'. These are:

1. On the making of Gold
2. The same
3. On the entire World
4. On that which is in actuality

5. On that part of the Divine Art which is in actuality
6. The same
7. The same.
8. On the division of the sacred art.
9. Teaching of the same.

In his translation of part of this work Sherwood Taylor (1897-1956) observed that Stephanos was certainly not a practical laboratory worker but one who viewed alchemy as a mental process. He did, however, give a full exposition of alchemy as was understood in the 7th century AD and his writings are of great importance in the history of the subject.[6,7]

Lecture I is a long declamation on the marvels of alchemy, summarising the concepts of the ancient writers and emphasising the importance of prayer and praise of God, the King of All, and his only begotten Son resplendent before the ages together with the Holy Spirit, before starting work in alchemy. In the words of Stephanos:

'To gather the fairest fruits of the work in hand of this very treatise, and we trust, to track down the truth'.

A typical passage from the first Lecture which has significance for modern ears reads:

'Who will wonder at the coral of gold perfected from thee?   From thee the whole mystery is fully brought to perfection, thou alone shalt have no fear of the knowledge of the same, on thee will be spread the radiant eastern cloud; thou shalt carry in thyself as a guest of the multiform images of Aphrodite, the cupbearer again serving the fire-throwing bearer of coals (then carrying such brightness from afar, in bridal fashion you veil the same, you receive the undefiled mystery of nature.)

The coral of gold must surely refer to the 'Golden Herb' or 'Philosopher's Tree' of Paracelsus, later demonstrated in the work of Archibald Cockren. The 'radiant eastern cloud' can only be the Rising Dawn, the beginning of all wisdom. Even the reference to the bride implies that the *mysterium conjunctionis* is not far from the writer's mind.

The second Lecture has equally suggestive phrases:

'O wisdom of teaching of such a preparation, displaying the work, O moon clad in white and vehemently shining abroad whiteness, let us learn what is the lunar radiance that we may not miss what is doubtful.   For the same is the whitening snow, the brilliant eye of whiteness, the bridal procession-robe of the management of the process, the stainless chiton, the mind-constructed beauty of fair form, the whitest composition of the perfection, the coagulated milk of fulfilment, the Moon-froth of the sea of dawn; the magnesia of Lydia, the Italian stibnite, the pyrites of Achaea, that of Albania, the many-named matter of the good work, that which lulls the All to sleep, that which bears the One which is the All, that which fulfils the wondrous work'.

E.J. Holmyard was of the opinion that:

'By the time of Stephanos, then, alchemy had very largely become a theme for rhetorical, poetical, and religious compositions, and the mere physical transmutation of base metals into gold was used as a symbol of man's regeneration and transformation to a nobler and more spiritual state'.[8]

The second Lecture does, however, give some practical instruction in the *marvellous making of gold* despite the emphasis by the author that as *the wise man speaks in riddles as completely as possible.*

'After the cleaning of the copper, and how is one to clean the copper yet bearing all its ios? How? I will tell you the accurate meaning of the phrase – Aphrodite walking through a cloud. 'After the cleaning of the copper', that is a trituration well managed, a consideration well taught beforehand; "After the attenuation of the copper", that is a finer condition of trituration, he also speaks of the blackness placed upon it and following upon these for the purpose of the later whitening; then is the solid yellowing.   For when it shall spurn the blackness of the wrinkled crust, it is transformed to whiteness; then the moon of shining light shall send forth the rays; then to the later whitening, when you shall see the white compound.'

The word *ios* is an important concept in Greek alchemy and Zosimos used it almost in the sense of a chemical compound, the potency of the metal being concealed in its *ios*. The word is used amongst the symbols in the *Chrysopoea* of Cleopatra.

The Third Lecture continues:

'So copper, like a man, has both soul and spirit.   For these melted and metallic bodies when they are reduced to ashes, being joined to the fire, are again made spirits, the fire giving freely to them its spirit ...   So also copper, being burnt and restored with oil of roses and being expelled, after it has undergone this many times, becomes white without stain, better than gold.'

The Dialogue of Cleopatra and the Philosophers

This work which was cited by Zosimos in his writings is not thought to be composed by Cleopatra, Queen of Egypt, but by an alchemist of the same name. Cleopatra was an ancient Archaian name meaning 'famously descended'. The alembic or still was ascribed to this Cleopatra and was an apparatus common to the laboratories of all early alchemists. Professor John Read was of the opinion that the reference in the *Emerald Tablet* to 'things above and things below' was probably derived from the use of the kerotakis and other forms of circulatory stills. One such still is illustrated in the *Chrysopoeia* of Cleopatra.   This is a particularly interesting document which consists of a single page of symbolic drawings shown in figure 8.3.[9]

At the top left-hand side of the page are two concentric circles enclosing the symbols for gold, silver and mercury. Within the inner circle is the Greek legend which may be translated 'One is the serpent which has its poison according to two compositions' and in the outer circle 'One is All and through it is All and by it is All and if you have not All, All is Nothing'. To the right of these concentric circles is a serpent's tail and a series of symbols of unknown meaning although apparently related to the formula of the 'Crab' in some way. Below the circles is the familiar serpent of eternity the 'Ouroboros' who eats his tail and encloses the Greek words 'en to pan' (One is All). To the right of the serpent is a diagrammatic representation of a still sitting on a sand bath. The small diagrams above the serpent seem to represent reaction vessels and receivers of various kinds.

Zosimos of Panopolis

About 300 AD an author named Zosimos and his sister Theosebeia (his *soror mystica*?) compiled an encyclopaedia (The *Cheirokmeta*) on alchemy in twenty-eight books. Some passages in the work are apparently original, but, according to Holmyard, a large part of the work is a compilation from earlier texts now lost.

Zosimos is considered to be the most important of the Graeco-Egyptian alchemists. His extant works which included his *Authentic Memoirs, On the Evaporation of the Divine Water that fixes Mercury*, and a *Treatise on Instruments and Furnaces* were published by Bertholet and Ruelle in 1887-8. Zosimos lived at a time when the basic concepts of alchemy were emerging as a result of '... the fusion of Egyptian metallurgical and other arts with the mystical philosophies of the Neo-Platonists and Gnostics'.[6] It would seem, however, that the Neo-Platonists regarded matter as the principle of unreality or evil, from which the disciple should attempt to detach himself, while the Gnostics cared little for the phenomena of the sensible world.Zosimos was himself a Gnostic and an outstanding alchemical author of his time. His works made reference to such contemporary workers in the field as Cleopatra, Mary the Jewess and several other alchemists, real or imaginary. He gave a description of the Still and the Kerotakis and is thought to be responsible for the famous 'Formula of the Crab'.

The Formula of the Crab

This mystic formula, said to embody the secret of transmutation, is thought by Read to have been a cipher used by Egyptian craftsmen engaged in aurifictive practices.[7] It has been analysed at length by Marcelin Bertholet and the symbols ascribed as follows:

[1] is equivalent to 'The message begins'
[2] a contraction of the Greek 'to pan' (the all) possibly a
        lead-copper alloy
[3] verdigris (rust of copper)
[4] the double symbol of copper perhaps indicating lead
[5] the crab, or crayfish, the symbol of fixation
[6] perhaps enclose in a vessel
[7] perhaps meaning divide into parts
[8] thought to symbolise weight
[9] signifies '14' relating to the weight
[10] thought to refer to the Philosopher's Egg
[11] repeats the last words of the previous sign
[12] translated as the Greek 'titanos'
[13] again the sign for copper or its oxide/acetate (verdigris)
[14] blessed is he who gets understanding!

It is said that the formula conveys to the initiated a method of colouring base metals by the use of copper compounds, so as to make them resemble gold. The Formula of the Crab reminds us that the ancient symbolism was quite meaningless unless explained in an oral tradition passed down from one alchemist to another.

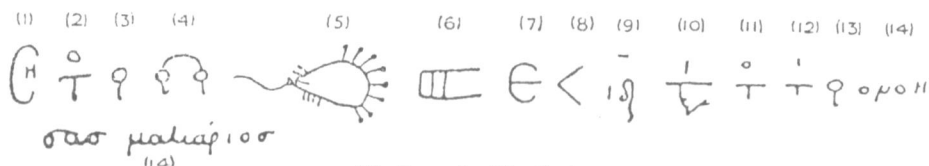

The Formula of the Crab

Zosimos is also responsible for an enigmatic description of the Philosopher's Stone:

'In speaking of the Philosopher's Stone, receive this stone which is not a stone, a precious thing that has no value, a thing of many shapes which has no shape, this unknown which is known of all'

A typical expression of alchemical paradox!

Zosimos on Egyptian Alchemy

Zosimos begins one of his books, which appears to have been dedicated to Theosebeia, his sister, or 'soror mystica' in the following manner:

> 'Herein is established the book of the truth. Zosimos to Theosebeia greetings! The whole of the kingdom of Egypt, lady, depends on these two arts, that of seasonable things and that of minerals. For that which is called the Divine Art, whether its dogmatic and philosophic aspect or its phenomena in general, was given to its wardens for their support; and not only this art, but also those that are called the four liberal arts and the technical manipulations, for their creative capacity is the property of kings. So that, if the kings permitted it, one who had received the knowledge as an inheritance from his ancestors would interpret it, whether from oral tradition or from the inscribed columns. But he who had the knowledge of these things in full did not himself practice the Art, for he would have been punished. In the same way, under Egyptian kings the workers of the chemical operations and those who had the knowledge of the procedure did not work for themselves, but served Egyptian kings, working to fill their treasuries. For they had special masters set over them and a strict supervision was kept, not only upon the chemical operations, but also upon the gold-mines. For if anyone in mining found anything, it was a law among the Egyptians that it should be handed in for entry in the public register'.

The Letters to Theosebeia

The published letters of Zosimos form part of the correspondence with Theosebeia who Zosimos had met during his studies in Constantinople and on his return to Alexandria continued the acquaintanceship as his 'soror mystica'. Unfortunately, it is only the letters from Zosimos to Theosebeia that have survived, her replies to him and other correspondence appears to have been lost. Those letters that do remain give a confused but understandable picture of alchemy as practiced in Alexandria at that time. This would seem to have been largely concerned with producing the appearance of gold rather than with any attempt of transmutation of base metals. Zosimos writes:

> 'Yes I know the great mystery of transmutation. The conversion of baser metal into gold is known to the sages of Alexandria.'

> 'The divine work is accomplished by the arts of the makers by means of the metal-creating stone in Egypt, in Cyprus and in Thrace; but principally at Alexandria and at Memphis. Silver is gained in the temples of Hephaistos-Ptah by whitening with cadmia, and gold by yellowing with cinnabar.'

> 'For the gilding of the statues of the gods in temples they use a solution of gold in quicksilver. This amalgam is called 'sun-water', 'concentrated rays of the sun' and also 'dissolved sulphur'. This is regarded as a great secret, and a supernatural achievement!'

> 'As philosophy is falsified by babblers, so quicksilver is adulterated by greedy merchants. They read chemical writings and then multiply. When they buy, they know many tests of purity: when they sell they swear upon their heads that they have never heard of such tests.'

> 'But thou, who knowest that the great work can be completed only by our own meditation, hold those who would learn from thee aloof from such mystery-mongering. Instruct openly, and bind no one by oath to silence. Thou sayest that the Book of Hermes can only be obtained in secret. That is true, but it should not be so: on the contrary, everyone should possess such a

book in freedom from all obligations to secrecy. Only from those just books, the old ones and those composed by me, can truth be gained. It demands, however, patience and attentiveness, industrious study and meditation, and eagerness for practical experiments. To him who fulfils these conditions, the great work is child's play. But many who are not called, but who are dazzled by silver and gold, have used up all the quicksilver of Phrygia and Iberia and have died without having discovered or even comprehended the truth.'

'The nature of the alchemical art is like that of creation. The soul of copper is purified until it receives the sheen of gold and turns into the royal sun-metal. The great mystery consists of a carrier of the right qualities, axerion, a strewing powder, which dyes, penetrates and fixes. Of a powder that gives a gold dye first superficially and then within, so makes permanently into gold.'

'He who will devote himself to the great work must be free from selfishness and greed and filled with piety and goodwill. He must know the true times of the planets, the magic formulae and processes, and the magic substances. Fruitless are all efforts of the unlearned and deceitful, who strive not after knowledge but after gold – after the curing of the incurable malady of poverty, a curing which might have attained by other means, as by marrying a rich wife with a great dowry.'[9]

Cbe Visions of Zosimos

In *The Treatise of Zosimos the Divine concerning the Art* the Visions are described following an introduction which begins:

> 'The composition of waters, the movement, growth, removal and restitution of corporeal nature, the separation of the spirit from the body are not due to foreign natures, but to one single nature reacting on itself, a single species, such as the hard bodies of metals and the moist juices of plants.'

The 'Visions', recorded as 'Lessons', have much of the character of actual dreams and are unique among the writings on Greek alchemy as they possess an elaborate allegorical character. They have been translated from the Greek by F. Sherwood Taylor[10] and form the subject of a lecture given to the Eranos Conference in 1937 by C.J. Jung.[11]

Each 'Lesson' is in two parts, the first being an introduction and the second an account of the vision or dream. Lesson 1. reads as follows:

> '1. The composition of waters, the movement, growth, removal and restitution of corporeal nature, the separation of the spirit from the body, and the fixation of the spirit on the body are not due to foreign natures, but to one single nature reacting on itself, a single species, such as the hard bodies of metals and the moist juices of plants.'

> 'And in this system, single and of many colours, is comprised a research, multiple and varied, subordinated to lunar influences and to the measure of time, which rule the end and the increase according to which nature transforms itself.'

> '2. Saying these things I went to sleep, and I saw a sacrificing priest standing before me at the top of an altar in the form of a bowl. This altar had fifteen steps leading up to it. Then the priest stood up and I heard a voice from above saying to me, 'I have accomplished the descent of the fifteen steps of darkness and the ascent of the steps of light and it is he who sacrifices, that renews me, casting away the coarseness of the body; and being consecrated priest by necessity, I became a spirit'. And having heard the voice of him who stood on the bowl-shaped altar, I questioned him, wishing to find out who he was. He answered me in a weak voice saying, 'I am Ion, the priest of the sanctuary, and I have survived intolerable violence. For one who came headlong in the morning, dismembering me with a sword, and tearing me asunder according to the rigour of harmony. And flaying my head with the sword which he held fast, he

mingled my bones with my flesh and burned them in the fire of the treatment, until I learnt by the transformation of the body to become a spirit'.

'And while yet he spoke these words to me, I forced him to speak of it, his eyes became as blood and he vomited up all his flesh. And I saw him as a mutilated little image of a man, tearing himself with his own teeth and falling away.'

'And being afraid I awoke and thought, 'Is this not the situation of the waters?' I believed that I had understood it well, and I fell asleep anew. And I saw the same altar in the form of a bowl and at the top the water bubbling and many people in it endlessly. And there was no one outside the altar whom I could ask. I then went up towards the altar to view the spectacle. And I saw a little man, a barber, whitened by years, who said to me 'What are you looking at? I answered him that I marvelled at the boiling of the water and the men, burnt yet living. And he answered me saying, 'It is the place of the exercise called preserving (embalming). For spirits fleeing from the body'. Therefore I said to him 'Are you a spirit?' And he answered and said, 'A spirit and a guardian of spirits'.

'And while he told me these things, and while the boiling increased and the people wailed, I saw a man of copper having in his hand a writing tablet of lead. And he spoke aloud, looking at the tablet, 'I counsel those under punishment to calm themselves, and each to take in his hand a leaden writing tablet and to write with their own hands. I counsel them to keep their faces upwards and their mouths open until your grapes be grown'. The act followed the word and the master of the house said to me, 'You have seen. You have stretched your neck on high and you have seen what is done'. And I said that I saw, and I said to myself, 'This man of copper you have seen is the sacrificing priest and the sacrifice, and he that vomited out his own flesh. And authority over this water and the men under punishment was given to him.'

'And having had this vision I awoke again and I said to myself 'What is the occasion of this vision? Is not this the white and yellow water, boiling, divine [sulphurous]? And I found that I understood it well. And I said that it was fair to speak and fair to listen, and fair to give and fair to receive, and fair to be poor and fair to be rich. For how does nature learn to give and to receive?'

'The copper man gives and the watery stone receives; the metal gives and the plant receives; the sky gives and the earth receives; the thunderclaps give the fire that darts from them. For all things are interwoven and separate afresh, and all things are mingled and all things combine, all things are mixed and all unmixed, all things are moistened and all things dried and all things flower and blossom in the altar shaped like a bowl. For each, it is by method, by measure and weight of the four elements, that the interlacing and dissociation of all is accomplished. No bond can be made without method. It is a natural method, breathing in and breathing out, keeping the arrangements of the method, increasing or decreasing them. When all things, in a word, come to harmony by division and union, without the methods being neglected in any way, the nature is transformed; and it is the nature and the bond of the virtue of the whole world.'

'And that I may not write many things to you, my friend, build a temple of one stone, like ceruse in appearance, like alabaster, like marble of Proconnesus, having neither beginning nor end in its construction. Let it have within it a spring of pure water glittering like the sun. Notice on which side is the entry of the temple and taking your sword in hand, so seek for the entry. For narrow is the place at which the temple opens. A serpent lies before the entry guarding the temple; separate his parts; thus reuniting the members with the bones at the entry of the temple, make of them a stepping stone, mount thereon, and enter. For the priest, the man of copper, whom you see seated in the spring and gathering his colour, do not regard him as a man of copper; for he has changed the colour of his nature and become a man of silver. If you wish, after a little time you will have him as a man of gold.'

Lesson 2

'1. Again I wished to ascend the seven steps and look upon the seven punishments, and, as it happened, on only one of the days did I effect the ascent. Retracing my steps I then went up many times. And then on returning I could not find the way and fell into a deep discouragement, not seeing how to get out, and fell asleep.'

'And I saw in my sleep a little man, a barber, clad in a red robe and royal dress, standing outside the place of the punishments, and he said to me 'Man, what are you doing?' And I said to him, 'I stand here because, having missed every road, I find myself at a loss'. And he said to me 'Follow me'. And I went out and followed him. And being near to the place of the punishments, I saw the little barber who was leading me cast into the place of punishment, and all his body was consumed by fire.'

'2. On seeing this I fled and trembled with fear, and awoke and said to myself 'What is it that I have seen?' And again I reasoned, and perceiving that the little barber is a man of copper clothed in red raiment, I said 'I have understood well; this is the man of copper; one must first cast him into the place of punishment'.

'Again my soul desired to ascend the third step also. And again I went along the road, and as I came near to the punishment again I lost my way, losing sight of the path, wandering in despair. And again in the same way I saw a white-haired old man of such whiteness as to dazzle the eyes. His name was Agathodaemon, and the white old man turned and looked at me for a full hour. And I asked of him, 'Show me the right way'. But he did not turn towards me, but hastened to follow the right route. And going and coming thence, he quickly gained the altar. As I went up to the altar I saw the whitened old man and he was cast into the punishment. O gods of heavenly natures! Immediately he was embraced entirely by the flames. What a terrible story my brother! For from the great strength of the punishment his eyes became full of blood. And I asked him saying, 'Why do you lie there?' But he opened his mouth and said to me 'I am the man of lead and I am undergoing intolerable violence'. And so I awoke in great fear and I sought in me the reason of this fact. I reflected and said 'I clearly understand that thus one must cast out the lead, and indeed the vision is one of the combustion of liquids.'

Lesson 3

'1. And again I saw the same divine and sacred bowl-shaped altar, and I saw a priest clothed in white celebrating these fearful mysteries, and I said 'Who is this?' And, answering, he said to me 'This is the priest of the sanctuary. He wishes to put blood into bodies, to make clear the eyes, and to raise up the dead'.

'And so, falling again, I fell asleep another little while, I mounted the fourth step and I saw, coming from the East, one who had in his hand a sword. And I saw another behind him, bearing a round white shining object beautiful to behold, of which the name was the Meridian of the Sun, and as I drew near to the place of punishments, he that bore the sword told me, 'Cut off his head and sacrifice his meat and his muscles by parts, to the end that his flesh may be boiled according to method and that he may then undergo the punishment'. And so, awakening again, I said, 'Well do I understand that these things concern the liquids of the art of the metals'. And again, he that bore the sword said 'You have fulfilled the seven steps beneath'. And the other said at the same time as the casting out of the lead by all liquids, 'The work is completed'.

The Interpretation of the Visions of Zosimos

Carl Jung has given a detailed explanation of the 'Visions' in terms of modern psychology. To the student of alchemy, however, there are a number of themes in these visions that are familiar from, for example, the *Splendor Solis* of Salomon Trismosin, the sixth parable of which describes the dismemberment of a white body with a golden head into parts that represent the four elements and the quintessence. The vision makes much of the solution of metals and their reduction into their elements.

The alchemical process hinted at must be one of reduction of the metal into its elements followed by the subsequent restoration to the metallic condition, but with the ratios of the elements so adjusted that the base metal is advanced to the status of silver or gold. There is no concept of any universal catalyst or reagent such as the Philosopher's Stone being used to bring about this change. It was only later this concept became a factor in medieval and subsequent alchemy.

The work has the flavour of apocalyptic writing such as that of St. John the Divine and would, therefore, appear to be concerned with both metallic and spiritual regeneration. Sherwood Taylor was of the opinion that Zosimos lived and wrote during the 3rd century AD as he cites both Democritos and Africanus, who died in 232 AD and also mentions the Serapeum, destroyed in 390 AD, as if it was still in being. The Visions of Zosimos are unique among Greek alchemical writings because of their allegorical character which is, to some extent, summarised in the opening paragraph on the composition of waters. The dream or vision is largely concerned with an allegory of the alchemical process as understood at that time. Metals in order to be reduced to their elements in solution must be subjected to that which in those days would have appeared to be intolerable violence or punishment. The survivor of such treatment is appropriately named 'Ion' for, in modern terms, the metal would assume an ionic form in solution! The separation of the spirit from the body must relate to the observed volatilisation of the solution and its condensation on the cooler parts of the reaction vessel.

The advice to build a temple of one stone with an enclosed spring of water may refer to the reaction vessel itself enclosed in the furnace. The concept of the fifteen steps and the seven steps of Lesson 2, must mark the progress of the alchemist in his experimental work, the choice of the right path and the use of the right reagent the *corpus mysticum* or universal spirit. It is thus that the Visions of Zosimos contain the essential features of alchemy, secrecy, symbolism and the correspondence of the operations in the reaction vessel to those in the outer spiritual world. There is no mention of the Philosopher's Stone, unknown at that time, although the transformation is accomplished by means of a 'universal spirit'. The object of the work of the Greek alchemists was the production of gold, aurifaction, and, if that failed, aurifiction!

The Emerald Tablet of Hermes Trismegistus

In the first book of the *Geheime Figuren der Rosenkreuzer*, published in 1785, there is a poem associated with an emblematic engraving and a version of the Emerald Tablet in German. The poem claims to interpret and explain the Tablet, although on examination the poem is more concerned with the engraving.

The *Tabula Smaragdina* is held to be one of the principal statements of alchemy which has survived from antiquity. It is claimed that this was composed by the legendary Hermes Trismegistus, the Egyptian god Thoth, being found, in due course, by Sarah, the wife of Abraham, in the hands of the mummified body of Hermes in a tomb near Hebron. E.J. Holmyard mentions other embellishing details of the legend; such as the writing being in Phoenician characters. This discovery has also been ascribed to Alexander the Great, or to Apollonius of Tyana who flourished in the first century AD. As Mrs. Atwood comments in her *Suggestive Enquiry into the Hermetic Mystery*: 'This Emerald Table, unique and authentic as it may be regarded, is all that remains to us from Egypt of her sacred art'. The Emerald Table has been the subject of much comment despite its general air of incomprehensibility. It is valued as a chart with which many of the alchemists started on their voyage of discovery, initially with the object of finding a route to the preparation of the Philosopher's Stone, but ultimately with the discovery of themselves.

Ferguson states that the earliest version of the *Tabula Smaragdina* is contained in a work *De Alchemia*, published in 1541. The Table was apparently known over 300 years earlier as Hortulanus, who wrote a commentary on it, flourished in the 10th century AD. In 1923 Holmyard discovered an Arabic text in Jabir's *Second Book of the Element of the Foundation*. Jabir was born in Tus in 721 or 722 which gives a still earlier date of its composition, receding into the mists of history.

Alchemy has always been known as 'the Hermetic Art'. It is not surprising that the *Tabula* with its likelihood of ancient origin and which, in addition, bore the name of Hermes and appeared to have a profound esoteric meaning, would be revered and cherished by the alchemists. The *Tabula* has all these qualifications. It is certainly not modern, has always been ascribed to Hermes and its significance is not easily apparent. As Ferguson remarked 'The man that runs cannot read it, nor, for that matter the man who sits. It is as profound a mystery, as great a puzzle as the mysterious *Aelia Laelia Crispis* itself'.

In the *Geheime Figuren* there is an emblematic figure known as the 'Vitriol Acrostic' which is relevant to the *Tabula*. The figure is of particular interest and is circular in form with the words VISITA INTERIORA TERRAE RECTIFICANDO INVENIES OCCULTUM LAPIDEM (Search the interior of the earth and by rectifying thou shalt find the hidden stone) written around the circle. The initial letters of these Latin words 'VITRIOL' form an acrostic indicating that the searcher should use vitriol in his quest: not any common vitriol but that of the sages. In early chemistry the term vitriol was used to designate any glistening crystalline body. Thus copper acetate is a vitriol which may be prepared as follows:

The Vitriol Acrostic

> Take some pounds of verdigris, extract its tincture with distilled vinegar, let it shoot, then you have a glorious Vitriol.

This acrostic can also be found in the *Viridarium Chymicum* of Daniel Stolcius. It is given in full in his *Second Treatise of the Sulphur, Vitriol and Magnet of the Philosophers* with the additional words 'VERAM MEDICINAM'. An epigram in the *Viridarium* states that:

> '... the wandering planets are held in the sky; earth is equal to them with her yield of metals. The Sun is father to the Stone, wandering Cynthia (the moon) its mother, Wind bore the child in its womb, Earth gave it food.'

Again, a direct reference to the *Tabula*.

Returning to the emblematical figure, within the encircling acrostic are a series of symbols, later to be described in verse. The planets shown in the firmament are exercising their influence on the conjunction of the masculine and feminine principles, represented by the Sun and the Moon. The chalice, a symbol of fruit-fullness, which is not unlike the Holy Grail, is the symbolic source of both physical and spiritual life. The Chalice rests upon the alchemical symbol for

Mercury, while the two-headed Eagle and the Lion are thought to represent Wind and Earth. The interlinked Star with seven points symbolises the seven metals of Earth corresponding to the seven planets of the firmament. Two hands point from the clouds with the fingers extended in blessing. The symbols of the other four elements, copper, tin, lead and iron, are laid clockwise around the Chalice. At the centre of the circle is a double ring to which is attached the three shields by means of a chain. The ring is expressive of the dual nature of the forces at work in producing metals and the Philosopher's Stone, while the emblems on the three shields, a Green Lion, a Seven-pointed Star and a Double-headed Eagle, represent not only the *Tria Prima* of Salt, Sulphur and Mercury but also the Body, Soul and Spirit, all three of which are essential to the Art of Alchemy. The Eagle may also represent the King, or Gold, being the King of Birds, while the Green Lion portrays the 'Secret Fire', the agent to bring about the reaction surrounding the conjunction. Within the area formed by the links of the chain is an Imperial Orb and between the shields, celestial and terrestrial globes; the former depicting the Four Elements and the latter Heaven and Earth.

A poem draws attention to the importance of the number seven from the seven-fold Star, the seven planetary metals and the seven words of the acrostic, each of which, it is claimed, stands for a city with but one gate:

> Seven letters and seven words,
> Seven cities and seven gates,
> Seven times and seven metals
> Seven days and seven ciphers.
> Whereby I mean seven herbs
> Also seven arts and seven stones.
> Therein stands every lasting art.
> Well for him who findeth this.

In the instructions which follow for 'the water that does not make wet', importance is given to the seven ways for the Art none of which must be neglected.

An Interpretation of the Smagarine Table

'True it is, without falsehood, certain and most true'. The *Tabula* begins with the assurance that what follows, however obscure, contains nothing but the truth. It may relate to 'the truth' which was Our Lord Jesus Christ as, clearly, there must be a spiritual counterpart to each section of the Table.

'That which is above is like to that which is below, and that which is below is like to that which is above, to accomplish the miracles of One Thing'. Hortulanus draws attention to the Philosopher's Stone being divided into two parts by the Magisterium. There is a superior part which rises above and an inferior part which remains below. The inferior part is the Earth which is called nurse or ferment and the superior part is the soul, which vivifies and resuscitates the whole Stone. For this the separation is made, the conjunction celebrated and many miracles come to be perpetuated and done within the secret work of Nature. Alchemy being both esoteric and exoteric, spiritual and material, work on either plane results in changes on the other. This is the only way in which the Stone may be prepared.

'And as all things come into being by the contemplation of One, so all things arose from this one thing by a single act of creative adoption.' As at the Creation God created all things, perhaps by separating matter from anti-matter, so will He be invoked to assist the formation of the Philosopher's Stone from the *prima materia* by a contemplative process associated with the Great Work.

'**The father thereof is the Sun, the mother the Moon. The Wind carried it in its womb, the Earth is the nurse thereof.**' The Stone is prepared from the *tria prima* of Sulphur, Mercury and Salt, the Body, Mind and Soul or the King, Queen and Secret Fire. The first is well purified, the second properly prepared, then the two are conjoined to form a compound known as the 'rebus' or 'hermaphrodite'. The process continues with the rebus being reduced to a fine powder and placed in a sealed vessel with the Secret Fire. This vessel is the Philosopher's Egg in which the colour changes, from the black of the *Nigredo* to the red of the *Rubedo,* take place under the correct heating conditions and over a pre-determined period of time.

'**It is the father of all works of wonder throughout the whole world. The power thereof is perfect.**' This could be a reference to the 'power' of 'Our Sulphur', generally considered to be purified Gold, for as Eirenaeus Philalethes wrote '... so let Gold be the subject on which you work, and none other.'

'**If it be cast on to Earth, it will separate the element of Earth from that of Fire, the subtle element from the gross matter.**' The theory of alchemy implies that all matter is compounded from the four 'elements' Earth, Air, Water and Fire, in various proportions to achieve the qualities which determined the particular matter under consideration. The elements were regarded not as different kinds of matter but rather as different forms of the one original matter, by which it manifested its different properties. The elements were based on four qualities: the dry, the wet, the warm and the cold. Hence Earth was cold and dry, Fire hot and dry, Air hot and wet and Water cold and wet.

'**With great sagacity it doth ascend gently from Earth to heaven. Again it doth descend to Earth.**' In a closed reaction vessel the volatile component of the mixture will, on gentle heating, be vaporised and will condense on the cooler parts of the apparatus from whence it will run down back into the mixture. This reflux action was in fact encouraged in certain types of alchemical apparatus. The Pelican is a case in point. It was described as a pot-bellied, two-armed circulatory vessel for use in continuous distillation. The vapour from one vessel condensed on the upper part or helm and ran into the second vessel from where it was distilled back into the first.

'**Then uniteth in itself the force from things above to the force from things below.**' This section refers to the influence of the planets in the astrological sense, on the reaction taking place in the vessel. It has always been held that the operations of the Great Work should be carried out at the right time of year. In the *Splendor Solis* of Salomon Trismosin, the various regimens are associated with corresponding astrological signs of the Zodiac and with the seven planets. Operations should be started in the Spring under the sign of Aries. This is again a reference to the hermetic doctrine of 'as above so below'.

'**Thus shalt thou possess the glory of the brightness of the whole world, and all obscurity will fly from thee. This thing is the might and power of all strength.**' This is surely a reference to the 'Vision of the Kingdom' later portrayed in the *Parabola* of Hinricus Madathanus. It gives a glimpse of the perfection which the possession of the Stone will induce. With this will come true enlightenment.

'**It will overcome every subtle thing and has the power to penetrate every solid substance.**' This could be a reference to the action of the Stone on metals during the transmutation process or to the elusive nature of the Stone during its preparation. Far from being a stable substance, the Stone's rather ethereal qualities apparently make it very difficult to retain within the reaction vessel.

'**Thus was the world created.**' The Great Work of Alchemy is both spiritual and material. It must be carried out with complete devotion on the part of the alchemist to the Creator of all things. The work has all the holiness of the Blessed Sacrament, which is itself a wonderful example of the power of transmutation on the spiritual plane. The achievement of the Creator in forming the Universe is of a similar nature but on a different scale to that of the achievement of the Philosopher's Stone. All achievement has a wonderful spiritual content resulting in a feeling of exhilaration and satisfaction. The sublime happiness of the successful alchemist can only be described as 'out of this world'!

'Hence will there be marvellous adaptions achieved, of which the manner is this.' Here the writer alludes not only to the transmuting power of the Stone by which means any imperfect Earthly matter could be brought to its utmost degree of perfection. For example, base metals could be changed into gold or silver, or flints into all manner of precious stones such as rubies, sapphires, emeralds and diamonds. It may also refer to its other properties such as the production of longevity and to its healing nature when taken in small doses.

'For this reason I am called Hermes Trismegistus, because I hold three parts of the wisdom of the whole world.' Hermes has always been regarded as the Greek equivalent of the Ibis-headed Moon God, Thoth, who, in turn has sometimes been identified with the deified Athosis of 3400 BC, or with Imhotep, both of whom excelled in the art of healing. The term 'Trismegistus' derives from the Egyptian superlative obtained through repetition (Hermes appears as 'Great, Great, Great' on the Rosetta stone) which is later simplified through the substitution of the prefix *tris* in the Roman period.

'That which I had to say about the operation of Sol is completed.'

The Interpretation and Explanation of the *Tabula* from the *Geheime Figuren*

This explanation is contained in a long octosyllabic poem, in German, which begins:

> *Dis Smahl angstehen schlecht und ting,*
> *heft in sich gross und wichtig ding.*

> This picture, plain and insignificant in appearance,
> Concealeth a great and important thing.
> Yea, it containeth a secret of the kind
> That is the greatest treasure in the world.

In the first thirty lines of the poem reference is made to the Three Shields, the Imperial Globe, Heaven and Earth and to the symbols of the metals between the outstretched hands of the emblem, all intended to illustrate the text of the *Tabula*. Reference is also made to the seven words of the Vitriol acrostic. The next section of the poem (lines 31-61) identify the shields with the *tria prima* of Salt, Sulphur and Mercury associating these with the Body, Soul and Spirit of Power:

> This is the meaning of the Art:
> The body giveth form and constancy,
> The soul doth dye and tinge it,
> The spirit maketh it fluid and penetrateth it.

Each word of the Vitriol acrostic is shown to stand both for a metal and for a city (each of which has but one gate.):

> The first signifieth gold, is intentionally yellow.
> The second for fair white silver.
> The third, Mercurius, is likewise grey.
> The fourth for tin, is heaven blue.

> The fifth for iron, is blood-red.
> > The sixth for copper, is true green.
> The seventh for lead, is black as coal.
> > Mark what I mean, understand me well!

If these gates are reflections of the twelve gates postulated by Sir George Ripley in his *Compound of Alchymie*, they represent seven rather than twelve regimens of the Great Work. The poem goes on to list seven words, seven cities, seven gates, seven times, seven metals, seven days, seven ciphers or herbs, seven arts and seven stones, not forgetting the seven colours of the rainbow. In the 120th line there is a reference to:

> There is a water which does not make wet.
> > From it the metals are produced.
> It is frozen as hard as ice.

The method of preparation of this water is next described:

> There are seven ways for this art,
> If thou neglectest any of them thou workest in vain.
> But thou must, before all things else, know
> > Thou hast to succeed in purification.
> And, although this be twofold,
> > Thou art in need of one alone.
> The first work is freely done by it
> > Without any other addition,
> Without distilling something in it,
> > Simply through putrefaction.
> From all of its earthliness
> > Is everything afterwards prepared.

Reference is then made to two paths, one involving strong fire while the other *extendeth through the strength of fire* to treasure and to gain. The fire art comes to an end on line 150 and is followed by a moderate warming of the reactants in the sun or in warm dung. The elements are separated by distillation (the sublimation of the Wise Men) and attention is drawn to:

> The earth on the ground has mislead many,
> > Having been deemed a worthless thing,
> Although all the power lieth in it.
> > Some know how to separate it
> From their Corticibus, therefore they fail.

Corticibus is thought to have been an extract derived from tree bark. After this purification the poem is interrupted with three pointing hands following line 183

> And thereupon followeth the mixture, observe!
> > And so it cometh to a wondrous strength,
> The finished figures with the unfinished.
> > And if the fire be likewise rightly controlled,
> It will be entirely perfect
> > In much less time than a year.
> Now thou hast the entire way in its length
> > On which are not more than two paths.

These are the Mercurius alone, while the other is called a Vinegar (which will not attack philosophical iron but will attack copper). Might this not be nitric acid?

> Only two things more are to be chosen
> > Which thou wilt find by now
> If thou dost follow the right way
> > And attend carefully to thy work.
> The composition is the one
> > Which the Wise men kept secret.
> The nature of the fire also hath hidden craft;
> > Therefore its order is another.
> With that, one should, not deal too much
> > Or else all execution is lost.
> One cannot be too subtle with it.
> > As the hen hatcheth out the chick
> So also shall it be in the beginning,
> > And time itself will prove it.
> For just as the fire is regulated
> > Will this treasure itself be produced.
> Be industrious, constant, peaceful, and pious,
> > And also ask God for His help:
> If thou dost obtain that, then always remember
> > The poor and their needs.

As far as heating the reaction vessel is concerned, it has always been held that too high a temperature will injure the nature of the materials involved. Metals and chemical compounds can no longer be regarded as just assemblies of atoms and molecules but as entities forming part of God's creation, animal, mineral and vegetable. All such creatures, having a spiritual component, may be altered by the treatment to which they are subjected.

The 251 lines of the poem, in the original German consisting of rhyming couplets, say much about the emblematic figure with the vitriol acrostic but little about the Emerald Tablet. To the alchemist who knows, the hidden content may well be apparent, but to the interested reader much study is required before any understanding is obtained or even any benefit felt!

Chapter 9

Aspects of Chinese Alchemy

Introduction

In the West alchemy was usually concerned with the transmutation of base metals into gold, a process which has been termed gold making or aurifaction. Associated with this was a more profound spiritual transformation in which the alchemist increased his own spiritual awareness and moved towards a state of grace. Longevity and immortality were not the principal goals of the Western alchemist before the time of Roger Bacon (c. 1214 - 1294). Although some alchemists such as Nicholas Flamel and the Compte de St. Germain were reputed to have achieved a shadowy immortality, there is little reliable evidence for any prolongation of their lives beyond the normal span.

In the East, however, the objectives appear to have been rather different. Emphasis was placed on longevity and immortality achieved by the preparation and ingestion of suitable elixirs. Aurifaction, the transmutation of base metals into gold, and indeed aurifiction or gold faking, the treatment of the surface of base metals to make them appear like gold, were practiced to a small extent. This was called 'WAI TAN' as distinct from the 'NEI TAN' of spiritual alchemy. NEI TAN was based on yogic and tantric spiritual and meditational practices to aid vitality and induce longevity.

While accounts of Chinese alchemy have been given in publications by Eric Holmyard, John Read, Sherwood Taylor and others, the most comprehensive and detailed study has been that of Joseph Needham (1900 - 1995) given in parts of the fifth volume of his monumental work **'Science and Civilisation in China'**. It is this latter work that has proved such a rich source of information for the present study.[1]

Immortality and Transmutation

Needham was of the opinion that the two concepts of alchemy, macrobiotics, the production of an elixir for the prolongation of life, and aurifaction, first came together in the minds of Chinese alchemists from the time of Tsou Yen in the 4th century BC. This is further confirmed by the fact that in 144 BC an Imperial edict was issued forbidding the unauthorised private minting and the making of 'false yellow gold'.

The Chinese alchemists believed that it was possible to prepare drugs or elixirs (*tan*) which would prolong human life beyond old age. These could rejuvenate the body and its spiritual parts so that the adept would endure through centuries of longevity, finally attaining the status of eternal life and assume the etherealised body of a true Immortal.

This concept derived from the religion of Taoism dominant at that time. The alchemical process involved was referred to by Needham as 'macrobiotics' and was one of the routes to immortality – the incorporation into the ranks of the invisible bureaucracy of the universe as a Heavenly Immortal (*thien hsein*),

> '... purified, ethereal and free, able to spend the rest of eternity wandering as a kind of wraith through the mountains and forests, enjoying the company of similar enlightened spirits'.

It has been argued that the three key operational concepts of alchemy were gold-faking (aurifiction), gold-making (aurifaction) and the preparation of the Elixir (drug of deathlessness).

In the West, the third of these concepts was missing until the 13th century when Roger Bacon referred to it in his writings as *longaevitas prolongata per multos annos per experientas secretas*, but in China all three were present at the beginning from the time of the philosopher Tsou Yen. The twin concepts of alchemy and macrobiotics were linked by the belief that longevity would also be obtained by eating and drinking from vessels prepared from alchemical gold or even by the ingestion of such gold in a 'potable' form.

The three fundamental concepts of Chinese alchemy which are of importance in formulating any theoretical background are:

the Five Elements (*wu hsing*),
the Two Fundamental Forces(*yin* and *yang*) and
the Book of Changes (*I Ching*).

The Theory of the five Elements

This theory was first postulated by the alchemist Tsou Yen who lived some time between 350 and 270 BC. A classical text, dating from the 3rd century BC, comments as follows:

'As for the five elements, the first is called water, the second fire, the third wood, the fourth metal, and the fifth earth. Water (is that quality in Nature) which we describe as soaking and descending. Fire (is the quality in Nature) which we describe as blazing and uprising. Wood (is that quality in Nature) which permits of curved surfaces or straight edges. Metal (is that quality in Nature) which can follow (the form of the mould) and then become hard. Earth (is that quality in Nature) which permits of sowing, (growth), and reaping.'

These elements differ widely from those of the Western tradition - earth, air, water and fire, with the fifth element of the Quintessence or Ether - but are similar in describing qualities of matter in terms of combinations of the qualities of hot, cold, moist and dry.

The Two fundamental forces

The concept of the two fundamental forces again dates from the beginning of the 4th century BC. The characters *yin* and *yang* are concerned with darkness and light respectively. The Chinese character *yin* involves pictographs for a hill (shadows) and clouds; while the character *yang* depicts slanting sunrays or a flag fluttering in the sunshine. The *yin-yang* concept can also be represented as a circular binary concept with a black pole in a white vortex and a white pole in a black vortex. This indicates that the passive is present in the active, and the active in the passive, just as man contains the nature of woman and woman the nature of man to a greater or lesser extent. In terms of alchemical theory, this *yin-yang* symbol represents the Sulphur – Mercury binary. Sulphur is the active essence or spirit and Mercury corresponds to the receptive and passive role of the soul itself. It is the Eastern equivalent of the Western '*mysterium conionctionis*'. The *yin-yang* dipole emphasises that there are only these two fundamental forces or operations in the universe, now one dominating, now the other, in a wave-like succession.

The legendary Fu His is said to have stated his philosophy as follows:

'*... The Illimitable produced the Great Extreme; the Great Extreme produced the Two Principles and the Two Principles produced the Four Figures.*'

From the Four Figures the Eight Trigrams of the *Book of Changes*' were produced:

The Two Principles

The Four Figures

The Eight Trigrams

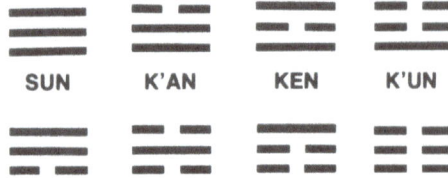

CH'IEN   TUI   LI   CHEN

SUN   K'AN   KEN   K'UN

The Great Extreme was depicted in tradition by the circle of YANG and YIN. It is said to be merely the immaterial principle. It is found in the male and female principles in Nature, in the five elements and in all things. From the time that the Great Extreme came into operation, all things were produced by transformation. The Great Extreme has neither residence, or form, nor any place to which it can be assigned. If it is spoken of before its development, then, previous to that emanation, it was perfect stillness. Motion and rest with the male and female principles of Nature (force and matter) are only the descent and embodiment of this principle. It is the immaterial principle of the two Powers, the four Forms, and the eight Changes of Nature. It cannot be said that it does not exist and yet no form or corporeity can be ascribed to it. It produced one male and one female principle of Nature, which was called the Dual Powers.

It was the combination of the eight trigrams that gave rise to the sixty-four hexagrams of the *I Ching*.[2]

The Book of Changes (*I Ching*)

The symbolic hexagrams of the *I Ching* are composed of six lines, whole or broken, corresponding to the *yang* and *yin* respectively. The Great primal Beginning of all that exists the *t'ai chi*, which originally meant a 'ridgepole' is represented as a line. With this line, which itself represents oneness, duality comes into the world, for according to Wilhelm, the line at the same time posits an 'above' and a 'below', right and left, front and back – the world of opposites. These opposites became known under the names of *yin* and *yang*.

陰 陽

YIN   YANG

In the alchemical treatise *Tshan Thung Chhi* of Wei Po-Yang, the hexagrams of the *I Ching* are thought to have been used to indicate times of day during which heating operations were carried out, and the trigrams the days of the lunar month when these took place. This use of the hexagrams and trigrams is a classic example of the function of the *Book of Changes* as a universal concept-repository to which any natural phenomenon could be referred. As Needham comments:

'The sixty-four symbols in the system provided a set of abstract conceptions capable of subsuming a large number of the events and processes which any investigation is bound to find in the phenomena of the natural world.'

The Legend of Wei Po-Yang (*c.*AD 120)

It is said Wei Po-Yang entered the mountain region to make efficacious medicines. With him were three disciples, two of whom he thought were lacking in complete faith. He was also accompanied by his dog. When the medicine was made, he tested the disciples. He said:

'The gold medicine is made but it ought to be tested first on the dog, we may then take it ourselves; but if the dog dies of it, we ought not to take it.' (Now Po-Yang had brought a white dog along with him to the mountain. If the number of treatments of the medicine had not been sufficient or if harmonious compounding had not reached the required standard, it would contain a little poison and would cause temporary death.)'

'Po-Yang fed the medicine to the dog, and the dog died an instantaneous death. Whereupon he said, "The medicine is not yet done. The dog has died of it. Doesn't this show the divine light has not been attained. If we take it ourselves, I am afraid we shall go the same way as the dog. What is to be done?" The disciples asked "Would you take it yourself, Sir?" To this Po-Yang replied, "I have abandoned the worldly route and forsaken my home to come here. I should be ashamed to return if I could not attain the *hsein* [immortal]. So, to live without taking the medicine would be just the same as to die of the medicine. I must take it". With these final words he put the medicine into his mouth and died instantly.'

'On seeing this, one of the disciples said, "Our teacher was no common person. He took the medicine and died of it. He must have done that with special intention." The disciple also took the medicine and died. Then the other two disciples said to one another, "The purpose of making medicine is to attempt at attaining longevity. Now the taking of this medicine has caused death. It would be better not to take the medicine and so be able to live a few decades longer." They left the mountain together without taking the medicine, intending to get burial supplies for their teacher and their fellow disciple.'

Wei Po-Yang, his dog, and his disciple, Yu, with a furnace

'After the departure of the two disciples, Po-Yang revived. He placed some of the well-concocted medicine in the mouth of the disciple and in the mouth of the dog. In a few moments they both revived. He took the disciple, whose name was Yu, and the dog, and went the way of the immortals. By a wood-cutter whom they met, he sent a letter of thanks to the two disciples. The two disciples were filled with regret when they read the letter.'[3]

Wei Po-Yang and the Earliest Book on Alchemical Theory

Very little is known about Wei Po-Yang, who was called by the Chinese 'The Father of Alchemy', save that he was a member of the Kaomen clan, coming from the region of Wu, the modern Chekiang in Kiangsu Province. He was born about 100 AD and, avoiding government service, eventually studied alchemy under a contemporary Master, Yin Chhang-Sheng (120-210 AD). His great contribution to the beginnings of alchemical literature the *Tshan Thung Chhi* (The Kinship of the Three) has been dated about the year 142 AD and Wei's name is found concealed in a cryptogram in the last paragraph of the epilogue. The earliest edition of the work is a Ming block-printed copy in the Beijing Library, but legend has it that an ancient manuscript of the work was found in a rock chamber – a legend reminiscent of the *Emerald Tablet* of *Hermes Trismegistus*!

The *Tshan Thung Chhi* has been translated by Wu and Davis,[4] but on the surface it appears to be even less intelligible than many Western alchemical texts. For example one passage reads:

'At the first double hour of the day, which corresponds to the *Fu kua*, the *Yang Chhi* [positive ether] begins to operate and at once appears to be slightly strong. At this time when the *Huang-chung* coincides with the ordinary *tzu*, a promising beginning flourishes forth. Let there be warmness and all will be well.

When the furnace is worked with sticks (*kua Lin*), room is made for the propagation of light. With the increase in brilliance the day becomes longer. This corresponds to the ordinary *chhou* and to the *Ta-lu*. Appropriateness is now realised.

Face upward to attain the *Thai* [greatness]. *Kang* and *Jou* [hardness and softness] both come to have sway. *Yin* and *yang* [negativeness and positiveness] are in contact with one another. Undesirable things give place to desirable ones. Activity centres at this, the ordinal *yin*, when fortune is at its high tide ...'

Apart from the obscurity of the text Wu and Davis deduced that this passage referred to the cyclical heating and cooling of the chemical reactants. In the text there is reference to three hexagrams of the *I Ching* – *Fu*, *Lin* and *Thai* – which give information as to the times of heating of the reaction vessel.

Interest in alchemy was stimulated by the Emperor Hsuan Ti (who ruled 73-48 BC) who encouraged the alchemist Lui Hsang by putting him in charge of the Imperial Workshops with the object of making alchemical gold. Unfortunately, Lui Hsang was not very successful and the Emperor had him imprisoned and sentenced to death. Lui was reprieved following the offer of a substantial ransom by his brother, the Marquis of Yang-chheng. Needham describes the failure of Lui Hsang's alchemical process to the proximity of capable metallurgists who understood the technique of cupellation. He suggests this was why Ko Hung and so many other Taoist alchemists emphasised the importance of carrying out alchemical operations at a site remote from the turmoil of worldly life and also from the proximity of capable assayers!

Ko Hung — The Greatest Chinese Alchemical Writer

Ko Hung was not only an alchemist with a profound knowledge of astronomy, meteorology and mineralogy but he also achieved a considerable reputation as a physician, comparable with his attainments as an alchemist. He produced an autobiography which, unfortunately, was not particularly informative about his scientific work.

Ko Hung was born in Nanyang in Chiang Su Province in 282 AD, the son of the Governor of Shao-ping. Although destined for the army, he studied alchemy under the guidance of the alchemists Cheng Yin and Paso Ching. Later, as an army officer, he was involved in the suppression of the rebellions of 303 AD and in 306 AD joined his friend Chi Han, the Governor of Kuang-chou, as military advisor while still indulging his interest in collecting books relevant to his alchemical studies. Chi Han was, in fact, an eminent botanist who produced a celebrated work *Record of Plants and Trees of the Southern Regions* (*Nan Fang Tshao Mu Chuang*). After the assassination of Chi Han the Governor, Ko Hung remained in the South for many years before becoming Magistrate of Kou-lou, in

Ko Hung (*c*.281-361 AD), Taoist philosopher and alchemist

in the cinnabar producing region of Chiao-chih. He lived for many years in the Lo-fou Shan mountains where he carried out his alchemical studies and where he wrote his celebrated treatise *The Pao Phu Tzu Book*'. He died in 343 AD.

The Book of the Preservation-of-Solidarity Master

Ko Hung's main contribution to alchemy is contained in the many chapters of this book, whose title in Chinese is *Pao Phu Tzu*, the esoteric chapters being known as *Nei Phien* and the exoteric chapters as *Wai Phien*.

Chapter 4 of the Nei Phien is concerned with the preparation of various forms of the Elixir including a magical elixir called the 'Elixir Flower' or 'tan hua':

> 'The first elixir is called 'elixir flower'. One should first prepare the 'mysterious yellow' (lead-mercury amalgam). Add to it a solution of realgar (arsenic disulphide) and a solution of alum. Take several dozen pounds each of rough Kansu salt, crude alkaline salt, alum, oyster shells, red bole clay, soapstone and lead carbonate; and with these make the sixty-one lute. After 36 days heating the elixir will be completed, and anyone who takes it continuously for 7 days will become an immortal. Now if this elixir is made into pills with 'mysterious fat' and placed upon a fierce fire, it will very quickly turn into gold. Gold can also be made by taking 240 *chu* (10 oz.) of this elixir and adding it to 100 *catties* (1 lb.) of mercury, then upon heating, it will all turn to gold. If this works we know that the elixir is right. If it does not, reseal the constituents and heat for as long as before. This never fails.'[5]

An elixir that would render a mortal immortal in three days was the *Thai Chhing tan* or 'Grand Purity Elixir'. This elixir underwent cyclical changes becoming more efficacious with each transformation. While exact preparative details are not disclosed, a number of chemical substances such as cinnabar, realgar, potash alum, malachite and magnetite appear repeatedly in the recipes. Ko Hung himself stressed that oral instruction from a teacher was imperative. Studies have been carried out by Chinese scholars on the possible chemistry of the reactions involved by these various substances with vinegar and natural fruit juices with the possibility of the formation of soluble gold, as the cyanide, or its colloidal suspension during the preparation of 'potable gold'.

Ko Hung gives several accounts of the transmutation of base metals into gold:

> 'First take any desired amount, but not less than five *catties*, of realgar obtained from from *Wu-tu*, vermillion in colour like a cock's comb, lustrous and free from bits of rock. This is pounded to powder and mixed with ox bile and heated until dry. Take a red clay pot with a capacity of one peck, spreading crude *Kansu* salt and blue vitriol in powder form all over the inside to a thickness of three-tenths of an inch. Then put in the realgar powder, to a thickness of five-tenths of an inch, and placing more of the salt over it until it is completely covered. Next spread on the top of this a layer of pieces of charcoal, about the size of jujube-date stones, two inches thick. The pot must be smeared all over outside with a *lute* made from the earth of earth worms and crude salt. Another pot is then inverted over and all the outside smeared with *lute* to a thickness of three inches so that there can be no leaks. After allowing the whole to dry in the shade for a month it is heated in a fire of burning horse-dung or three days and three nights. When cool remove the contents then work the bellows to liquefy the copper and it will flow like newly smelted copper or iron. This copper substance is then cast into the shape of a cylindrical container and filled with an aqueous solution of cinnabar. This is again to be heated in a horse-dung fire for thirty days and then taken out, pounded and smelted. Two parts of this with one part of crude cinnabar added to mercury will immediately solidify it into gold. It will be bright and shining with a beautiful colour, fit for making into *ting*.'[6]

It is thought the result of this complex operation will be a copper-arsenic alloy and the various processes through which it passes will result in a gold-coloured metal. A classic case of aurifiction rather than transmutation!

Thao Hung-Ching – Alchemist of the 'Golden Age'

**Thao Hung Ching (456-536 AD),
Alchemist listening to music**

During the next 400 years alchemy was actively encouraged by the Chinese Emperors to the extent that in the 3rd year of the Thei-Hsing reign period a Professor of Macrobiotics or Alchemist Royal was appointed and an Imperial Laboratory built at Phing-chheng in Shansi province. Provision was made for adequate supplies of fuel for the furnaces and convicted felons were provided for field tests on the elixirs produced, often with fatal results! The most celebrated alchemist to practice during the 5th and 6th centuries was Thao Hung-Ching who lived from 456 to 536 AD. As a physician and pharmaceutical naturalist he followed in the tradition of Ko-Hung and, as a friend of the Emperor Liang Wu Ti, was supplied with gold, cinnabar, copper sulphate, realgar and other necessary materials. Thao Hung Ching prepared several elixirs, which had the appearance of frost and snow. These had the effect of making the body feel lighter and also enhanced the respect that Liang Wu Ti had for his alchemist. It is said in the Official History of the Southern Dynasties (Nan Shih) that:

'His respect grew so great that he (Thao Hung Wu Ti) burnt incense whenever he received a letter from him!'

Thao Hung Ching was involved with the Taoist Church and the latter part of his life was spent in association with an important group at Mao Shan near Nanking, a centre for magical and liturgical activities. While no alchemical manuscripts by Thao have survived, it is thought that a mass of his writings on minerals and chemical substances as well as on plants is contained in the pharmaceutical natural histories compiled by other writers. For example that on mercury gives:

'There are two sorts of mercury, crude and refined. That which comes from the earthy plain of Fu-ling is obtained from cinnabar, but the ore is found in pale sandy places. The best way to powder it is to roast it. The colour is white, and not as impure as the crude product. Mercury is able to soften and change gold and silver with the formation of a paste. People use it for the plating of objects. It can be reclaimed and converted back into *tan* (elixir or cinnabar), so the Manual of the Immortals say; and they add that if taken with wine warmed in the sun it will give longevity and immortality. On being heated it volatilises and a kind of ash sticks to the top of the reaction vessel; this is called *hung fen*, or popularly *shui-yin hui*. This is excellent for getting rid of fleas and lice.'[7]

Sun Ssu—Mo Physician and Alchemist

Sun Ssu-Mo, the next alchemist of the stature of Ko Hung and Thao Hung-Ching, was born about 581 AD. Very little is known of the life of this sage save he was the author of more than a dozen medical treatises and that he died in the year 682 AD. His alchemical work is known through a work entitled *Essentials of the Elixir Manuals for Oral Transmission* (*Thai-Chhing Tan Ching Yao Chueh*), written about 640 AD. and published in 1022.

Sun Ssu-Mo gives lucid descriptions of preparations that he had found successful, details of the essential alchemical apparatus, the furnace, the reaction vessel and of a suitable lute for hermetically sealing the joints in the vessel. His procedures are described in simple language without undue concealment of the nature of the materials used. Two substances alone are described by the trigrams *Li* and *Tui* and these materials have been interpreted as cinnabar and lead respectively. As an example, pills containing 'a minor cyclically-transformed elixir' were prepared by grinding separately the following components prior to mixing and making into pills:

mercury 1 lb.
sulphur 4 oz.
cinnabar 3 oz.
rhinoceros horn 4 oz.
musk 2 oz.

Sun Ssu-Mo (*d.*682 AD)
physician and alchemist
with scrolls

Meng Shen and Counterfeit Gold

A disciple of Sun Ssu-Mo, Meng Shen (621-718 AD) also achieved a reputation as an alchemist and as an outstanding pharmacist. On one occasion he upset the Empress by detecting the counterfeit gold which she used to give away to her civil servants as awards for their service:

> 'From early youth onwards Meng Shen was fond of alchemy. On one occasion, he visited the home of Liu Wei-Chih, vice president of the Department of the Imperial Grand Secretariat, and saw some gold given as rewards. Whereupon he said "This is alchemical gold, and if you submit it to the fire, coloured vapours will be seen above it." Afterwards the matter was put to the test, and he was proved right, but when Tse Thein, the Empress, heard about it she was not amused. So she found a pretext for having Meng re-posted to Thaichow, away from the capital.'[7]

Pai Chu-I Poet and Alchemist

Needham has pointed out that during the 8th century the poetry and the literature of the Thang dynasty were saturated with the ideas of alchemy and immortality. There were two outstanding Chinese poets at this time, Li Pai (701-762 AD) and Pai Chu-I (722-846 AD). Both had friends among the alchemists and, in particular, the poet Pai Chu-I was guided in his studies by the Taoist alchemist Kuo Hsu-Chou. Pai set up his own laboratory and carried out experiments, based on an alchemical manuscript the *Tshan Thung Chhi* loaned to him by Kuo on the preparation of the elixir. Feeling that he had failed in the operation, Pai confided his reactions to the following lines:

> 'I read it, and day by day the meaning grew clearer
> 'Till no doubt was left in my mind at all.
> The Yellow Sprout, yes, and the Purple Carriage
> Seemed to be perfectly easy things to produce …
> I bade a lofty farewell to the world of men;
> All my hopes were set on the silence of the hills
> My platform of brick was accurately squared,
> Compasses showed that my aludel was round.
> At the very first motion of the furnace bellows
> A red glow augured that all was well;
> I purified my heart and sat in solitary awe.
> In the middle of the night I stole a furtive glance,
> The *Yin* and *Yang* ingredients were in conjunction
> Manifesting an aspect I had not forseen,
> Locked together in the posture of man and wife
> Intertwined like dragons coil upon coil …
> The bell sounded from the *Chein-Chi Kuan*,
> Dawn was breaking on the Peak of Purple Mist.
> It seems that the dust had not yet washed from my heart;
> The stages of the firing had gone all astray.
> A pinch of elixir would have meant eternal life;
> A hair's-breadth wrong, and all my labours lost!
> The Master snapped his fingers and rose to go;
> The Elegant Girl flew up with the smoke to the sky …'

'Then I knew at last' continued Pai, '*… that on the plane of Assembled Occasions one cannot escape from the secret laws of predestination*'. With these words, worthy of Ernest Brahmah's Kai Lung, he dismantled the furnace and on the following day received the news that he had been made Governor of Chung-Chou!

The Decline of Alchemy in China

If alchemy in China can be said to have had a Golden Age (400 to 800 A.D.) when the Elixir of Life became attractive to the emperors who actively supported the work of their alchemists, it should be remembered that Taoism was the predominant religion of the country at that time, and had been since the 4th century BC. The theoretical basis of the Art was based on the Tao theories of the Five Elements, the Two Fundamental Forces and the Trigrams and Hexagrams of the *I Ching*.

Of course, elixir poisoning was a continuing problem. No less than six emperors of the Thang dynasty succumbed, and an unknown number of court officials died from the same cause.

It is uncertain if, like Wei Po-Yang, they went off and joined the immortals! As a result there was a change in emphasis from *Wei Tan* or laboratory alchemy to *Nei Tan* or physiological alchemy. When the empire fell into the hands of the Mongols at the end of the 13th century, Taoism lost favour and was replaced by Buddhism. With this the alchemists went to ground. There appears to have been a slight revival under the Ming emperors (1368-1644) although the predominant Neo-Confucianism orthodoxy did little to encourage either the theory or the practice of alchemy.

As in the West, alchemy had been associated with the change to chemistry. Its relevant practical techniques were gradually transferred into the industrial channels of metallurgy and pharmacy and like its Western counterpart, alchemy developed into pure and applied chemistry.

From the 9th century onwards writings on alchemy became more obscure in style and were more in the nature of compilations of the work of earlier alchemists. Those with outstanding personalities such as Teacher Keng, the daughter of Keng Chhien, made names for themselves. She was a court alchemist who had mastered the 'art of the yellow and the white'. It is reported that on one occasion she managed to transmute mercury into silver by a low temperature process in which a sealed package of mercury enveloped in several layers of beaten bark-cloth was placed 'in her bosom'. After a long time '... *there suddenly came a sound like the tearing of a piece of silk*'. On opening the sealed package it was found that the mercury had all turned into silver!

There were many legendary stories of the transmutation of base metals into gold or silver as well as the attainment of longevity and the preservation of youth. However, the former were sometimes associated with deliberate deception. Emperors, however, continued to express interest in elixirs, often with fatal results. Li Sheng, the founder of the Southern Thang kingdom, died from elixir poisoning. An alchemical laboratory was maintained for seventy years in the Imperial Academy during the time of the Emperor Chen Tsung (997-1022). This laboratory was known as the 'Hall of the Golden Elixir' and it is recorded it was supplied with five loads of charcoal daily for many years.

It was during this time that a merchant known as Wang Chieh, was taught the arts of alchemy by an un-named Taoist, and developed magical and hypnotic powers. Wang subsequently worked his way up until finally he achieved the confidence of the Emperor when he produced for him a quantity of alchemical gold and silver of considerable value. Wang spent much on the poor and needy and, it is said, built the Khai-Yuan temple at Tingchow. He lived a peaceful and respected life at the court and died regretted in the odour of sanctity. He was given the posthumous title of 'Legate of Chen-nan', a unique honour. An assistant, who worked with Wang Chieh, called Pi Sheng, was the inventor of movable-type printing.

Cbe Secret of tbe Golden flower (*T'ai I Chin Hua Tsung Chih*)

The title of this well known text has been translated by Joseph Needham as *Principles of the (inner) Radiance of the Metallous (Enchymoma)* and is generally held to be a Taoist *nei tan* treatise on meditation and tantric techniques with a Buddhist influence. The work became known in the West following its translation by Richard Wilhelm, with the Commentary by Carl Jung,[8] which was published in German in 1929. The work was thought to have been transmitted orally since about the 8th century AD, the T'ang period, and it was attributed to the Taoist adept Lu Yen (Lu Tung-Pin) and his school. Such was the longevity of Lu Tung-Pin who was born in the year 755 that he was canonised in popular folklore as a leading member of the Eight Immortals. Votive temples and shrines in his memory were erected and are, apparently, still to be found all over China. The *Chin Hua* was first printed during the late 17th century and then reprinted as a limited edition in Peking in 1920.

Although ostensibly compiled by a Taoist adept, the work contains much that is derived from Mahayana Buddhism. It is apparent from the text that the form of Confucianism which is based on the *I Ching* is also introduced.[9]   The eight trigrams or fundamental signs – *Ch'ien, K'un, Chen, K'an, Ken, Sun, Li,* and *Tui* - are used as symbols for certain inner processes. As Richard Wilhelm writes:

'The sign *Chen*, thunder, the arouser, is life which breaks out of the depths of the earth; it is the beginning of all movement. The sign Sun, wind, wood, gentleness, characterises the streaming of the reality forces into the form of the idea. Just as wind pervades all places, so the principle for which *Sun* stands is all-penetrating, and breaths realisation. The sign *Li*, sun, fire, the lucid plays a great role in this religion of light. It dwells in the eyes, forms the protecting circle and effects the rebirth. The sign *Kun*, earth, the receptive, is one of the two primordial principles, namely the *Yin* principle which is made real in the forces of the earth. It is the Earth which, as a tilled field, takes up the seed of Heaven and gives it form.   The sign *Tui*, lake, mist, serenity is an end condition on the *Yin* side, and therefore belongs to the autumn. The sign *Ch'ien*, Heaven, the creative, the strong is the reality form of the *Yang principle* which fertilizes *K'un* the receptive. The sign *K'an*, water, the abysmal is the opposite of *Li*, as is shown in its structure. It represents the region of Eros, while *Li* stands for Logos. *Li* is the Sun, *K'an* the Moon. The marriage of *K'an* and *Li* is the secret magical process which produces the child, the New Man. The sign *Ken*, mountain, quietness represents meditation, which by keeping external things quiescent, quickens the inner world.   Therefore *Ken* is the place where death and life meet, where *Stirb und Werde* is consummated.'[10]

The *Chin Hua* consists of eight chapters or sections leading the adept on his path to achieving enlightenment and longevity. The Golden Flower, the elixir of life, is the Light and the secret of the magic of life consists in using action in order to achieve non-action. It is stated at the beginning of the treatise that 'The work on the circulation of the Light depends entirely on the backward-flowing movement so that the thoughts are gathered together (the place of Heavenly Consciousness, the Heavenly Heart). In fact the first section of the work is entitled just that: 'Heavenly Consciousness'. In Taoism the achievement of material immortality as a *Hsien* or a 'True Man'. This required that the Adept underwent a considerable amount of training in such practices as respiratory, gymnastic, sexual, dietary, alchemical, pharmaceutical and helio-theraputic techniques.

Meditation, stage 1: Gathering the Light

Meditation, stage 2: Origin of a new being in the place of power

From the understanding of the tantric practices used in Indian alchemy and the general

acceptance of the Yin-Yang theories in China it might appear that the key to immortality may well reside in the prevailing concept of human sexuality against a cosmic background.   As Needham has pointed out:

> 'The Taoists considered that sex, far from being an obstacle to the attainment of *hsein*ship, could be made to aid it in important ways.'

Techniques practiced in private were called 'the method of nourishing life by means of the Yin and the Yang'. Their basic aim was to conserve as much as possible of the seminal essence (*ching*) and the divine element (*shen*), especially by 'causing the ching to return'. At the same time, the two great forces, as incarnated in separate individuals, were to act as indispensable nourishment, the one for the other.   In practice, one technique for the attainment of *hsein*ship involved frequent *coitus reservatus*, carried out under an elaborate system of prohibitions depending upon the seasons, phases of the moon, the weather and the astrological situation. Another, a physiological technique, which involved diverting the seminal fluid into the bladder, was called 'making the *ching* return'.[11]

Returning to the *Chin Hua*, it is pointed out in the first section that if a man can be absolutely quiet then the Heavenly Heart will manifest itself. If the Light is set in circulation then the Elixir of Life will be created - this is the conscious action that will give rise to the desired non-action!

The second section, entitled 'The Primordial Spirit and the Conscious Spirit' describes the role played by these entities in the making of the human body. The use of the primordial spirit during life is reflected on its behaviour after death. If the life has been generally concerned with doing good, the spirit emerges from the upper openings of mouth and nose, but if the life has been concerned with avarice, folly, desire, lust and other sins, the power of the spirit is confused and passes through the lower openings, sinks down into Hell and becomes a demon. The first stage of meditation is called the 'gathering of the Light'.

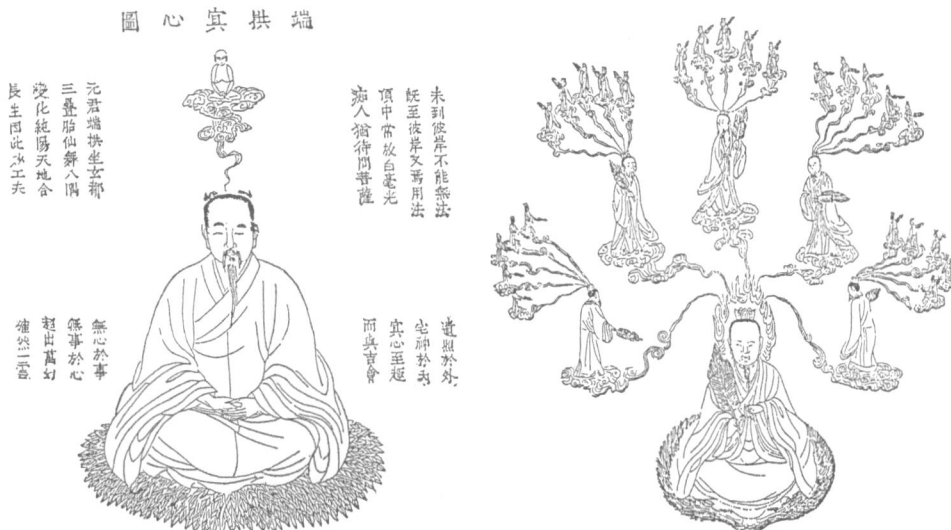

Meditation, stage 3: Separation of the spirit-body for independent existence

Meditation, stage 4: The centre in the midst of the conditions

The third section is on the circulation of the Light and the Protection of the Centre. The concept of a separation of a spiritual body which is likened to a non-being in the middle of one's being and in the middle of the non-being, a being. This idea was derived from the *Yin-Yang* binary concept. The emanation and dissemination of this spiritual consciousness is chiefly brought about by the power that flows downwards and, again, the meaning of the Golden Flower depends upon the backward-flowing method. Further instruction is given on the art of meditation with the lids of both eyes lowered but not closed and the breathing rhythmical. Further instruction is given on the art of meditation in section 4.

In section 5 attention is paid to the possibilities for error and to the importance of providing the right conditions and the right place for meditation. The school of Lu Tzu has confirmatory signs for each step of the way and these are set out in section 6:

> 'If, when there is quiet, the spirit has continuously and uninterruptedly a sense of great gaiety as if intoxicated or freshly bathed, it is a sign that the Light principle in the whole body is harmonious; then the Golden Flower begins to bud. When, furthermore, all openings are quiet, and the silver moon stands in the middle of Heaven, and one has the feeling that the great Earth is a world of light and brilliancy, that is a sign that the body of the heart opens itself to clarity. It is a sign that the Golden Flower is opening.'[12]

In section 7 it is emphasised that when there is gradual success in producing the circulation of the Light, a man must not give up his ordinary occupation. It is recommended that meditation should occupy two to four hours each morning and then, after two or three months, '... *all the perfected Ones come from Heaven and sanctify such behaviour.*'

Section 8 contains a magic spell for the Far Journey. This spell, left by Yu Ch'ing, reads as follows:

> 'Four words crystallize the spirit in the place of power.
> In the sixth month the white snow is suddenly seen to fly
> At the third watch the disc of the sun sends out shining rays.
> On the water blows the wind of gentleness
> Wandering in Heaven, one eats the spirit power of the receptive.
> And the deeper secret within the secret;
> The land that is nowhere, that is the true home ...'

This mysterious spell is interpreted by four words, these being 'non-action in action', the former preventing man from becoming entangled in form and image and the latter stopping him from sinking into numbing emptiness and a dead nothingness.

> 'When the right man makes the use of wrong means, the wrong means work in the right way!'

The *Chin Hua* while guiding the adept in his way of life towards the attainment of longevity is not concerned with the practical study of the art of alchemy as such, although it may have played a part in the spiritual development of individual alchemists.

Chapter 10

Aspects of Indian Alchemy

It has been said in the East the Adept, working mainly in the realm of the spirit and in search of the ultimate reality, attained a number of super-human abilities. One of these was the ability to prepare the Philosopher's Stone and to transmute base metals into gold. These powers were known as SIDDHAS. There is a story of a monk who by means of a great effort had become an ARHAT, that is, one who could be considered a saint - an enlightened one. Standing upon a high mountain he looked upon the people toiling in the valley below and thought how good it would be if he turned the mountain of stone into one of iron. This he did and the people were delighted. He then thought how good it would be if he turned the mountain of iron into one of gold; and this he did and the people were beside themselves with joy. However, he then saw their potential greed and avarice and so, for their own good, he turned the mountain back into a mountain of stone.

Yet one is forced to the conclusion that even if such powers were attained by a small number of individuals they were of no significance to them because they were but stepping stones on the way, their eyes being set on vistas far beyond. He who truly seeks the Philosopher's Stone, the Quintessence, the Tingeing Stone of the Wise, does so in the realm of the Spirit. The philosophic mercury of the alchemist becomes the creative force of the higher consciousness, which must be purged from the gross elements of matter, in order to attain the state of perfect purity and radiance - the state of enlightenment. Then does the mind of the Adept become the precious jewel, the Stone of the Philosophers.

It is no coincidence that what is known of the *Vajrayana* school of Buddhism, which is Tantric in essence, has been translated as the 'Diamond Vehicle', and whose most important symbol is the *Vajra*, the highest spiritual power, irresistible and invincible, which, like a diamond, is capable of abrading all substances and deterrents on the way. It was during the Tantric Period that the practice of alchemy reached its highest development in India.

The work for the sincere alchemist both in the East and in the West became a Way, a Quest, as in the doctrine of Tantra, a journey to a greater knowledge and higher values of achievement. The experiments and formulae of the alchemist remained only a form of material chemistry until his mind was projected into the operation. There would appear to be ample evidence that something of this kind must have happened. The journey with all its difficulties and complex stages led to the attainment of what was thought to be the final goal, only to find the work was still not complete and that still further stages remained to be accomplished. With his successive purifications, slowly, infinitely slowly, the alchemist came to realise that the *prima materia* was within himself - the light without flame - the centre of the rose. Perhaps by then he had even attained the ability to carry out the physical act of transmutation of metals, but it would now be of little value or interest. He would have reached the ultimate stage and have himself become the very Gold of the Philosophers.[1]

As Elias Ashmole expressed it in his *Theatrum Chemicum Britannicum*:

> 'Certainly he to whom the whole course of nature lies open, rejoiceth not that he can make gold or silver or the devils to become subject to him, as that he sees the heavens open, the Angels of God ascending and descending and that his own name is fairly written in the Book of Life.'[2]

Alchemy had existed in the East from the earliest times as a dual purpose art: the synthesis of gold and the prolongation of human life. The Philosopher's Stone or Elixir itself had a dual role – on the material plane of transmuting base metals into gold, the perfect metal, and on the spiritual one of perfecting the base nature of the alchemist himself, of preserving man's eternal youth and bringing him to a state of enlightenment. In Western alchemy more emphasis was placed on the achievement of the Philosopher's Stone, while in the East the search for longevity – the Elixir of Life – predominated. The Indian scholar, Dr. S. Mahdihassan, argues for a link between Elixir alchemy and the ancient Vedic worship of *soma*.

The *Rig Veda*, the oldest of the *Vedas*, which was composed between 1700 and 1000 BC, describes the *soma rasa*, or the juice of the *soma* plant, as *amrita*, which is thought to have been of the same nature as the Greek Ambrosia, the Nectar of the Gods. The exact nature of *soma* is uncertain. It must, however, have been an extremely potent euphoriant. Scholars are of the opinion it was the juice of a milky climbing plant, most probably a species of *Asclepias* or *Ephedra*, both of which are known in herbal pharmacy. The juice was ritually extracted by crushing the plant between two stones or by pounding in a mortar, the extracted liquid being filtered through sheep's wool and then mixed with milk, butter or honey.

Vedic texts give an account of its action on the body as no less than 'the roar of a bull'! *Soma*, when brewed or fermented became an intoxicating drink, and when prepared with the appropriate mystic rites and incantations could confer immortality, cure all diseases of body and mind, act as an inexhaustible source of strength and vitality, increased sexual energy and stimulated speech!

The *soma* plant was an object of particular adoration and the Vedic worshippers achieved ecstasy with the exhilarating effects of the fermented juices expressed from it. In the *soma rasa* (juice of the soma plant) and its attributes lies the dawn of Hindu alchemy. Plants were also recognised as helpful agents in the treatment of diseases. Their use was, however, invariably associated with the employment of charms, spells and incantations. The *artharva-veda*, the latest of the four *Vedas*, (c.850-500 BC), deals chiefly with sorcery, witchcraft and demonology and in this respect is similar to the ancient Egyptian Leyden Papyrus. It was in India at this time, however, that alchemistical notions gathered round the metals gold, mercury and lead.

The Vedic hymns clearly imply ecstatic visions in those who participated in the rites associated with the brewing of *soma* and consumed the orange-coloured juice *haoma-soma*. Recently it has been suggested that this juice might well have been the fluid extracted from the fungus *amanita muscaria* (fly-agaric) which is a powerful hallucinogen. This possibility is particularly plausible in the light of a statement by Joseph Needham (1900-1995) that such psychotropic substances can pass through the human body unchanged (up to five times!) while the accompanying nauseous principles are metabolised and lost. It is known that in the Shamanic ceremonies the urine of the officiating priest was frequently drunk as a libation. It is inferred from a study of the texts that this also occurred in the Vedic rites.[3]

Be this as it may, the Vedas state that:

> '*Soma* strengthens the weak, prolongs life, gives divine power to the gods, especially Indra and the moon, Chandra. Participating worshippers were healed and made imperishable as they repeated the refrain "We have drunk the soma, we have become immortal, we have entered into the light, we have known the gods." '

Gold and Soma

In ancient India the metal gold was intimately bound up with the soma sacrifice itself. This is confirmed in the *Satapatha-Brahmana* compiled some time between the 8th and 4th centuries BC. Gold (*suvarna*) is continually mentioned in this manuscript and held to be the semen of the fire-god Agni, the purest of earthly things, a sacramental symbol of light, fire and immortality. Gold played a part in the ritual buying of the *soma* as is shown in the following extract:

> 'Clarified butter being a thunderbolt and the *soma* being a seed he washes his hands

lest he should injure the seed, *soma*, with the thunderbolt the *ghee*. Thereupon he ties the piece of gold to the finger. Now twofold is this, there is no third, truth there is and untruth; the gods are the truth and men are the untruth. Gold having sprung from Agni's seed, he ties the gold to this in order that he may touch the stalks with the truth, and handle it by means of the truth …'He then makes (the sacrificer) touch the gold and say "Thee, the pure, I buy with the pure." For he indeed buys the pure with the pure, when (he exchanges) god for *soma*, "… the brilliant with the brilliant, the immortal with the immortal …" '

Gold threads were woven into the strainers for the *soma* juice and the metal itself also played an important part in the ritual.

Soma as a Hallucinogen

An entirely new light has been thrown on the nature of *soma* by the discoveries of R.G. Wasson and his collaborators. During the sixties of the last century they studied a number of hallucinogenic plants which included the cactus, the Mexican peyotl [*mescal*] and the fungus, fly-agaric [*amanita muscaria*].   The hallucinogens are substances which, as their principal and primary action, provoke transient mental perceptual and behavioural changes resembling the symptoms of psychotic disease. In appropriate doses they produce changes in the perceptual functions, especially the visual system with changes in colour perception, illusions, and hallucinations. There are also personality changes and alterations of the sense of time and space. Apparently, consciousness and awareness are not altered and memory is unimpaired. The principal substances included among the hallucinogens are lysergide, bufotenine, harmaline, mescaline, ololiuqui, psilocin and psilocybin.[4,5]

Alchemical Practice

Alchemical knowledge was widely cultivated in ancient India but reached its zenith in the Tantric renaissance period of 700-1300 AD. Ajit Mookerjee was of the opinion that the Tantrikas, those who followed the Tantric way, intimately understood the body and its cosmic affinities and were also well aware of the various techniques for attuning the body by the use of mercury, various medicaments, breathing exercises and helio-therapeutic meditation.[6]

Alchemical experiments were concerned mainly with the reduction or 'killing' of elements and their use in their primary forms.

'There is no such elephant of a metal that cannot be killed by the lion of Sulphur'[6]

The whole universe had, it was believed, been formed from a type of *prima materia* and this could be reduced or precipitated as powder – the ash form of matter. Reduction of a substance into ashes was therefore regarded as a form of purification. Such ash was the basic cohesive element for the preparation of the *elixir* vitae. Further purification was achieved when the ash was dissolved in a still more elemental substance called *rasa*, or liquid. *Rasa*, a term which included liquid mercury, was linked with the cosmic sea and might also have been related to dew or *ros*.

The Elements of Indian Alchemy

It was generally held, by the Indian philosophers, that the creation of the world occurred in a series of stages. In the beginning the world existed as an unmanifested state of pure consciousness. From this state the vibrations of the first sound, the soundless *Aum* began to manifest and it was from this vibration that the akasa, the element space or ether started to develop. Movement of the ether gave rise to air (*vayu*) and produced friction from which light or fire (*tejas*) was formed. The fire dissolved or liquefied parts of the ether giving rise to water (jala) which then solidified to form the element earth (prthivi). These five elements could be perceived by the five senses of hearing, touch, sight, taste and smell.

The physical characteristics of these five subtle elements, the *pancha mahabhutas*, recognised in Indian alchemy, and in the healing art of *ayurveda* were as follows:

> *Prthivi* (earth) is heavy, rough, hard, inert, dense, opaque; exciting the sense of smell.
> *Jala* (water) is liquid, viscous, cold, soft, slippery and fluid; exciting the sense of taste.
> *Tejas* (fire) is hot, penetrative, subtle, light, dry, clear; rarefied and luminous to the sense of sight.
> *Vayu* (air) is light, cold, dry, transparent, and rarefied; impingent on the sense of touch.
> *Akasa* (ether) is imponderable or light, rarefied, and elastic; capable of sound (vibrations) for the sense of hearing.

Indian alchemists believed that the most fundamental, even the quintessence, of all substances was mercury (*rasa*) and in all Tantric alchemical treatises the term *rasa* or juice refers to the metal mercury. They recognised about seven metals and two alloys but associated them with the heavenly or celestial bodies in a different order from that adopted by Western alchemy. Gold, the semen of the god Agni, was associated with Jupiter. Silver, the tears of Siva, was associated with the Moon. Copper, the semen of Kartikeya, came under the Sun, Iron, from the bodies of the Lomikas demons, with Saturn, Tin with Venus, Lead, the semen of the Snake God Vasuki, with the planet mercury, and Quicksilver (mercury) with the planet Mars.

An important Tantric treatise, the Sanscrit alchemical work entitled the *Rasaratnakara* which was thought by the Indian scholar Acharya Prafulla Chandra Ray (1861-1944) to be falsely ascribed to the philosopher Nagarjuna (*c*.800 AD), provides information on the various preparations of mercury and on its extraction, together with zinc, from their ores. Recent studies by Wujastyk[7] would confirm the ascription of the *Rasaratnakara* to the alchemist Nityanatha. Nagajuna was, however, responsible for an obscure alchemical work the *Rasendramangala*, and was a great Buddhist philosopher associated with the beginnings of Tantrism.

It is still uncertain whether it was this Nagarjuna who was the author of the various alchemical texts that have appeared under his name, Indeed, a number of authors of the same name are scattered through Indian history from the 2nd to the 8th centuries. Dr. Ray in his *History of Chemistry in Ancient and Medieval India* dated the *Rasaratnakara* from the 7th or 8th centuries based on internal evidence, while others believe it was an earlier work dating from the 3rd or 4th centuries. It has even been suggested that Nagarjuna was alive throughout the whole period having achieved longevity through his alchemical studies on the Elixir of Life!

A Chinese pilgrim, Hsuan-Chuang, writing sometime before 646 AD records:

> 'The Bodhisattva Nagarjuna was deeply versed in the techniques of pharmacy, and by eating certain preparations he attained a longevity of several hundreds of years, without

decay either in mind or body. Sadvaha Raja had also partaken of these mysterious medicines and had likewise reached an age of several centuries.'

A few pages later he continues:

'The Bodhisattva Nagarjuna scattered some drops of a numinous and wonderful liquid over certain stones, whereupon they all turned into gold.'

The Raja(h) Sadvaha is thought to have been the historical Satavahana who established a south Indian dynasty in the 2nd century. It is interesting to note that the *Rasaratnakara* contains dialogues with a Rajah named Salivahana or Sadvahana who might well be the same ruler.

The Importance of Mercury in Indian Alchemy

Mercury was held to be the quintessence of all substances and Hindu medicinal preparations were largely based on this metal and its compounds. The view was held that mercurial compounds should be prepared by amalgamating mercury with other components such as air, blood, semen and the ashes of various materials. Ingredients such as mica, sulphur, orpiment, pyrites, cinnabar, acids, alkalies, bitumen, gold, silver and arsenic were sometimes also incorporated. These preparations are described in the *Rasaratnakara* and also in an earlier tantric treatise the *Rasanava* (*c.*1200 AD). The range of apparatus and equipment used in alchemical laboratories is described in the 13th century iatrochemical treatise the *Rasaratna-Samuccaya*.

A typical recipe given for the preparation of a mercury-based elixir reads as follows:

'Mercury is to be rubbed with its equal weight of gold and then further admixed with sulphur, borax etc. The mixture is then to be transferred to a crucible and its lid put on, and then submitted to gentle roasting. By partaking of this elixir the devotee acquires a body not liable to decay ...'

In his *Travels* Marco Polo mentions that there are yogis [the *Chugghi*] who have a life span of 150 or 200 years:

'These people make use of a very strange beverage, for they make a potion of sulphur and quicksilver mixed together and this they drink twice every month. They say this gives them long life and it is a potion they are used to take from their early childhood.'

An Ayurvedic-Tantric treatise concerned with the internal and external applications of mercury gives details of the preparation of a potion prepared by pounding sulphur and copper pyrites together with mercury and roasting the mixture in a closed crucible. The product was administered mixed with honey.

Alchemical Apparatus

In Book IX of the *Rasaratnasamuchchaya* a text of the Iatrochemical period (1300 to about 1550 AD) a description is given of the apparatus, called *yantras*, used in alchemy and early chemistry. These include the *dola yantram* for liquid extraction of drugs, the *patana yantram* for sublimation and distillation, *valuka yantram* or sand bath, for the heating of vessels, *tiryakpatana yantram* for distillation of liquids, *vidyadhara yantram* for the extraction of mercury from cinnabar and

dhupa yantram for the fumigation of gold leaf with vapours from a mixture of sulphur, realgar, or piment. Detailed descriptions are also given of such common items of laboratory apparatus as crucibles, mortars and pestles.

Dola Yantram   **Patatna Yantram**

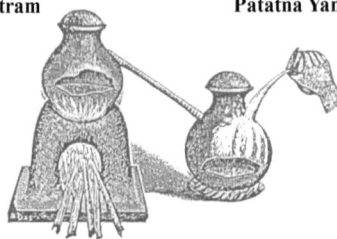

Tiryakpatana Yantram

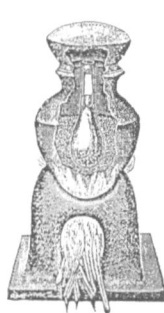

Valuka Yantram   Vidyadhara Yantram   Dhupa Yantram

Alchemy and Tantra

The pupil had to be initiated into the secrets of alchemy and mercurial lore by a guru devoted to *Siva* and his consort *Parvati*. It is recorded that the pupil should be full of reverence for his teacher, well-behaved, truthful, hard-working, obedient, free from pride, conceit and strong in faith. Such chemical operations that he undertook should be performed under the auspices of a god-fearing ruler who also worships *Siva* and *Parvati*. The laboratory should be located in the depths of the forest, containing an abundance of medicinal herbs and wells, should have four doors and be adorned with portraits of the gods. An artificial phallus made from an amalgam of gold and mercury should be placed in the east of the laboratory, for worship in due form, while the furnaces should be located in the south-east, washing operations in the west and drying carried out in the north-east.

It should be remembered that ritual or even ritual magic was closely associated with all preparative arts. From the 7th to the 8th centuries of this era the links of alchemy with tantric rituals and with *Hatha* Yoga became closer. Tantra is not a theoretical study but a practical discipline involving such disciplines as meditation, mantra (sounds and vibrations [*aum*]), mandalas (*yantra*), *mudras* (gestures and postures) and yoga (*Hatha* Yoga).

Shri Yantra

Central to the Tantric teaching was the concept that Reality is Unity as one indivisible whole. The Western *mysterium conjunctionis* of *Siva – Sakti* – the passive male and the active female – the Cosmic Consciousness in which *Siva* with his creative power and *Sakti* are eternally conjoined in union such that one cannot be differentiated from the other, and cosmic consciousness is thus '*... endowed with the essential potential of self-evolution and self-involution*'.

In Western alchemy the *Siva – Sakti* conjunction is represented by the Sulphur – Mercury, King and Queen, the sulphur being the *Pingala Nadi* and the mercury the *Ida Nadi*, which are the Sun and Moon breaths respectively. The Salt or 'secret fire' of the *Tria Prima* is associated with the tantric concept of *Kundalini* in which the 'coiled serpent', located at the base of the spine, flows upwards through the *Sushuma*, or central channel, by way of the seven *Chakras*, the energy centres of the body, to produce a state of enlightenment. It is said that *Kundalini* can be identified with *Sakti*:

> 'She is beautiful like a chain of lightning, and fine like a lotus fibre, and shines in the minds of the sages. She is extremely subtle, the awakener of pure knowledge, the embodiment of all bliss, whose nature is pure consciousness.'[8]

According to Mookerjee[9] the *Kundalini* is the microcosmic form of universal energy or, more simply, the vast storehouse of static, potential psychic energy that exists in latent form in every being. It is the most powerful manifestation of creative force in the human body. The concept of *Kundalini* is not peculiar to the tantras but forms the basis of all yogic practices, and every genuine spiritual experience is considered to be an ascent of this power. The *Kundalini* is described as lying coiled, or in trance sleep, at the base of the spine, technically called the *Muladhara Chakra* or root centre, blocking the opening of the passage that leads to the cosmic consciousness in the brain centre.

Throughout the alchemical literature of the West there are a number of references to the alchemist being assisted in his operations by his wife or female companion, his *soror mystica*, the *Sakti* to his *Siva*. The classical example is that of Nicholas Flamel and his wife Perrenelle who made projection upon half a pound of mercury on 17th January, 1392, 'Perrenelle only being present'. The *Mutus Liber* includes the alchemist and his wife in many of its plates and it would appear that her participation was an integral part of the process. Thomas Vaughan (1621-1666) mentions the participation of his wife in his *Aqua Vitae*.

A more modern case is that of Armand Barbault who reports in *Gold of a Thousand Mornings*:

> 'On Sunday, August 3rd, 1947 I went for the first time with my female collaborator to the site in order to inspect the ground and fix the exact point at which, from that day on, intensification of certain currents should take place. These currents were destined to preside over the most daring attempt we could ever make: to seize the *First Matter* and capture the etheric forces, which would then gradually increase in intensity during the irradiation of the *Matter*.'

There is no doubt that the forces inherent in the various ramifications of the instincts for survival and reproduction are considerable. Might it not be the case that the basis of the alchemical secret is dependent on these forces, which are of a psycho-sexual nature so powerful that even nuclear transformations may be brought about? This would certainly be suggested by the concept of the *mysterium conjunctionis*.

Chapter 11

Aspects of Islamic Alchemy

n the name of God, the Merciful, the Compassionate! Vouchsafe Thy Help, O Gracious One! Thus begins one of the lost alchemical treatises of Ar-Razi, the first of an encyclopaedia of twelve books on the subject compiled by that author. Ar-Razi was the third of the great alchemists of Islam during the last four centuries of the first millennium. The others were Khalid ibn Yazid (*c.*650 AD), Jabir ibn Hayyan (720-813 AD) and Abu ibn Sina (Avicenna) (980-1036 AD).

The foundation of the religion of Islam and the establishment of the Empire of the Caliphs that spread during the 7th to the 10th centuries to Asia Minor, Syria, Persia, Egypt, Africa and Spain, had a decisive effect upon the development of alchemy. At first, the attitude of the Arabs to the learning of the Western infidels was markedly hostile. After about 750 AD, however, under the Abbasid caliphs of Baghdad and such enlightened rulers as Harun al-Rashid (764-809) and Al-Ma'mun (786-833) they developed a thirst for learning and became more tolerant of the West. Greek and Indian works of science, philosophy and mathematics were eagerly translated into Arabic, for the most part by Syriac-speaking Nestorians based on the great university of Jundi-shapur in South Western Persia. There appears to have been an insatiable desire for this knowledge. These translations were facilitated by the manufacture of paper at Baghdad, the capital of the Islamic empire, about 795 AD. The Arabs established universities like those in Baghdad, Damascus and Cordova that became great centres of learning, attracting students from all over Europe, thereby having an important influence on the spread of scientific knowledge.

In the West with the decline and fall of the Roman Empire and the destruction of the great library at Alexandria at the end of the 4th century AD which resulted in a loss of many of the works of scientific value, alchemy had reached its nadir. The lamp of learning grew dim and was only to be re-kindled in the 12th and 13th centuries by the translation of, largely pirated, Arabic works back into Latin by such scholars as Robert of Chester and Gerard of Cremona. It was in this way that the Western alchemical tradition was regained in an enriched form with spiritual and other overtones from Eastern sources. There still remain many Arab manuscripts relevant to alchemy which have yet to be translated and studied by Western scholars.

A valuable source listing early Arabic literature on alchemy is the Tenth Discourse in *The Book of the Catalogue* (*Kitab al-Fihrist*) compiled by Abu-L-Fara Muhammad ibn Ishaq An-Nadim in 987 AD. Ibn An-Nadim was a bookseller in Baghdad who became familiar with nearly all the books in Arabic on alchemy to be found in the book-market at that time. The *Fihrist* gives details of the writings of such pre-Islamic alchemists as Hermes, Ostanes and Zosimos. It includes a list of some fifty-two alchemists from Hermes to Al-Azqiri.[1]

Prince Khalid Ibn Yazid

`Khalid ibn Yazid ibn Mu'awiya ibn Abi Sufyan, an Umayyad prince who died about 704 AD, was the son of Caliph Yazid I, who on the death of his elder brother, Muawiya II, should have succeeded to the Caliphate but, being a minor his cousin, Marwan, was made Caliph until he became old enough to take over. Marwan, however, reneged on this agreement by appointing his own son Abdul-Malik as his successor and, to make matters worse, accused Khalid's mother of immorality. The lady retaliated by having Marwan put to death. All this so disgusted Khalid that he withdrew from the court to devote himself to a study of the sciences which, at that time, included alchemy. When the remark was made to him 'You have wasted the best part of your labour in search of the Art', Khalid replied: 'I only did so to enrich my friends. I tried for the

Caliphate, but I failed. I have not found any compensation for it, save in reaching the highest degree of this Art, so that nobody who knows me, or whom I know, may be compelled to stand at the gate of a ruler in want, or in fear.' And it is said – but Allah knows best – that he was successful in the Art.

According to An Nadim, Khalid wrote a number of treatises and books on the subject of alchemy and composed many alchemical poems comprising some 500 folios. He also compiled *The Book of Amulets, The Great Book of the Scroll, The Small Book of the Scroll and The Book of the Testament to his Son on the Art*. There was another treatise in the library of the Nawab of Rampur written by Khalid that described how the author had learned from the monk Stephanos (of Alexandria) and the knowledge that he acquired from him. A second part gives an account of a process for manufacturing gold. This commences:

> 'The beginning of the work is like its end, and the commencement is like the finishing. What spoils it is the very thing that mends it. The work is two-fold. Let them begin by the second before the first. Everyone of them has referred to the way of entering into the work by mystical words. The finishing of the work in its most perfect degree comprises four operations in four years.'

To scholars in the West Khalid is best known for his 'dialogues' with the Palestinian hermit Morienus. He was, according to An-Nadim, the first Muslim for whom medical, astronomical and chemical writings were translated into Arabic.

The Dialogues of Khalid with Morienus

The *Dialogues* was to be known as *The Revelations of Morienus, Ancient Adept and Hermit of Jerusalem to Khalid ibn Yazid ibn Mu'awiyya, King of the Arabs of the Divine Secrets of the Magisterium and Accomplishment of the Alchemical Art* and was one of the first Arabic works on alchemy to be translated into Latin by Robert of Chester for Western consumption. The reference to Khalid as 'King of the Arabs' would perhaps indicate that despite his seclusion from the court Khalid had, in the end, inherited his rightful title following the demise of Marwan.

The Legend of Prince Khalid and Morienus

Prince Khalid had spent many years in searching for someone who could enhance his alchemical studies by revealing the secrets of *The Book of Hermes* to him. According to Holmyard:

> '... the story goes that Khalid had previously surrounded himself with self-styled experts in the art but was invariably disappointed at their failure to effect transmutation. Morienus, who was leading the life of a hermit in Jerusalem, heard of Khalid's great interest in learning and resolved to pay him a visit in the hope of converting him from the religion of Islam to Christianity. He was received with courtesy, and finding that Khalid desired above all things to witness the alchemical production of gold he asked for a room and equipment and there and then performed a successful transmutation. When Khalid had gazed at the alchemical gold, with true Oriental despotism he ordered the execution of the fraudulent alchemists, but in the resulting commotion Morienus vanished.'

A search for Morienus was set in hand, but all to no avail.[2] Eventually, news of the search reached a citizen of the town of Dirmanam who knew of the recluse Morienus and where he lived in the mountains of Jerusalem. Accompanied by Ghalib, Khalid's faithful servant, this citizen journeyed to Jerusalem and encouraged Morienus to return to the court of Khalid and to instruct him in the secrets of the Hermetic Art.

The Text of the Revelations of Morienus

In the version published by Lee Stavenhagen[3] the text begins:

'In the name of the Lord, holy and compassionate: this is the story of how Khalid ibn Yazid ibn My'awiyya came into possession of the spiritual riches handed down from Stephanos of Alexandria to Morienus, the aged recluse, as is written in the book of Ghalib, a bondsman of Yazid ibn Mu'awiyya.

The things in which the entire accomplishment of this operation consists are the red vapour (orpiment), the yellow vapour (sulphur), the white vapour (quicksilver), the green lion (green vitriol), ochre, the imputities of the dead and of the stones, blood (yellow arsenic sulphide), eudica (glass) and foul earth (sulphur).

Then begin in the Creator's name and with his vapour take the whiteness from the white vapour (quicksilver). Now pour out these things and set them about, taking an equal weight of each. Then they are mixed together. Pack the mixture into a vessel that will just hold the full charge. This is the best way to handle these things for they contain vapours which escape unless sealed in a container. Now stop the mouth of this vessel well with the luting cement which the authorities use, adding a little salt to it to make it stronger, that it may better withstand the heat of the fire. Let the furnace be heated with sheep dung or olive leaves and the vessel with its contents put into it to sublime. The sublimation should be done after sunset and the vessel allowed to stand until the cool of the evening. Then unstop the vessel and shatter it, examining that which is thus extracted, and if you find it well consolidated, like a stone, grind it thoroughly. Sift it and put it into another vessel the bottom of which is rounded. Construct for it a philosopher's furnace and let a philosopher's fire be started in it and kept going for a space of twenty-four days. After this time you should withdraw the vessel from the furnace and dry that which you find in it. Join one part of this with four parts of the purified earth and then with one part of the un- purified, proceeding in this order one after the other until all are mixed together, from which the elixir may then be made. Then examine its interior, and if you find the whiteness retained, well fixed and calcined, not driven off by the heat of the fire even after much heating, you have accomplished both natures of this operation. But know that if you had given your entire kingdom and brought all your subjects together at once to project the whiteness into the purified matter, they would have been powerless to do so by such means. For the spirit readily enters its own body, but if you try to put it into some foreign body, you will waste your efforts. The truth is perfectly clear.'

Morienus stresses that:

'The whole key to the accomplishment of this operation is in the fire, with which the minerals are prepared and the bad spirits held back, and with which the spirit and body are joined. Fire is the true test of this entire matter.'

Jabir ibn Hayyan (721-815 AD)

Abu Musa Jabir ibn Hayyan Al-Azdi Al-Kufi Al-Tusi Al-Sufi was the son of a druggist belonging to the famous South Arabian tribe of Al-Azd which was based at or had settled on the River Euphrates in Iraq shortly after the Muhammedan conquest in the 7th century AD. It was here that the druggist Hayyan had his pharmacy. Being interested in politics, he became an ardent supporter of the powerful Abbasid family. The object of the Abbasids was to overthrow the Umayyad dynasty and supplant the reigning Caliph. In furtherance of this objective the Iman, Muhammad ibn Ali, sent Hayyan to stir up discontent in Persia and thus prepare the way for revolution. It was while he was on this mission that he visited the town of Tus where his son Jabir was born in 721 or 722 AD. It was shortly after this nativity that Hayyan fell into the hands of the Caliph's men and suffered death by beheading and impalement. The young Jabir was sent to Arabia where in due time he studied the *Koran*, theology and a range of other subjects with a teacher named Harbi al-Himyari.

Jabir, as a young man, is thought to have studied occultism with the famous religious leader and Sufi teacher the Iman Ja'far al-Sadiq, himself an alchemist. It is probable it was at this time that Jabir turned his attention to alchemy – the Sophic or Sufic art. Sufism is an ascetic system of mysticism within Islam and was so-called from the Arabic suf (wool), because, it was thought, the early adherents of the sect used to wear a coarse garment made of that material. Partly a reaction from the extravagance and licence of the court, it soon became a movement whose members sought ecstatic union with God, practiced rigid austerity and gave themselves to contemplation and other religious exercises. It is said many of its tenets were similar to those of Neo-Platonism.

Because of his father's efforts in the Abbasid cause, which ultimately resulted in the overthrow of the Umayya house, Jabir was welcomed at the court of the fifth Abbasid Caliph, Harun al-Rashid in Baghdad. He was soon established there as an alchemist and became a personal friend of the sixth Shi'ite Imam, Ja'far al-Sadiq. Jabir also became a friend and physician to the powerful ministers, the Barmecides, eventually sharing in their expulsion from the Court in 803 AD. when he fled to his father's town of Kufa. He remained there, in seclusion, for the rest of his life. There is some evidence he returned to the Court for a short while in 813 when al-Ma'mun succeeded Harun al-Raschid as Caliph. Tradition has it, however, that Jabir died in Tus in 815 AD, with the manuscript of one of his works *The Book of Mercy* (*Kitab ar-Rahma*) under his pillow.

An-Nadim in his *The Life of Jabir ibn Hayyan As-Sufi and the titles of his works* makes the following observations:

'He is Abu Abbdallah Jabir ibn Hayyan B Abdullah Al-Kufi, known as As-Sufi. People differ about him for the Shi'ites say he is one of their great men and one of the Gates (spiritual guide), and they assert that he was the companion of Ja'far As-Sadiq -Peace be upon him! - and that he was a Kufan. Some philosophers, however, maintain that he was one of themselves, and that he composed books on logic and philosophy, whereas the seekers after the Philosopher's Stone assert, that the leading position in this art in his days was held by him, but that he lived in concealment. They maintain that he kept roaming about the countries without settling in a place because he feared that the government would attempt his life. It is also said, however, that he belonged to the circle of the Barmecides, was devoted to them and showed respect to Ja'far B. Yaha; for those who maintain this, say that he means, 'by his master Ja'far, this very Barmecide, whilst the Shi'ites assert that he means by this Ja'far As-Sadiq - Peace be upon him! A reliable man who practiced the Art that he used to live in the street of the Syrian Gate, in a lane known as 'Gold Lane'. Now this man told me that Jabir for the most part lived at Kufa and, owing to the city's good air, prepared there the elixir. When at Kufa, the cellar was discovered in which a golden mortar weighing 200 riftl was found, the place – said this man – where they found it was the actual house of Jabir b. Hayyan, but they found nothing else in the cellar except the mortar and a place for carrying out Solution and Fixation. This happened in the days of 'Izz Ad-Daula ibn Mu'izz Ad-Daula (967-977). The Chamberlain Abu Sabuktagin told me that he himself went to receive this.

Many scholars and elders of the Booksellers' Corporation say that this man, I mean Jabir, did not exist at all, whilst some say that, if he really did exist, he composed nothing but the *Book of Mercy* (*Kitab ar-Rahma*), and that those books were written by other people and then ascribed to him. But I say that for an eminent man to sit down, and weary himself out with the composition of a book comprising 2000 folios, fatiguing his talents and thoughts in composing it and his hand and body in writing it down, and then to assign it to some other person either real or imaginary - this, I say, is a kind of foolishness. Such a thing no one would endure; nor would anyone who has busied himself with science even for a moment embark on it; for what kind of profit would there be in it and what sort of advantage? The man really existed; his circumstances are too clear and well known and his writings too important and numerous.'

As Holmyard has observed:

'The life and works of this master of the art have presented a very difficult historical problem, which, however, research carried out during the last fifteen years (1940 to 1955) has gone far to solve. There seems no reason to doubt that at the court of Harun al-Rashid there was an alchemist named Jabir ibn Hayyan al-Sufi, the son of a druggist of Kufa. As a young man he enjoyed the patronage of the 6th Iman Ja'far al-Sadiq ... he was probably born in 721 or 722, he is said to have survived the banishment of the Barmecides in 803. Difficulties begin to arise when consideration is given to the numerous treatises on alchemy that pass under his name. How much of the alchemical operations and theory described in it is of the 8th century and how much of the 10th must remain a conjecture.'[4]

An-Nadim lists a very large number of books said to be written by Jabir although Paul Kraus, a Jabirian scholar, was of the opinion the entire Arabic Jabirian corpus was the compilation of a Muslim religious sect, the Ism'illiya or Brethren of Purity. The writings that survive and form the bases for the Latin ritings attributed to Geber, such as the *Summa Perfectionis*, were composed in the 10th century.[5]

More recently, William Newman has shown that the Latin *Summa Perfectionis* was definitely based upon manuscripts of Jabirian translations already in circulation and was probably the work of one Paulus de Taranto, an obscure Franciscan monk, known as the author of a number of other alchemical treatises.[6]

Abu Bakr Muhammed Ben Zakeriyah er-Rasi

According to Ferguson[7] er-Rasi or Rhazes was the son of a merchant of Ray in Iraq, Chorassan, where he was born about 850 to 860 AD. Holmyard gives his life span from the years 825 or 826 to 925.[8] As a young man he studied philosophy, logic, metaphysics and poetry, and developed a talent for music becoming an accomplished lute-player. At the age of thirty he visited Baghdad where he came into contact with an apothecary who introduced him to the study of medicine which Rhazes followed with considerable success. Indeed, his books on the subject were known throughout Islam. Later, when translated into Latin, his fame spread to Western Europe, where they became prescribed reading in universities in the Netherlands as late as the 17th century.

Rhazes was made chief physician at the hospital at Rayy and, later, physician-in-chief to the great hospital at Baghdad. His works on medicine included a treatise on smallpox and measles, two diseases which he was the first to describe in any detail. He was also the first to make any attempt to classify drugs and chemical substances and to group them into the well-known categories of animal, vegetable and mineral. The minerals he sub-divided into six classes: spirits, bodies, stones, vitriols, boraces and salts.

His classification was as follows:

> Spirits included substances such as:
> sulphur, arsenic sulphide, sal ammoniac and mercury
> Metals or metallic bodies:
> tin, lead and iron

Stones:
   magnesia, marcasite, tuttia, lapis lazuli, alum, antimony sulphide, talc, gypsum and glass.
Vitriols were classified according to their colour:
   black, red, white, green and yellow.
Boraces included:
   borax, natron, bone-ash and tinkar.
Salts comprised:
   cooking salt, sweet salt, quail (crude sodium carbonate) and salt of ashes (crude potassium carbonate).

The Book of the Secret of Secrets (Kitab Sirr al—Asrar)

The *Secret of Secrets* is one of the more important writings of Rhazes on the subject of alchemy. It conveys the impression, however, that he was more interested in practical chemistry than in theoretical alchemy. The *Secret of Secrets* is very much a laboratory manual and, although difficult to interpret, is thought to describe experiments that Rhazes carried out himself. As Holmyard remarks, by reversing the relative importance of experiment and speculation he brought about a revolution in alchemy. Earlier adepts had obscured experimentation in floods of unsupported hypotheses. Rhazes emphasised the importance of work in the laboratory as opposed to burning the midnight oil in the study. By adopting a practical approach to alchemy he transformed the subject from a speculative philosophy into a new and strictly scientific system. It was this scientific approach that attracted the attention of the greatest minds of the West for over 700 years.

The Character of Al—Razi

A description of Rhazes, the teacher, was given by a native of the city of Rayy and was quoted by Gerard Heym (1888 - 1972) who was a founder of the Society for the Study of Alchemy and Early Chemistry.   The description reads:

> 'Asked by a stranger about al-Razi, a native of Rayy gave the following description of the great man. Al-Razi, an old man now, was surrounded by his pupils in the lecture hall, and behind these sat their pupils and behind these there were more pupils of pupils. Al-Razi had a large stomach and running eyes, the latter having been caused by over-indulgence in eating garlic (or beans), which later resulted in blindness; he was friendly and courteous, receiving all strangers and attending to their maladies to the best of his abilities. He never left his desk and his books; he was always either writing a work of his own or copying that of another author.'[9]

There is a tradition that Rhazes became blind towards the end of his life due to cataract or to his close application to his alchemical work. It has been suggested his blindness resulted not from over-consumption of garlic but from a blow received from some great personage following a refusal of Rhazes to apply his knowledge of alchemy to the actual production of gold.

Al—Razi's Theory of Matter

The system of thought employed at the time when the Islamic mind had attained its greatest consummation was based on a belief in the interaction of the macrocosm with the microcosm.

This was united with a scheme for social betterment, a source of conflict between the Isma'ilite sect and established authority. Furthermore there was a belief in strict secrecy (to divulge the esoteric doctrine of the sect was considered equivalent to adultery) and also to a belief in the presence on earth of the concealed Iman. The Isma'ilies took great pride in the transmission and preservation of 'Greek' science – this in spite of their immersion in other speculative and often fantastic ideas. They were, in one sense,'humanistic', as were the Manicheans of the day, who believed in free-learning as opposed to the obscurantism of the orthodox party. For this reason al-Razi was also called a Manichean.

Al-Razi not only possessed a great interest in empirical science as demonstrated by his works on medicine, but also had a mind influenced by a system of abstruse and symbolical metaphysics. The Isma'ili philosophy is probably one of the keys to the understanding of the more obscure parts of al-Razi's alchemical works.

Gerard Heym describes Rhazes theory of matter in the following words:

'He taught that there were five eternal principals: the Creator, the soul, matter, time, and space. Bodies were composed of individual elements and of the empty space that lay between them. These atoms or elements were eternal and possessed a certain size.

The characteristics of the four elements earth, air, water and fire, that is their lightness and their heaviness, their transparency and colour, their softness and hardness, these characteristics were determined by the density of the elements, in other words by the measure of the emptiness of the spaces between them.These spaces determined the natural motion of the elements: water and earth moved downwards air and fire upwards.'[10]

It would appear from his writings on alchemy that Rhazes did not regard the transmutation of metals as its chief aim. He emphasised the importance of the knowledge of chemical substances as applied to medicine, a form of iatrochemistry. The object of experimental work was to attempt to improve the base metals by transforming lead, tin, copper and iron into silver or gold, and also to improve the quality of ordinary mineral stones, such as flints and glass, into more exotic materials such as red yaqut, green emerald and other precious stones. The means by which these transformations were to be accomplished involved a powder or liquid prepared by means of an arduous series of operations. This powder or liquid behaved like a strong medicine (or virulent poison) when it penetrated the base metals and pulverised stones to transform them into the desired improved product.

The theory on which the possibility of achieving a substance which had these miraculous properties was based on the hypothesis that all forms of matter contain within them a number of specific attributes, which are capable of being increased in potency to the highest degree of effectiveness, or, on the other hand, capable of being weakened or destroyed. The method by which this might be accomplished consisted in adding other forms of matter with similar or opposite specific attributes to the substances being experimented upon. Al-Razi also describes certain processes for this transformation which are difficult to understand!

He stressed the importance of colours and colour changes in a substance when heated or when mixed with something else. The admixture of mercury, for example, removed moisture, that of ammonium chloride removed earthiness, sulphur and arsenic produced whiteness. By removing the sulphurous characteristics from these substances the red colour was reduced, while mercury additions resulted in hardening, ammonium chloride in sublimation and distillation, whereas sulphur and arsenic were involved in distillation, washing, roasting and boiling.

Calcination was defined by Rhazes as the deprivation of bodies of their association with one another and the removal of any sulphur or oil they may contain to transform them into a white substance called *Nuqra*.

Al—Razi and the Philosopher's Stone

There is no reference in Rhazes work to any substance comparable with the Philosopher's Stone. His use of the term *hajar* relates only to organic substances from which elixirs are produced. The equivalent of the 'Stone' is , of course, the elixir or *iksir*, also referred to as *durur* and *haba* (fine powders). Al-Razi also used the term *jauhar*, meaning 'essence', for a substance which can transform base metals into gold and pebbles into precious stones. He gave a definition of the term 'elixir' in his work the *Kitab al-Asrar*. Elixir apparently meant a medicine which when fed into a molten body converts it into silver or gold, or changes its colour to whiteness or yellowness. The technique of preparing the elixir is summarised in four stages and forms an excellent summary of those methods employed by alchemists in Iraq and Persia during the 10th and 11th centuries. The four stages are:

1. A red tincture is extracted from a spirit (sulphur) by heating with a solution of sulphur in caustic soda (Strong Water). The extract is evaporated to a solid and kept separately with the residual 'dregs'.

2. Gold or copper, but preferably gold is calcined with sulphur.

3. Similarly, an amalgam of mercury with either tin or lead is treated with sulphur.

4. The various products after separately undergoing sublimation and conversion into wax-like incombustible substances are dissolved, mixed together and the whole coagulated into what was believed to be the desired elixir.

Another typical process for bringing about the desired transmutation of base metals into silver or gold as described by Al-Razi began with the purification of the material by some suitable treatment such as distillation, calcinations or amalgamation.

This *materia prima* being freed from its impurities, the next stage was to reduce it into an easily fusible condition by means of ceration. After this the product was to be further reduced by solution in 'Sharp Waters'. These were generally not acids but usually alkaline and ammoniacal liquids, although lemon juice and sour milk, which are weakly acidic, were sometimes employed.

The solutions of various substances, chosen to give the appropriate combination of properties, were then combined and subjected to a process of coagulation or solidification. If the experiment was successful, the resulting product would be the elixir.

Abu ibn Sina (Avicenna)

Avicenna or Abu ibn Sina was born at Afshana near Bukhara in Persia in 980 AD. Such was the ability of this Islamic genius that he became in time known as 'the Prince of Physicians' and he was even called 'the Aristotle of the Arabians'. After the birth of Avicenna's younger brother the family moved to Bukhara where a tutor was engaged to instruct Abu in the *Koran* and in Arabic poetry. Such was his progress in these subjects that additional tuition was soon necessary. He was therefore taught arithmetic by a local greengrocer, law by an ascetic named Ibrahim and Euclid by a wandering scholar called Natili. Avicenna also set himself a considerable programme of private study in a range of subjects. He made such progress in medicine that by the age of sixteen he had advanced so far adult qualified physicians came to learn new methods of treatment from him. He was soon appointed physician to a Persian prince and at the age of seventeen held many important posts, later rising to the rank of Grand Vizier (Prime Minister) to Shams al-Daula at Hamadhan. After living for a period at Ispahan he returned to Hamadhan where he died in 1036 or 1037.

In the fifty-six or so years of his life Avicenna accomplished a considerable amount of literary, medical, philosophical and scientific work. He achieved the reputation as an almost legendary hero in Islam and his fame spread to medieval Europe. He is said to have written over a hundred books and his celebrated *Canon of Medicine* contains over a million words.

Contemporary Islamic Views on Alchemy

At that time most of Avicenna's contemporaries certainly believed the transmutation of base metals into gold or silver was possible. Indeed, many of them held that it had already been achieved. Some, however, while accepting the theoretical possibility of transmutation held grave doubts it had ever been successfully carried out. Avicenna's views on the subject are to be found in his work the *Kitab al-Shifa* (Book of the Remedy).

Following the teaching of Jabir he ascribes the basic constituents of all metals to be mercury and sulphur or to substances that closely resembled them. If the mercury is pure and if it is commingled with the virtue of a white sulphur that neither induces combustion nor is impure, but on the contrary is more excellent than that prepared by the alchemists, then the product is silver. If, however, the sulphur besides being pure, is even better than just described, and whiter, and if, in addition, it possesses a tinctorial, fiery, subtle and non-combustive virtue, it will solidify the mercury into gold. If the mercury is of good substance, but the sulphur that solidifies it is impure, possessing the property of combustibility, the product will not be gold but copper. If the mercury is corrupt, unclean, lacking in cohesion and earthy, and if the sulphur also is impure, the product will be iron. As for tin, it is probable its mercury is good but its sulphur is corrupt, and that the commingling of the two is not firm but has taken place, so to speak, layer by layer, for which reason the metal 'shrieks'. Lead is probably formed from an impure, fetid, and feeble sulphur, for which reason its solidification has not been thorough.

It might be thought that Avicenna expressing these views on metallic structure, which were generally acceptable to contemporary alchemists, might well have accepted the possibility of converting one metal into another. As a young man Avicenna wrote two treatises on the quest for the tincture and certainly believed in the possibility of transmutation. He cannot have been successful in his quest, however, for towards the end of his life he expressed complete incredulity concerning the subject of transmutation.

'There is little doubt that the alchemists can contrive to make solids in which the qualities of metals are perceptible to the senses, though the alchemical substances are not identical in principle or in perfection with the natural ones, but merely bear a resemblance to them. Hence the belief arises that their natural formation takes place in this way or in some similar way, though alchemy falls short of nature in this respect and, in spite of great effort, cannot overtake her.

As to the claims made by the alchemists, it must be clearly understood that it is not in their power to bring about any true change of the metallic species. They can, however, produce excellent imitations, whitening a red metal so that it closely resembles silver, or tinting it yellow so that it closely resembles gold. They can also colour a white metal in such a way as to make it resemble gold or copper, and they can free lead and tin from most of their defects and impurities. Yet in metals so treated the essential nature remains unchanged; they are merely so dominated by induced qualities that errors may be made concerning their real nature. I do not deny that such a degree of accuracy in imitation may be reached as to deceive even the shrewdest, but the possibility of transmutation has never been clear to me. Indeed I regard it as impossible, since there is no way of splitting up one metallic combination into another. Those properties that are perceived by the senses are probably not the differences which distinguish one metallic species from another, but rather accidents or consequences, the essential specific differences being unknown. And if a thing is unknown, how is it possible for anyone to endeavour to produce or destroy it?

It is likely that the properties of the elements which enter into a composition of the essential substance of each of the metals enumerated is different from that of any other. If this is so, one metal cannot be converted into another unless the compound is broken up and converted into the composition of that into which its transformation is desired. This, however, cannot be effected by fusion, which maintains the union and merely causes the introduction of some foreign substance or virtue.

There is much I could have said upon this subject if I had so desired, but there is little profit in it nor is there any necessity for it here.'[10]

This apposite quotation of the words of Avicenna concludes this account of Islamic alchemy. Although there is no certain evidence that any of these Islamic alchemists solved the problem of metallic transmutation, their work stimulated similar activities in the West and resulted in the valuable contributions of such great minds as Roger Bacon and Albertus Magnus. Alchemy with all its mystery was still the science of its day and made an immense contribution to the sum of human knowledge.[11]

Chapter 12

Aspects of European Alchemy

St. Thomas Aquinas - Doctor Angelicus

It has always been held that Thomas Aquinas had a deep interest in alchemy. He was a pupil of Albertus Magnus in Cologne at a time when Albertus was making an intensive study of alchemy and other occult matters. It may well have been at this time that Thomas acquired an interest in the sublime art.[1]

Thomas Aquinas was born in Rocca Secca near Naples about the year 1225 of Norman descent and was the son of Count Landulf of Aquinas and of Theodora, Countess of Teano. Thomas was the youngest of four sons and four or five daughters. As a small boy of five years old he was placed in the care of his uncle Landulf Sinibaldo, the Abbot of the nearby monastery at Monte Cassino. Taken away from the monastery in about 1235 because of the disturbed state of the times and general political unrest, he was sent in 1239 to the University of Naples to study the arts and sciences. It was here he began to coach other students and became attracted to the Order of Preachers. While at Monte Cassino he had become an oblate of the Benedictine Order but he now wished to become a Dominican. After waiting some three years and at the age of nineteen, Thomas was received and clothed in the habit of the Order of Preachers and achieved his ambition.

His parents on receiving the news of his reception expressed considerable indignation. They had no objection to his joining a religious community and becoming a Benedictine as he might then eventually have followed in his uncle's footsteps as the Abbot of the monastery at Monte Cassino. However he had elected to join a mendicant Order which, like the Franciscans, practiced not merely individual but corporate poverty and this fuelled their indignation. The Order had no possessions except its actual houses and churches. Its members lived by begging.

Thomas's family determined to frustrate his life as a Dominican despite the friars taking him to their convent of Santa Sabina in Rome and then on to Bologna. It was while Thomas was resting by the roadside at Aquapendente he was overtaken by his brothers accompanied by a troop of soldiers and brought back, first to Rocca Secca and then to the nearby castle of Monte San Giovanni. Here he was held in close confinement, only his worldly-minded sister Marotta being allowed to visit him. It was during this 'captivity' that he studied the *Sentences* of Peter Lombard and learned by heart the greater part of the Psalms, Gospels and Letters of St. Paul.

His family continued in their efforts to undermine his determination to enter upon the religious life and even attempted to seduce him by sending a beautiful courtesan to visit him in his prison. Thomas is said to have chased her away with a burning brand from the hearth. Legend has it that he immediately fell into a deep sleep in which he was visited by two angels who girded him around the waist with a cord emblematic of chastity. From this time on he is said to have abhorred the sight of women.

In 1245, however, possibly due to the influence of the new Pope (Innocent IV) the family relented and Thomas was allowed to return to his Dominican convent. To complete his studies he travelled to Paris to study under the greatest teacher of that time Albertus Magnus (St. Albert the Great). Albertus, who was engaged in an extensive study of alchemy and other occult arts, both practically and theoretically, enhanced his knowledge by visits to known alchemists in their laboratories throughout the country.

In the year 1246 Thomas accompanied Albertus to the newly established Dominican *Studium Generale* at Cologne. It was here that Thomas received the nickname of 'the dumb Sicilian ox' due to his reticent silence at disputations and his short bulky stature. Quiet he may have been, but he possessed an acuity of mind and was able to elucidate the most difficult passages of prose clearly and correctly. As Albertus said on one occasion: 'We call Brother Thomas 'the dumb ox', but I tell you that he will yet make his lowing heard to the uttermost parts of the earth'! It was from Albertus that Thomas was inducted into all the burning intellectual and spiritual problems of the age and introduced to the works of Aristotle and Avicenna.

In 1252, at the request of Albertus and Cardinal Hugh of Saint-Cher, Thomas returned to Paris to become a lecturer at the Dominican convent of St. Jacques. It was here that he wrote his defence of the mendicant Orders *Contra Impugnantes Dei Cultum.* In 1256 he was awarded the degree of Master of Theology, received his Doctor's chair and was sent by the Order into Italy where he taught as *Lector Curiae* at Agnagni and Orvieto (1259-65), at Santa Sabina and the Dominican *Studium Generale* at Rome (1265-67) and at Viterbo (1267-69). He was recalled to Paris in 1269.

In 1272 he was moved back to Italy and appointed Regent of the study-house in Naples. It was here that he compiled his great but unfinished work the *Summa Theologica.* On his way to the Council of Lyons, Thomas Aquinas died on 7 March 1274, at the Cistercian monastery of *Fossa Nuova.* He was canonized in 1323, when Pope Pius V conferred upon him the title of 'Doctor of the Church'.

It is reported Thomas's learning was only exceeded by his piety and that after he had been ordained priest his union with God seemed closer than ever. He spent many hours in prayer, both day and night, and, according to his biographer William de Tocco, when consecrating at Mass '... he would be overcome by such intensity of devotion as to be dissolved in tears, utterly absorbed in its mysteries and nourished with its fruits'.

Thomas Aquinas and Alchemy

Thomas's views on the subject of alchemy have been summarised by Lynn Thorndike. Aquinas declared that alchemy was a true although difficult art. He accounted for the efficacy of its operations by its utilisation of occult forces of celestial virtue. [*Unde etiam ipse Alchemistae per veram artem alchimiae sed tamen diffficilem propter occultas operations virtutis coelestis ...*]

The Aurora Consurgens

A treatise on alchemy attributed to Thomas Aquinas is the *Aurora Consurgens* (The Rising Dawn), sometimes also known as *De Alchimia* and *Liber Trinitatis*. This work is of particular significance as it relates to the problem of opposites in alchemy. Despite opinions to the contrary it is thought by some scholars this treatise may well have been a transcript of St. Thomas's last seminar given while he was in a state of ecstasy and lying on his deathbed. Marie-Louise von Franz, the editor of the work, believed that the *Aurora* is indeed a transcript of this seminar. She wrote:

> 'The last precious words of St. Thomas were not preserved officially as one might certainly have expected, but were transmitted as an "apocryphon" for the very good reason that they reveal the other, unconscious, personality of the Saint, which had overpowered him on several occasions before in his trance-like states and now manifested itself again in the mystic marriage at the moment of his death'.

The *Aurora Consurgens*, originally written in Latin, is a comparatively short work and comprises twelve sections as follows:

I. The Aurora or *Aurea Hora* of Blessed Thomas Aquinas. Here beginneth the Treatise entitled *Aurora Consurgens*.

In this introduction the author cites the Wisdom of Solomon from the Apocrypha, together with the works of the alchemical writers Alphidus and Senior, on the value of wisdom in learning and understanding the 'Science of God'. The alchemical paradox of that which is beyond price being passed by daily in the streets and trodden into the mire by beasts of burden and cattle is introduced, and the stage set for the revelation of the secrets of the Great Art.

II. What Wisdom is.

The writer continues with a description of Wisdom, but he is now the 'wise man' who reveals the secrets of symbols to the elect. Wisdom is not only a source of knowledge, but also has a divine and mysterious nature containing within herself the secret of immortal life.

III. Of Them Who Know Not and Deny This Science.

The section starts with a quotation from the Book of Proverbs '*Fools scorn wisdom and discipline*' and goes on to imply that a degree of ability and understanding is necessary for those who will succeed in this science.

IV. Of the Name and Title of This Book.

This chapter is devoted to the reason behind the choice of *Aurora Consurgens* (The Rising Dawn) as the title for this work. Four reasons are given. First of all, the dawn is the 'Golden Hour' and for those who rightly perform the work there will be a golden end. Secondly, the dawn is midway between night and day, shining with twofold hues, namely yellow and red (*flavus et rebedo*), the two colours of the work, midway between white and black (*albedo et nigredo*).   Thirdly, at dawn those who work in darkness are relieved and see the light and 'all evil odours and vapours that infect the mind of the laborant fade away' replaced, as the Psalmist says, 'by gladness'.

V. Of the Provocation of the Foolish.

This section serves as an introduction to the seven parables which follow in the text as it cites a quotation from the Book of Proverbs with the words 'Understand, ye foolish ones, and mark the parable and the interpretation, the words of the wise and their mysterious sayings'. Then an important symbol is introduced as the writer equates Wisdom with the Queen of the South, like unto the morning rising, the apocalyptic woman 'clothed with the sun and the moon under her feet, and on her head a crown of twelve stars'. Is she not familiar as the symbol of the *mysterium coniunctionis* as depicted in the *Spendor Solis* of Solomon Trismosin and, indeed, of the Blessed Virgin Mary herself?

VI. The First Parable: Of the Black Earth, Wherein the Seven Planets took Root.

The parable begins: 'Beholding from afar off I saw a great cloud looming black over the whole earth'. This, it has been suggested, is as though the author is watching what was going on in the retort during the initial stages of the work. The blackness of the cloud is a reference to the primary stage of the process, the *nigredo*, the primary union of the opposites in the *materia prima*. The implication of the parable is that the author has identified himself with the reagents in the retort and the process through which the transformation is taking place applies to him also! This almost foreshadows the *Dark Night of the Soul* of St. John of the Cross (1542-1591) three centuries later. 'On a dark night, Kindled in love with yearnings – O happy chance! – I went forth without being observed, My house being now at rest.' Here the soul is setting out upon its journey on the spiritual road that leads to the attainment of the perfect union in love with God. So also is the *prima materia* being transformed into that perfect body – the Philosopher's Stone.

VII. The Second Parable. Of the Flood of Waters and of Death which the Woman both brought in and put to flight.

According to Thomas, the old and oft quoted saying 'The corruption of one is the generation of another' is a fundamental law of all change aiming at perfection. As he wrote in his *Summa Theologica* '... through many generations and corruptions we arrive at the ultimate substantial form.' (S.T. I, q. 118). The alchemical symbols used in the parable have the important function of uniting the opposites. At the beginning of the parable the reactants are again identified with the alchemist himself. The allusion to 'ten wise virgins' reflects the female power of Wisdom while the Massacre of the Innocents has deep alchemical connotations and was to find a place in the hieroglyphical figures of Nicholas Flammel (1330-1417): 'On the other side of the fifth leaf was a king armed with a falchion, who caused to be killed in his presence, by soldiers, a multitude of little children, whose mothers wept at the feet of the pitiless slayers. The blood of these infants was gathered up by other soldiers and put in a great vessel, wherein the Sun and Moon came to bathe themselves.' Just as the reactants in the Philosopher's Egg have passed through the darkness of the *nigredo* so the alchemist has begun his process of re-integration. The *albedo* is in sight!

VIII. The Third Parable: Of the Gate of Brass and Bar of Iron of the Babylonish Captivity.

The third parable ends with the words 'Manifold are the alternations of the seventy precepts' and makes reference to the Arabian alchemist Rhazi and the work of that name attributed to him. The opening of the parable has been interpreted as an allusion to the gradual extraction of the 'fluid soul' from the mineral body. There is some indication the size of the reaction vessel should be increased and that additions be made to the initial ingredients to continue the albification process. After standing for a period, their bulk is reduced by evaporation.

IX. The Fourth Parable: Of the Philosophic Faith, which consisteth in the Number Three.

Following an explanation of the Most Holy Trinity, Aquinas likens the Godhead to the concept of body, spirit and soul. This is a reference to the words of Senior in his *De Chemica*: 'Our ore is like to a man, having spirit, soul and body. Wherefore do the wise say: "Three and three are one". They say also: "In one are three" and "Spirit, soul and body are one and all things are of one".' This refers to the alchemical adage *'Omnia ab uno et in unum omnia'*. The generation of the Stone is likened to the nine months of human pregnancy, three months of preservation with water, three months of nourishment with air and three months of heating with fire. The Holy Spirit baptizes with water, in blood and in fire. The water is the saving Water of Salvation, the blood the Blessed Sacrament and the fire the gift of the Holy Spirit.

The sevenfold gifts of the spirit are next pursued in some detail – Righteousness, Wisdom, Knowledge, Fortitude, Understanding, Counsel and the Fear of the Lord. The alchemical process is also set out in seven stages.

The first of these is the *conjunction* in which the male is set on the female, the warm upon the cold (Senior) and the second is *ignition* in which change is brought about by allowing the reactants to ignite in the air. The third stage is *liquifaction* in which the reactants are brought again into solution and the fourth, *albification* in which the black is made white and the white red. The fifth is *separation* when the pure is separated from the impure and the sixth, *exaltation* in which the hidden is made manifest and, finally, the seventh, *inspiration* when the earthly body is made spiritual and by distilling seven times the Stone is set apart from corrupting humidity.

This Parable includes quotations from a number of established alchemists and alchemical texts: Calid the Less, Senior, Avicenna, Morienus, Alphidius and Rhazes, the *Turba Philosophorum*, the *Book of the Quintessence* and the *Emerald Tablet* of Hermes Trismegistus.

X. The Fifth Parable: Of the Treasure House which Wisdom built upon a Rock. 'To them that unlock this house shall be befitting holiness and also length of days'.

The parable has a quotation from Alphidus to the effect that this might also be achieved with four keys, which are the four elements and interpreted as the extraction of the water, the softening of the earthy body, its saturation and fixation. The parable refers to the 'fourteen corner stones of the House of Wisdom' which are: Health, Humility, Holiness, Chastity, Virtue, Victory, Faith, Hope, Charity, Goodness, Patience, Temperance, Spiritual Discipline or Understanding and Obedience. Thus clothed and subjected to the Will of God, the alchemist beholds, in the Holy of Holies, the *mysterium coniunctionis* of Sun and Moon. An interpretation of the reference to the 'fourteen Corner Stones' may be a reference to the different aspects of the Philosopher's Stone.

XI. The Sixth Parable: Of Heaven and Earth and the Arrangement of the Elements.

The concluding words of the Sixth Parable are 'When thou hast water from earth, air from water, fire from air, earth from fire, then thou shalt fully and perfectly possess our art'. This is a quotation from the *Secreta Secretorum* of the pseudo-Aristotle and is often quoted in alchemical literature. At first sight this could well be another of the paradoxes of alchemy. At the Creation, however, all things were made from one substance (*omnia ab uno*) and on the death of the body man returns to the ashes whereof he was made. These ashes are symbolised in the reaction vessel of the alchemist and when mixed with permanent water, which is the ferment of gold, yields up a divine water, which is the secret fire. A secret reserved for revelation by God to those especially worthy.

XII. **The Seventh Parable: Of the Confabulation of the Lover and the Beloved.**

The last and final parable is truly a celebration of the *coniunctio*. The Biblical quotations are largely from the *Song of Songs* with a final reference to Calid and his *Liber trium verborum*: 'For he had sowed his seed, that there might ripen thereof threefold fruit, which the author of the Three Words saith to be three precious words, wherein is hidden all the science, which is to be given to the pious, that is to the poor, from the first man even unto the last'. This must refer to the multiplication stage of the work following the preparation of the Red Stone, the *Rubedo*. In some ways the parable is one of recapitulation, for at the beginning it returns to the *Nigredo*, the bride being the *prima materia* of the work. Throughout the *Aurora* there is reference to the Lapis-Christus parallel. Just as the Philosopher's Stone will transmute base metals into gold so, also will Jesus Christ transmute our base natures that we may enter more fully into a state of grace. It is this *unio mystica* with God that is the consummation of the whole treatise.

The extent of St. Thomas's theological and philosophical writings is immense and characterised by a compactness of thought and expression. The vast output of over eight million words is the more remarkable in that he died at the early age of about fifty. He produced commentaries on the Gospels, upon Aristotelian treatises and upon the *Sentences* of Peter Lombard. He compiled two *Summae* of theology, the first *On the Truth of the Catholic Faith against the Gentiles*, known as the *Summa contra gentiles*, and the second, which is his chief masterpiece, the *Summa Theologiae*, which was left unfinished at his death.

The commentaries upon the treatises of Aristotle are particularly relevant in giving some idea of Thomas's views on alchemy that, as Sherwood Taylor has written, '... was sweeping like a fever over 13th century Europe'. Commenting on the Third Book of Aristotle's *Meteorologica* Thomas gives a long discussion on the views put forward as to the generation of metals and their ores under the earth from the reaction of a dry or smoky vapour and a moist vapour. While Aquinas accepts this view he adds something not found in Aristotle, namely that this combination requires a celestial virtue to bring it about. He writes:

> 'The remote material of such metallic bodies is the vapour included in the stony parts of the earth, but the immediate materials of metals are sulphur and mercury, as the alchemists say - so in the aforesaid stony places of the earth by the mineral virtue mercury and sulphur are first generated, and then from these are generated metals according to their various mixture. And so the alchemists, through the true art of alchemy, (but yet a difficult art, because of the occult operations of the celestial virtue, namely the mineral virtue, which because they are hidden, are imitated by us only with difficulty) – these alchemists, by the above principles or by principles laid down by themselves, sometimes make a true generation of metals, sometimes indeed from the aforesaid sulphur and mercury without the generation of the exhalation: but sometimes by making the aforesaid vaporous exhalation exude from certain bodies, by the application of a proportionate heat which is a natural agent' (Lectio IX ad finem)

It is said that the style and contents of the *Aurora* differ widely from St. Thomas's usual manner of writing being to some extent a pastiche of quotations from the Bible and from contemporary and earlier alchemical sources, mainly Latin translations of Arabic originals. It is very often found that it was the practice of alchemical authors to quote extensively from other authors, perhaps with the object of 'lending an air of verisimilitude to an otherwise bald and unconvincing narrative' but in the main to establish the credentials of the author as being well-read and instructed. St. Thomas, on the other hand, uses quotations from other relevant sources, principally the New and Old testaments and such established alchemists as Senior and Alphidus, to emphasise the points that he is making. This is also a characteristic in such works as *Summa Contra gentiles* where, for example, the views of Averroes and Aristotle are contrasted with frequent reference to their works.

The influence of Albertus Magnus, who introduced Thomas to all the burning intellectual and spiritual problems of the age and made him more closely acquainted with the writings of Aristotle and Avicenna and Arabic alchemical philosophy in general, may well have stimulated the young Thomas in his quest for a comprehensive concept of creation. The nature of the text of the *Aurora* might suggest a theoretical rather than a practical approach to alchemy, with greater emphasis on the transmutation of the soul rather than of the base metals themselves.

On the Feast of St. Nicholas in 1273, Thomas was celebrating Mass when he received a revelation which so affected him that he wrote and dictated no more and left his great work, the *Summa Theologiae*, unfinished. He is said to have observed:

> 'The end of my labours is come. All that I have written appears to be so much straw after the things that have been revealed to me.'

Two days before his death, following his last confession to Father Reginald of Priverno and on receiving the viaticum he concluded:

> 'I am receiving thee, price of my soul's redemption: all my studies, my vigils and my labours have been for love of thee. I have taught much and written much of the most sacred Body of Jesus Christ; I have taught and written in the faith of Jesus Christ and of the Holy Roman Church, to whose judgement I offer and submit everything.'

He died in the early hours of the seventh day of March, 1274, being about fifty years of age.

Basil Valentine

'Frater Basilus Valentinus of the Benedictine Order, monk and hermetic philosopher', so runs the inscription surrounding a portrait that was included as a frontispiece to the Hamburg 1677 edition of his collected writings (*Fr. Basilii Valentini Benedictiner Ordens Chymische Schriften*). The name Basil Valentine became one of considerable importance in alchemy. It is reputed that he was born in the year 1394 and, if so, he was a contemporary of an equally mysterious figure Christian Rosenkreutz who was said to have lived from 1374 to 1484.[2]

According to such autobiographical details as he saw fit to include in his works, Basil Valentine was a monk or canon of St. Peter's Priory in Erfurt in Saxony. He travelled widely during the course of a long life, visiting Belgium, Holland and England and, it is believed, he made a pilgrimage to the shrine of St. James at Compostella in northern Spain. This was a journey considered highly important by the alchemistical fraternity and notably by Nicholas Flamel. In his old age Valentine is thought also to have visited Egypt.

In his alchemical writings, his views were not unlike those of Paracelsus (1493-1541) in that the medical aspects of the Philosopher's Stone were to him paramount. This followed the doctrine that as imperfections in metals their falling short of the perfection of gold is akin to human sickness, so cures for the one should, therefore, be effective for the other. Also, like Paracelsus and Christian Rosenkreutz, he lived in a world where the unity of all knowledge was still apparent, and could be comprehended by a single individual as was the Divine Purpose of the Creator for his creation.

Very little is, in fact, known about Basil Valentine as a man. It is thought by many scholars that his name, the Latin and Greek for 'Mighty King', was assumed with the object of preserving his anonymity while enhancing his status as a alchemist. Even successful alchemists lived in fear of the extortion of their secrets under duress imposed by impecunious princes, so it could well be the assumption of a *nom de plume* was just self-protection and not deliberate deceit by a syndicate of literary forgers. Nevertheless, because of this device, the evidence for a monk of that name existing at all is somewhat tenuous. There are no surviving manuscripts of his works which were only published in the early 17th century, about the same time as the Rosicrucian manifestos. This uncertainty is to be regretted because Basil Valentine appears to have been a gifted chemist who made important contributions to that science and was apparently well ahead of his time if, in fact, he lived and worked in the three or four score years after 1394.

During the 17th century a number of tracts ascribed to Basil Valentine were published under the editorship of Johann Tholde of Hesse. Tholde was the part owner of a flourishing salt business in Franckenhausen in Thuringia. The following titles are those of the most important of Basil's tracts published under his name:

On the Great Stone (with the 12 Keys) (1599)
On Natural and Supernatural Things (1603)
On Occult Philosophy (1603)
The Triumphal Chariot of Antimony (1604)
On the Microcosm (1602)
The Manual Operations (1624)
The Last Will and Testament (1626)

One of the historians of the Rosicrucian movement, J.L. Jager (1728-1787), mentions that Tholde was encouraged to join the Order in 1624 after twenty years of its known existence so that it would be able to preserve a monopoly in the writing of Basil Valentine and prevent any further editions being published. Jager goes on to state that Tholde was to become the secretary of the Order, an office he held for a considerable period. It is noticeable that he edited no further works of Basil's after 1624 and published nothing of his own. Tholde's *Coelum Chemicum*, a collection of Rosicrucian arcana, which he preserved in manuscript, was not published until 1737, long after his death. In the preface to this work it is stated that it was known to the Rosicrucian brotherhood some hundred years earlier.

Returning to the published works of Basil Valentine, as far as is known the original 15th century manuscripts, from which the edited versions were produced, have disappeared completely. There is, apparently, no reference to them in the catalogues of the various Continental libraries. Their late publication has been explained as due to the fact (or legend) that on his death in the priory at Erfurt the manuscripts were enclosed, together with a portion of the 'powder of projection' under a marble tablet on the High Altar. This only came to light many years later during repairs to the Church building which had been damaged when it was struck by lightning. This miraculous disclosure of the 'Works' is well within the alchemical tradition, but the evidence for it is very dubious!

Apart from the Works, with their doubtful authenticity and which contain occasional autobiographical details, the first outside reference to Basil is given by an historian of the town of Erfurt writing in 1675, when he states:

'Basil Valentine lived about 1413 in St. Peter's Benedictine monastery at Erfurt, was a searching investigator into the wonders of medicine and science and of whom it was also said he was one of those who hoped to make gold but deceived himself.'

There have been a number of searches made into the records of the Benedictine Order in the hope some reference would appear to a monk of that name living then, or at any other time, and who was based at Erfurt or another priory in the region, but without any success. It is believed the Emperor Maximilian I (1459-1519) caused similar enquiries to be made in 1515, but was also unable to find out anything. As mentioned above, the simple explanation could well be the monk who called himself Basil Valentine wrote under a pseudonym to preserve his anonymity and to protect himself and, indeed, his priory from the persecution that might follow success in the field of alchemy – real or claimed. The choice of the name 'Basilius Valentinus', a composite of Greek and Latin signifying 'Mighty King', however, seems entirely appropriate for one engaged in the Royal or Sacerdotal Art of Holy Alchemy!

Under normal circumstances the works of the alchemists can be dated with reasonable certainty by identification of the various quotations and topical allusions incorporated by the writer. However, the works ascribed to Basil Valentine are strangely free from quotations, although they do contain a number of topical allusions that have thrown doubt on the dates on which the various tracts are claimed to have been written.

The Triumphal Chariot
of
Antmony

It has even been questioned whether they could have been written at all during the lifetime of one born in 1394. For example, mention is made of America, discovered by Columbus in 1492, the French sickness (syphillis) which swept across Europe during the closing decade of the 15th century and the use of metal type and tobacco which would date the manuscripts at the end of the 15th century when Basil would have been a very old man.

On the assumption these contemporary allusions were not introduced during the editing of the manuscripts for publication by Tholde, which is a distinct possibility as it would have enhanced the reputation of the author and the selling value of the tracts, there is nothing to suggest the monk who wrote under the name of Basil Valentine could not have enjoyed a considerable, if not enhanced, life span as an alchemist. There is no reason why he should not have lived well into the time of Paracelsus (1493-1541). In one of the most readable of his books, the *Triumphal Chariot of Antimony*, he refers to the time when he was a young man, which implies that at the time of writing he was well advanced in years.

The questionable situation with regard to his identity was summarised by John Ferguson, the bibliographer of Alchemy, who wrote the following words:

'Whether Basilus Valentinus was a real person or not, whether he was a Benedictine monk at Erfurt or at Walkenreid or not, whether he was a Benedictine monk at all or not, whether he was a native of Alsace or not, whether he flourished in 1413 or 1493 or in both or neither, whether his works had been hidden and were afterwards revealed by a flash of lightning or not, whether they were by him or by his editor Tholde or Tholden, whether they are all genuine or some by other writers, whether Paracelsus copied him or he Paracelsus, whether the works are not really by Paracelsus, whether the name Basilus Valentinus is not made up and may even denote the Alchemical Mystery itself, are questions which have been debated and some of which have been provisionally answered, but all of which are still open to discussion if only fresh data would come to light.'

This is as true now as it was in 1906 when Ferguson wrote it!

Basil Valentine's account of how he began his studies is of interest, partly on account of his style but mainly because of his conformity to the general beliefs of the alchemists of that time (whenever that was!) The following passage, taken from the preface to the *Great Stone of the Philosophers*, reads:

> 'Being possessed with human fear, I began to consider, out of the simplicity of Nature, the miseries of this world, and exceedingly lamented with myself the offences committed by our First Parents, and how little repentance there was throughout the world, and that men grew daily worse and worse, an eternal punishment without redemption hanging over the heads of such impenitents: Therefore made I haste to withdraw myself from sin, and bid farewell to the World, and addict myself to the Lord as his only servant.'

> Having lived some time in my Order, then also, after I had done my appointed devotions, meddling not with frivolous things, lest my vain thoughts through idleness should yield causes of greater evils; I took upon me diligently to searching into Nature, and thoroughly to Anatomise the Arcanae thereof, which I found to be the greatest pleasure next to Eternal things. Having found in our Monastery many books written by Philosophers of ancient time, who truly followed Nature in their Study and Search; this gave a greater encouragement to my mind, to learn those things they knew; and though it proved difficult to me in the beginning, yet at last it proved more easie. The Lord so granted (to whom I daily prayed) that I should see those things that others before me had seen.'

Apparently the original language of the bulk of the writings of Valentinus was not Latin but an Upper Saxony dialect of German and his style a mixture of pious mysticism, intolerable verbosity and sharp invective very reminiscent of Paracelsus but without the latter's undoubted ability to invent, and use, new words. Despite this Basil himself writes:

> 'I propose to set forth what I have to say in a few simple straightforward words, for I am no adept in the art of multiplying words; nor do I think that exuberance of language tends to clearness; on the contrary, I am convinced that it is many words that darken council.'

He goes on to say that:

> 'Although many are engaged in the search after the Stone, it is nevertheless found but by very few. For God never intended that it should become generally known. It is rather to be regarded as a gift which he reserves for those favoured few, who love the truth, hate falsehood, who study our Art earnestly by day and by night, and whose hearts are set upon God with an unfeigned affection.'

In order that the 'Disciple of the Spagyric Art' should know how to lay the requisite stable foundations Basil then sets out a rule of meditation for diligent consideration by '... those who are in possession of the wisdom of philosophy as by all who aspire after that wisdom which is attained by our Art.'

The five point rule runs somewhat as follows:

1. Invocation of God

2. Contemplation of Nature
3. True Preparation
4. The Way of Using
5. The Use and Profit

The preliminary act of invocation emphasises the spiritual side of alchemy, the acknowledgement by the creature of his dependence upon his Divine Creator, the striving for perfection that is the prerequisite for enlightenment. It emphasises that the Oratory is complementary to the Laboratory, the importance of the Hermetic doctrine 'as above so it is below' and the signification of the old Latin tag *'laborare est orare'* – to work is to pray.

To Basil Valentine the contemplation of Nature covered a wide field. He spoke of Spirits inhabiting Fire, Air, Water and Earth in much the same way as there are Salamanders, Sylphs, Undines and Gnomes all derived from the concepts of Paracelsus. He explained the macrocosm like the microcosm was created by God from nothing, first as primal matter that then took on form. Such form and matter is Earth and Water which were separated at the creation of everything else including animals and plants, Air and Fire. Life came from the form and matter through motion that originated in Air and was completed by warmth and heat the Sulphurous Hot Spirit.

Although Basil adopted the concept of the Four Elements, he laid much more stress on the Three Principles – Salt, Sulphur and Mercury – and in this he might well have forestalled Paracelsus:

'I have already indicated that all things are constituted of three essences, namely Mercury, Sulphur and Salt and herein I have taught what is true. But know that the Stone is composed out of One, Two, Three, Four and Five. Out of Five, that is the Quintessence of its own substance. Out of Four, by which we must understand the Four Elements. Out of Three, and these are the Three Principles of All Things. Out of Two for the Mercurial Substance is two-fold. Out of One, and this is the First Essence of everything which emanated from the final Fiat of Creation.'

Contemplation was two-fold, the contemplation of the possible and of the impossible. Basil further considered the former as being:

'Essential properties of a thing, the circumstance by which it is conditioned, its matter, its form, its operations and their source, whence it is infused and implanted, how it is generated by the Stars, formed by the Elements and produced and perfected by the Three Principles and likewise how it may be separated and by what processes or manual operations.'

Amongst these manual operations Basil listed calcination, sublimation, reverberation, circulation, putrefaction, digestion, distillation, cohobation and fixation.

Impossible contemplation, on the other hand, was concerned with such matters as the Eternity of God, the Infinite nature of the Godhead and the Sin against the Holy Ghost – the only sin for which forgiveness is not possible.

The next stage of preparation relies upon the use of manual operations and requires diligent application and knowledge based upon experience. It is designed to bring out the latent and hidden nature of the substance. Mary Anne Atwood, with her background interest in Mesmerism and a wide knowledge of the writings of the alchemists through the ages, was inclined to attribute to the manual operations the processes by which hypnotism is induced. Whether Basil Valentine is referring to this is uncertain, although it is held by some scholars that initiation into the Eleusinian Mysteries involved the use of hypnosis by skilled practitioners in the art with the object of communicating with the deeper levels of the personality and thus to bring about enlightenment.

After preparation, the fourth stage was the determination of proportions by weight or dosage. This could only be accomplished by experiment as, apparently, could also the final stage, the establishment of uses. As Basil comments:

> 'It may happen that a medicine properly prepared and given in proper quantity is nevertheless rather harmful than curative in certain diseases. So it is necessary to discover the conditions under which alone it is likely to be beneficial.'

So much for the basic regimen necessary during the preparation of the Stone. The two best known works of Basil Valentine will now be considered. These are the *Practica concerning the Great Stone of the Ancient Sages together with the Twelve Keys* with which 'We may open the doors of knowledge of the most ancient Stone and unseal the most secret fountain of health' and the *Triumphal Chariot of Antimony*. These two works are considered to be as authentic as any such works can be. They set out to describe the preparation of two versions of the Philosopher's Stone – the Stone itself, which is based on gold, and the Fire Stone that is based on the metal antimony. Both Stones would transmute base metals into gold, but the Fire Stone was not a Universal Tincture like the former, which is prepared from the essence of gold. While it would tinge silver into gold and would make tin and lead perfect, it would not transmute iron and copper. In addition, one part of the Fire Stone had no power to transmute more than five parts of any imperfect metal, unlike the true Stone that was almost unlimited in its effect.

In both tracts a brief account of the generation of metals is given. This tends to be rather Neoplatonic in origin and is to the effect that a vapour is extracted from the elementary earth by means of the celestial influence of the heavenly planets. The vapour then descends from above on to those things that are below and resolves into a kind of water or earth-like substance, giving rise to a kind of water which penetrates the ground to form the Three Principles – Mercury, Sulphur and Salt – by way of an internal Soul, an impalpable Spirit and a visible Body.

All metals are produced by the combination of these principles in various proportions. The metals may be fixed or not fixed, permanent and unchangeable or volatile and variable. These properties are those displayed by the seven metals (gold, silver, copper, iron, tin, lead and mercury) generally known at that time. The Three Principles will also give rise to the other minerals such as vitriol, antimony, marcasite, electrum, etc. Vitriol was any glassy mineral such as gypsum (calcium sulphate), blue vitriol (copper sulphate) or green vitriol (ferrous sulphate) or even sulphuric acid itself! Mention of the subject of the vitriols is a reminder that the famous Vitriol Acrostic has been attributed to Basil Valentine. The word VITRIOL stood for its expansion VISITA INTERIORA TERRA RECTIFICANDO INVENIES OCCULTUM LAPIDEM (Visit the inward parts of the earth, by rectifying thou shalt find the hidden stone). Antimony, or Hungarian antimony, was, in fact, not the metal of that name but the ore stibnite (antimony sulphide) used by the ladies of ancient Egypt as a form of eye shadow! Marcasite is, of course, crystallised iron pyrites (ferrous sulphide) or 'Fool's Gold', while electrum is an alloy of silver and gold used by the ancients.

The two tracts then set out to describe the preparation of their respective Stones – the *Practica with its Twelve Keys* by means of much allegory and the *Triumphal Chariot* with fairly down-to-earth chemical experiments. For example, the preparation of an important starting material, Glass of Antimony, is described by Basil as follows:

> 'Take Hungarian antimony (the sulphide) and powder it finely. Heat the powder in an earthenware dish in a furnace. When smoke rises, stir the powder with an iron spoon and continue stirring and heating until no more smoke is given off and the powder sticks together in small globules. Remove the dish, re-powder the antimony and repeat the process again until no further smoke is evolved. The powder is now calcined antimony. Fuse this calcined antimony in a crucible by itself to give a glass which is glass of antimony. This may also be prepared by fusing the antimony ore with borax when a red glass will be produced – the redness may be extracted with a special spirit of wine'

This is rather more helpful than instructions such as:

> 'Take a quantity of the best and finest gold, and separate it into its component parts by those media which Nature vouchsafes to those who are lovers of the Art, as an anatomist dissects the human body. Thus change your gold back into what it was before it became gold; and thou shalt find the seed, the beginning, the middle and the end from which OUR GOLD and its female principle are derived, viz. the pure and subtle spirit, the spotless soul, and the astral salt and balsam.'

The *Twelve Keys* are fascinating for they include examples of the typical quaint pictorial symbolism used by the alchemists for communicating inner truth. That they are symbolic is certain, but without the key to the *Keys* the interpretation can only be speculative.

Take the first key, which is fairly well known. Here is depicted the King and Queen with their castle in the background and two figures in the foreground. To the left is a wolf jumping over a crucible on a furnace, while to the right is a one-legged man with a scythe, representing Saturn or lead, standing over a small sealed vessel in a water bath (Bain Marie) on a fire. The text accompanying the Key is an example of Basil being obscure rather than lucid. He begins with the statement:

> 'No impure or spotted things are useful for our purpose, metallic substances must be purged as a physician would purge the inward parts of the body.'

The purification of gold (the King) is by fusing it with antimony (the Wolf), an operation that must be carried out thrice according to the three-flowered plant in the hand of the Queen. The silver (the Queen) is likewise purified with lead. The purified gold and silver become the sophic Sulphur and sophic Mercury respectively.

The First of the Twelve Keys

It might be expected the first Key would give some explanation of that which is essential to all operations of alchemy, the *materia prima* or starting material. Perhaps it does, but whether this is gold or antimony or the ore of antimony is left to the intuition of the reader.

In the **Triumphal Chariot of Antimony** Basil writes:

'... take, in the name of God, equal parts of the ore of antimony obtained after sunrise, and of saltpetre; pulverise finely, mix well, place over a gentle fire, bake dexterously (and the method of baking is the key to the whole work). There will remain a blackish substance. Out of this prepare glass, which pound, extract its red tincture with strong distilled vinegar (made of the same ore), remove the vinegar by distillation in the (water) bath. There remains a powder from which you should make a second extract with highly rectified spirit of wine. Let the sediment settle, when a beautiful sweet red extract of great medicinal value will be obtained. This is the Sulphur of antimony.

Take two pounds of this extract and four ounces of the salt of antimony. Pour over these the extract, circulate for at least one month in a closed vessel when the salt will unite with the extract of the sulphur. Remove the sediment, if any, extract the Spirit of Wine in St. Mary's bath, sublime the powder that remains and it will be distilled in the form of a many coloured, sweet, pellucid, reddish oil. Rectify this oil in St. Mary's bath so that the tenth part remains and it is then prepared.

Then take the living MERCURY of antimony, half the quantity of the saltpetre, melt and pour into a copper dish, purify with Tartar (potassium carbonate) and saltpetre until it becomes silvery white. Pour on to it the red oil of the Vitriol, made over iron and highly rectified, remove by distillation in sand (on a sand bath) the viscidity of the Mercury and you will have a precious precipitate of glorious colour.

Take equal parts of this precipitate with sweet oil of Antimony, put in a well-closed phial – if exposed to gentle heat, the precipitate will gradually be dissolved and fixed in the oil; for the fire consumes its viscidity and it becomes a red, dry, fixed and fluid powder which does not give out the slightest smoke. This is the FIRESTONE.'

Instructions such as these could not be clearer. However there are one or two problems. The salt of antimony is not specified and its preparation may involve the substance *Omphacium* about which nothing is known. Recent work attempting to reproduce this preparation in a modern chemical laboratory has shown that the results obtained depend upon the impurities contained in the 15th century materials. When this factor is taken into account similar results are obtained! (Lawrence Principe, Ambix, 1987, Vol. 34, pp. 21-30, ' *"Chemical Translation" and the role of impurities in alchemy: examples from Basil Valentine's Triumph-Wagen'*)

Basil's preparation of the Firestone is summarised as follows:

'The first preparation is calcination and liquifaction into glass, the second is digestion by which the extract is perfected, the third is coagulation, the fourth distillation into an oil, the subtle being separated from the gross, while the fifth is fixation by the last coagulation. This gives rise to the pellucid fiery stone which can operate on metals only when it is fermented and rendered penetrative.'

In alchemy the source and degree of heat is very important. There is much mention of the 'blood heat' such as was produced by bacterial action in horse manure, a favourite source of heat in the early stages of the preparation. The 'Philosopher's Egg' required gentle heating as does a real egg, no greater miracle being required in one than in the other. Basil Valentine defines five intensities of fire, increasing in intensity, as follows:

1. Fire of Heaven (which quickens our heart's love towards God our Father)
2. Elementary heat of the Sun (on the Earth's surface)
3. Corporeal fire by which we cook our food

4. That which one day will consume the visible world
5. Instrument of everlasting punishment (of Hell).

As a postscript to the *Practica* Basil writes:

> 'The regimen of fire is the only thing on which you need to bestow much attention. This is the sum and goal of our search. For our fire is a common fire, and our furnace a common furnace, and though some of my predecessors have left it in writing that our Fire is not common fire, I may tell you that it is only one of their devices for hiding the mysteries of our Art. For the material is common, and its treatment consists chiefly in the proper adjustment of the heat to which it is exposed.'
>
> The fire of the spirit lamp is useless for our purpose. Nor is there any profit in horse dung, nor in the other kinds of heat in the providing of which so much expense is incurred. Neither do we want many kinds of furnaces. Only our threefold furnace affords facilities for properly regulating the heat of the fire. Therefore do not let any babbling sophist induce you to set up a great variety of expensive furnaces. Our furnace is cheap, our fire is cheap, and our material is cheap, and he who has the material will also find a furnace in which to prepare it, just as he who has flour will not be at a loss for an oven in which it may be baked. It is unnecessary to write a special book concerning this part of the subject. You cannot go wrong, so long as you observe the proper degree of heat, which holds a middle place between hot and cold. If you discover this, you are in possession of the secret, and can practice the Art, for which the Creator of all Nature be praised, World without end. AMEN.'

The Firestone in the *Triumph-Wagen* should be prepared and matured by the corporal fire so the reference to the baking of bread might well be of significance.

Much more could be written of Basil Valentine, the alchemist. It may be he was successful in his quest for he apparently prolonged his life to a ripe old age. It seems fairly certain his choice of name was to conceal his real identity and there is a distinct possibility that the edited versions of his various tracts which were put before the public at the beginning of the 17th century and which attracted much attention, contained certain contributions from the editors themselves. Despite this, Basil Valentine appears to have had a good practical grasp of chemical operations and he has contributed much to the science of modern chemistry. He was the first to prepare hydrochloric acid from marine salt and oil of vitriol. He made brandy by the distillation of wine and beer with rectification on calcined tartar (potassium carbonate). He described the extraction of copper from its sulphide ore by oxidation to the sulphate and reaction with scrap iron to give metallic copper. He also described the preparation of diethyl ether from alcohol (spirits of wine) and Oil of Vitriol, he was acquainted with solutions of caustic alkali, the acetates of lead and copper and with fulminating gold. Finally, he attempted a description of the preparation of metallic antimony and its derivatives.

Despite Brother Basil's deliberate obscurity where he thought it necessary, he could be refreshingly precise. While readers of the *Triumphal Chariot of Antimony* may not achieve the preparation of the Firestone, they will at least feel that it is almost within their grasp!

Philip Aureolus Theophrastus von Hohenheim (Paracelsus)

> 'This is my vow: To perfect my medical art and never to swerve from it so long as God grants me my office and to oppose all false medicine and teachings. Then to love the sick, each and all of them, more than if my own body were at stake. Not to judge anything superficially, but by symptoms, nor to administer any medicine without understanding, nor to collect any money without earning it. Not to trust any apothecary, nor to do violence to any child. Not to guess but to know ...' (Paracelsus, Colmar, 1528)

Phillip Aureolus Bombast von Hohenheim was generally known as Theophrastus and is now called by his assumed name Paracelsus, which he adopted towards the end of his life, both to express his eminence as a physician (superior to Celsius) and as a Latin version of the family name of Hohenheim. He is mentioned in the Rosicrucian tract the *Fama* as one of those '... painful, worthy men, who broke with all force through darkness and barbarism and left us who succeeded to follow them. Such a one likewise hath Theophrastus been in vocation and callings although he was none of our fraternity'.

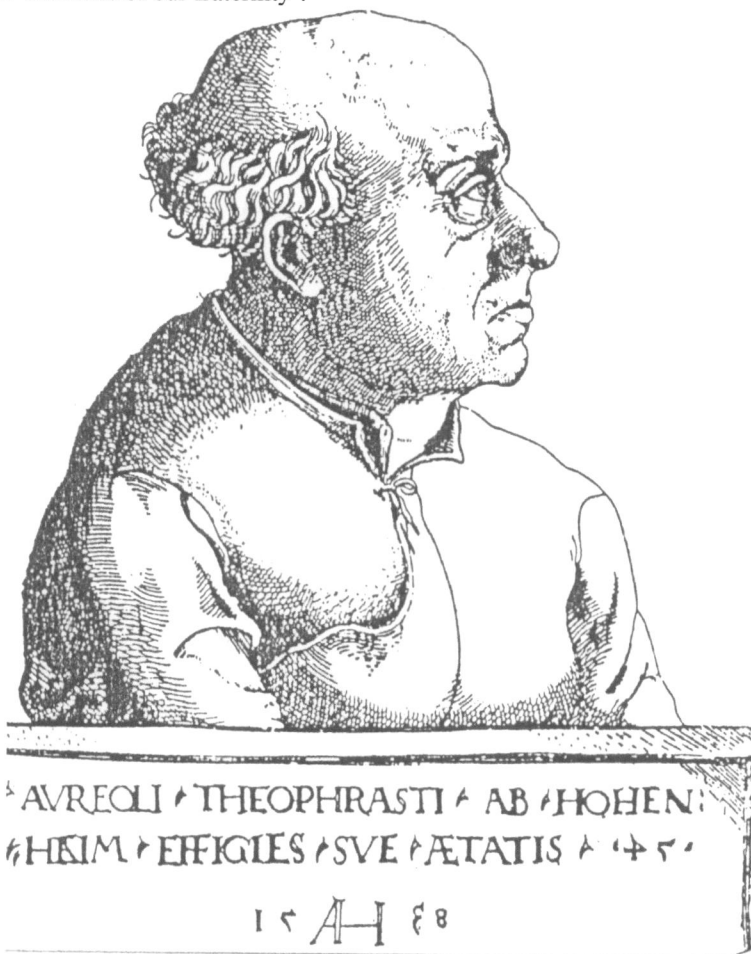

Phillip Aureolus Bombast von Hohenheim

Despite this, Paracelsus' teachings and writings and the particular brand of Neoplatonism which became associated with his name, produced a profound effect on future generations of workers in medicine and science in general and in the Rosicrucian movement in particular. Indeed, some later writers have believed that he was a member of that fraternity in spite of the denial of this in the *Fama*. Also, some points of similarity between the early travels of Paracelsus and those of Christian Rosenkreutz have led some scholars to associate one with the other. There is, however, no doubt concerning the real existence of Paracelsus, the itinerant physician.

In contrast with many of his predecessors, Paracelsus has been the subject of a considerable amount of erudition and speculation both by historians of chemistry and those of medicine. Despite this, however, very little is known about his personal life, although a number of portraits are available and the large bulk of his writings have been preserved. The Sudhoff-Mattheissen edition of the *Collected Works (Paracelsus: Samtliche Werke)* has 14 volumes in 'Part I: Medical, Scientific and Philosophical Writings', while 'Part II:Theological and Religious-Philosophical Writings' is still being compiled, surely evidence that he was a prolific writer with titanic energy!

Eliphas Levi, in his *Transcendental Magic* describes marvellous Paracelsus as a somewhat Rabelaisian character '... always drunk and always lucid ...' but this is not the whole story. A recent biographer, Joland Jacobi, is perhaps more sympathetic and describes him from his portrait as:

> '... a rather slightly built man, with sensitive nervous hands, a relatively large bald skull framed in unruly hair, blazing eyes deep-set and mysterious: a man of an easily irritable temper. Yet withal he was a sincere Christian with deeply held beliefs, an enthusiastic and contentious lecturer beloved of his students but scorned by traditional authority, a genius with an insatiable curiosity about the whole nature of man.'[1]

Paracelsus was born in 1493 (or 1494) in the little village of Einsiedeln, just South of Zurich in Switzerland. The Church and Abbey of Saint Maria of Einsiedeln was place of pilgrimage and the young Paracelsus, growing up in a family of the village doctor, must have met all sorts and conditions of men among the bands of pilgrims to whom, no doubt, his father must have ministered from time to time. His father, Wilhelm Bombast de Riett, was descended from a very old Swabian family from around Stuttgart and was the illegitimate son of George Bombast of Hohenheim (1453-1496), Knight of the Order of St. John. Paracelsus' mother, named Els Ochsner, was a bondswoman of the Benedictine Abbey at Einsiedeln who died when her son was quite young. It is thought her death occurred about the year 1502, at the time when the family moved to Villach near Klagenfurt in Carinthia, where Wilhelm continued to practice his calling as a physician until his death in 1534.

Little is known of Paracelsus' childhood, save that gathered from auto-biographical fragments in his writings:

> 'Ever since my childhood I have pursued these things (alchemy and medicine) and learned them from good teachers, who were thoroughly grounded in adepta philosophia and well versed in the arts. First, from Wilhelmus von Hohenheim, my father, who has never forsaken me, also from many writings of ancients and moderns of diverse lands, who laboured mightily.' (*Grossen Wundarznei*, 1536).

In 1509, at the age of fifteen, Paracelsus travelled to the University of Vienna or, as some believe, of Basle, to formalise his studies. A year or so later he is reputed to have journeyed to Wurzburg to study under Hans von Trittenheim (1462-1519), better known as Trithemius. Some scholars believe that his only contact with Trithemius was through his books, which were concerned with alchemy and the kabbalah, but from the little we know about the young Paracelsus it is more likely he would have made to journey to Wurtzburg to be able to study under so famous a teacher as the Abbot of Spondheim, then nearing the end of his life.

During his formative years Paracelsus absorbed as much as he could of what was known as medicine and related subjects which, in those days, covered the whole field of knowledge, of Man, of Nature and of God. At the present time it is difficult to appreciate the closeness of man to nature. The predominating forces today appear rather different from those of the 16th century yet, by a curious paradox, Paracelsus was also subjected to wars and rumours of wars. The Reformation was under way and he was a contemporary of Martin Luther (1483-1546).

Paracelsus was a non-conformist in many things and also an individualist both in his theological as well as in his naturalist and medical thinking. He submitted neither to the Church of Rome, which to his mind compromised with the social and political injustice of the times, nor with the Reformation which made concessions to the existing aristocratic and civic orders of society. As Walter Pagel, Paracelsus' most recent biographer, has written:

'Paracelsus agreed with the social reformers of his time; the pacifist and the belligerent, the communist and the charitable, the ascetic and the worldly in their criticism of the contemporary powers both ecclesiastical and secular. Yet he differed from them in many respects. On questions of dogma Paracelsus was anti-Baptist, within the limits of private property he recognised the rights of the individual and the family as the fundamental unit of society and of State and Church activity. He endeavoured to preserve the mediaeval idea of Christian Community life and his communism was charismatic rather than dogmatic. He practiced a truly Christian involvement with rich and poor alike and a genuine love for suffering humanity. Both property and poverty were for him, objects of religious rather than economic interpretation.'

After leaving the University of Vienna, Paracelsus journeyed south into Italy to continue his medical studies at the University of Ferrara before returning home to his father at Villach in 1514. Very soon after this he set out again to work for a period in the mines and metallurgical workshops of Sigmund Fueger at Schwaz in the Tyrol, near Innsbruck. Here he was to gain experience in the practical operations of metal extraction and alchemy.

In 1516, at the age of twenty-two, Paracelsus again set out on his travels, with the world at his feet but with very little money, to gain experience in his chosen profession and, as it turned out, to revolutionise medicine. As he was to write in the *Grossen Wundartzney* in 1536:

'... for many years I studied at the Universities of Germany, Italy and France, seeking to discover the foundations of medicine. However, I did not content myself with their teachings and writings and books, but continued my travels to Granada and Lisbon, through Spain and England, through Brandenburg, Prussia, Lithuania, Poland, Hungary, Wallachia, Transylvania, Croatia, the Wendian mark and yet other countries which there is no need to mention here, and wherever I went I eagerly and diligently investigated and sought after the tested and reliable arts of medicine. I went not only to the doctors, but also to the barbers, bath-keepers, learned physicians, women and magicians who pursue the art of healing; I went to alchemists, to monks in monasteries, to nobles and common folk, to experts and the simple ... I have oftentimes reflected that medicine is an uncertain and

The First Page of the *Grossen Wundartzney*, 1536

haphazard art scarcely honourable to practice, curing one and killing ten … Many times I abandoned medicine and followed other pursuits, but then again I was driven back to it. Then I remembered Christ's saying: "The healthy need not a physician, but only the sick." And so

I made a new resolve, interpreting Christ's words to mean that the art of medicine is true, just, certain, perfect and whole, and there is nothing in it that should be attributed to the deception of spirits or chance, but that it is an art tested in need, useful to all the sick and beneficial in restoring their health.'

There is a legend that while Paracelsus was in Russia he was made a prisoner of the Tartars and brought before the Grand Cham. Impressed by his knowledge and understanding, Paracelsus was retained by that monarch at the court and became a royal favourite. On the occasion of an ambassador being sent from China to Constantinople, Paracelsus accompanied the Grand Cham's son as friend and interpreter. Be this as it may, in Constantinople he met the celebrated alchemist Solomon Trismosin (or Pfeiffer) from whom, it is said, he learned the secret of the preparation of the Philosopher's Stone. Trismosin was the compiler of the beautifully illustrated *Splendor Solis* (Harley MS 3469).

It is thought that about this time Paracelsus wrote an early and fundamental treatise entitled the *Paramirum* (*c.*1520) on industrial medicine (*Medica Industria*) in which he set out the five principal causes of disease, which are related to their cures. The five types of disease were: astral, poison, natural, spiritual and divine. He also refers to five types of divination to be practiced by the physician in order to obtain insight into the patient and the nature of the disease. These were – prophets and sibyls from God, signs in Nature in humans and in spirits, and augury from animals and spirits.

Having acquired knowledge of the supreme secret of alchemy, the Universal Dissolvent, the *Azoth*, *Alcahest* or Secret Fire, he continued his travels and became involved in the Venetian wars, gaining further experience as an army surgeon. It was from this time on that legends and reports of the miraculous cures of patients began to circulate. Patients who would otherwise, in the medical opinion of the time, not have survived treatment for their sickness, rapidly recovered.

At the present time it would not appear strange for a physician to enlist as an army surgeon but in the 16th century there was a great gulf fixed between physicians and surgeons. The former were university graduates, doctors of medicine in fine clothes, accepted by local society, while the surgeons were of a more lowly status being classed with barbers and midwives. Indeed, as is well-known, the barber's pole, characteristic of their calling, demonstrated that they carried out the menial tasks of medicine, the letting of blood, the bandaging of wounds and such simple surgical operations as were deemed necessary to prolong rather than to preserve life. On the field of battle the opportunities for the surgeon were legion and Paracelsus must have gained a good working knowledge of both anatomy and surgery. This social dichotomy between physicians and surgeons was anathema to him and he did much to bring about a fusion of the two medical roles.

Paracelsus was, first and foremost, a dedicated physician, but to him the book learning beloved of the followers of Galen and Aristotle was irrelevant. The patient was the book for the physician to read and study, not the antiquated doctrines of the medical establishment! He was later to remark, and modesty was never his strong point in his public utterances:

'In this mid-century, Monarchy of all the Arts pertains to me, Theophrastus Paracelsus, Prince of Philosophy and Medicine. For to this aim I am chosen by God that I may extinguish all fantasies of all far-fetched, false and putative works and presumptuous words, be they of Aristotle, Galen, Avicenna or any of their adherents.'

In 1524, at the age of 30, Paracelsus returned from his travels, hoping to settle and practice medicine in Salzburg which was comparatively near to his home town of Villach, and of sufficient size to provide ample opportunity for the application of his skill and ability as a physician. Unfortunately, in the general unrest of those days of the Reformation, the peasants and serfs throughout Europe were rebelling against their lords. In one such uprising in Salzburg, Paracelsus sided with the peasants and narrowly escaped with his life, slipping away quietly into

Italy to continue his medical studies at Ferrara which resulted in his being awarded a doctorate of medicine in 1525.

The following year he returned to Germany and made application for the rights of citizenship at Strasbourg where he was inscribed in the City Register as a surgeon and, in accordance with the regulations in force at that time, joined the local Guild of Grain Merchants and Millers, which permitted him to practice in the city. Unfortunately, it was not long before he was in trouble. He entered into a public disputation on anatomy with the celebrated anatomist Wendelin Hock. Paracelsus won the argument but forfeited his favour in the city!

It so happened that about that time a prosperous printer and publisher in the City of Basle, one Johann Froben (or Frobenius), fell ill with a sickness which baffled the local medical practitioners. Frobenius appears to have suffered an apoplectic stroke and having heard that there was a new physician at Strasbourg he sent for Paracelsus, who was able to cure him completely. Staying with Frobenius at this time was the Dutch scholar Erasmus, regarded as the leader of Renaissance learning in Northern Europe. The house provided a centre for the gathering of other scholars which included Johann Heussgen (Oecolampadius), the professor of Theology at the University of Basle. The professor was much impressed with the personality and medical skill of his friend's physician and recommended him to the City authorities for the, then vacant, position of Medical Officer of Health (City Physician) and Professor of Medicine. The offer was at once accepted.

True to his nature, Paracelsus immediately proceeded to make both friends and enemies in his new appointment. His lectures at the University were well attended and he was popular with the students, but he upset the academic establishment. To begin with, he insisted, as Johann Sebastian Bach was to do later at Leipzig, in delivering his lectures in German rather than in the customary Latin. In a printed broadsheet he promised he would teach all parts of medicine in a different way to that which had been done by former physicians. An outline of his course of lectures for the Summer of 1527 has survived and reads as follows:

> 1. Propositions on internal diseases and their remedies – stomach and bowel troubles worms, epilepsy, consumption, rashes, gout, asthma, fevers, headaches, disorders of the womb, toothache, and ailments of the eyes and ears.

> 2. Lessons on general pharmaceutics and prescription of remedies, general introduction to special pathology and therapy.

> 3. On the treatment of external wounds, injuries, furoncles, ulcers, swellings and growths.

> 4. During the vacations, particularly dog days, special lectures on diagnosis by pulse and urine, on purges and blood letting, on interpretation of Hippocrates aphorisms and commentaries on Macer's *Herbal*.

To inaugurate this course of lectures Paracelsus is said to have burnt the works of Avicenna on the St. John's Day bonfire, (or in a brass pot with nitre!), thus publicly affronting the medical establishment in much the same manner as did his contemporary Martin Luther with the Church establishment, when he nailed his famous *Ninety-five Theses* to the door of the Schlosskirche at Wittenberg!

By attacking Galen and Avicenna, Paracelsus decried the virtues of their old-fashioned herbal remedies, which were then generally accepted and used by most physicians. In their place he extolled the power of medicines made from mineral substances.

This use of minerals and his critical approach to accepted practice was a significant contribution to the science of medicine. Paracelsus in his way did more for medicine than, perhaps, Luther ever did for religion. He made people think and in so doing, issued in the age of

Iatrochemistry and Chemotherapy.

In order to explain the basis for the introduction of mineral remedies it is necessary to consider Paracelsus' views on alchemy and what he thought it should achieve. He was basically by nature an empiricist. He believed in experiment and had little time for theoreticians. As a practicing alchemist, he appreciated the value of purification, for it was by constantly repeated purifications that the quintessence of a body might be prepared. As a neoplatonist, Paracelsus accepted the doctrine of the Four Elements: Earth, Air, Water and Fire, together with the Quintessence or Ether. In addition he postulated that all substances were mixtures of the three hypostatical principles: Sulphur, Mercury and Salt. These three fundamental states of matter stood for the combustible, the fluid and the solid states, corresponding, on the human level, to the Soul, Spirit and Body. Sulphur bestows oiliness, inflammability, viscosity and structure, Mercury wateriness, spirit, vapour and vivifying powers, while Salt gives bodies their rigidity, solidarity, dryness and earthiness. He believed in the Hermetic doctrine that the seven metals, the tinctures and even the Philosopher's Stone itself all derive from these three substances. This was in accordance with the essential unity of all things, animal, mineral and vegetable.

The correspondence between the seven metals and the planets, from which they derived their names, was obvious from astrological considerations. These planets ruled the various signs of the zodiac and hence the various parts of the body influenced by these signs. Disease, like impurity in metals, was the result of wrong proportions of the hypostatical principles, and treatment of a disease with the appropriate quintessence or arcana should bring relief. The extraction of the arcana of the appropriate metal was usually achieved by solution of the metal in a mineral acid followed by repeated distillation.

Paracelsus believed too much Sulphur in the body gave rise to fever and the plague, too little resulted in gout; too much Mercury gave paralysis and depression and an excess of Salt resulted in diarrhoea and dropsy. If the Sulphur distilled from one organ to another, the patient would suffer from delirium. Some diseases he regarded as entities in themselves, evil principles that enter the patient from outside and are identical with such forces as are immanent in the stars, planets and minerals. In the same manner as Robert Fludd's demons Azael gave rise to plague, smallpox and phthisis, Azazel to apoplexy and Mahazeal to melancholy, stupor, leprosy and scabies!

In his opposition to conventional remedies, Paracelsus soon became the object of hatred to all the druggists and apothecaries in Basle as well as to his fellow physicians. Although his students and the City authorities supported him, attempts were often made to disrupt his lectures. The situation became particularly unpleasant and came to a head when a prominent citizen, Canon Lichtenfels, fell sick and offered a fee of 100 *gulden* to any physician who would cure him. Paracelsus accepted the offer and restored the Canon to good health. The canon, however, refused to pay the agreed sum. As might be imagined, Paracelsus with all his other vexations was not the man tamely to submit to such treatment. He took the matter to Court. For some reason or other, possibly a legal quibble supported by the ill feeling of the local medical fraternity since he seems to have had right on his side, judgement was given against him. He was so infuriated by this that he was to surpass even his own virtuosity in his execration of the startled magistrates. Warned that this flow of scurrilous and slanderous epithets had rendered him liable to severe punishment for contempt of Court, he left the town secretly in great haste and set out on what was to be the second and last stage of his travels. February 1528 was to be the turning point of his life.

He journeyed first to Colmar where he stayed with his friend Lorenz Fries, to be joined shortly by a pupil Operinus who brought such of Paracelsus' belongings as remained to him. It is to this Operinus, later to become Professor of Greek at Basle University and publisher of the works of Versalius, that we owe a contemporary description of Paracelsus. This occurs in a reminiscence included in a letter written to Solenander in 1555. Walter Pagel regards this as an honest treatment of a gentle soul, shocked for life by the rough and irregular habits and jokes of a not-altogether-sane genius. Perhaps some of it should be taken with a pinch of one of the three Hypostatical Principles!

Operinus found fault with Paracelsus' lack of piety and scholarship. He never saw him pray, but often overheard his slighting remarks against both Luther and the Pope and all theologians, none of whom Paracelsus believed to have penetrated to the heart of the Scriptures. Above all, Operinus was repelled by his addiction to drink, yet continues by observing rather charitably that:

> '... that which Paracelsus dictated late after a nocturnal drinking bout and while to all appearances still inebriated, made complete sense and could not be improved upon by a perfectly sober person!'

At this stage of his life, during his travels he never undressed but threw himself upon his bed with his long-sword, a gift from the town hangman, still girded about him. Suddenly he would jump up brandishing the sword, behaving like a madman and generally frightening Operinus to death or distraction. All day he was busy at his furnace producing violent fumes, which once overpowered the assistant when made to sniff the contents of one of the alembics. As has been mentioned earlier, there is some evidence from his written formulations that Paracelsus may have prepared diethyl ether by his treatment of alcohol with sulphuric acid and used its narcotic properties in his surgery. Could the overpowering of the laboratory assistant have been due to an early example of anaesthesia?

Operinus was of the opinion that Paracelsus lived luxuriously. No doubt in his earlier days in medical practice he was never short of money, spending it on new clothes of which he was very fond. The portraits surviving from the first Alsatian period (1520-1530) would seem to bear this out. He tried to give his old clothes away, but they were so dirty that no one would accept them. He was said to have worked miracles on gastric ulcers without restricting the diet of his patients but feasting with them, so he cured them with a full stomach! Up to his twenty-fifth year Paracelsus had been adverse to drink, but later acquired a taste for it and frequently challenged peasants to drinking contests from which he emerged victorious. He was not interested in women, except, perhaps as patients.

From Colmar he journeyed to Esslungen where he wrote one of his several tracts on syphilis. This disease, recently introduced into Europe by, it is thought, sailors returning from the New World, had reached epidemic proportions and was spreading rapidly. Then he went on to Nurenburg and Beratzhausen – travelling South to St. Gallen where he had completed his work on the *Paramirum* in 1531/32. In the following year he passed through Appenzell to his native village of Einsiedeln where he seems to have undergone his deepest spiritual experience. It was at Einsiedeln that the way of the hermit took possession of him to the depths of his being, never again to relinquish him. With this period are associated more than 100 religious treatises, filled with yearning for God, and in which he approached and considered the ultimate metaphysical problems of life and death.

He was not long to remain in this region, however, for he was soon off across Austria to the mining district of Hall and later Schwarz, outside Innsbruck, where he had served his apprenticeship with Sigmund Fueger. His work in these mining areas had resulted in the first treatise on occupational diseases (*On the Miner's Disease*) in which he drew attention to chronic arsenic and mercury poisoning. Still he journeyed on, exercising his calling as he went and spreading the Gospel of the New Medicine, south into Italy where he stayed at Sterzing for a short period before moving on to St. Moritz and then north again to the mineral baths at Oberschuzz.

The year 1536 saw the completion of his great treatise on surgery the *Grosse Wundartzney* and its author riding impetuously through Kempten, Menningen, Ulm and Augsberg to arrive at Munich in 1537. Thence he travelled on to Passau and into Bohemia and then to Vienna. His appearance at this time is given in the Hirschvogel portrait which is dated 1538 and which is held to be the best likeness of all the portraits. Either this portrait, or one by the same artist, dated 1540 was modified by Franz Hoogenburgh and used to illustrate the first edition of the *Philosophia Magna* published by Burchmann in 1567. An interesting feature of this portrait, mentioned by Pagel, is the representation of unmistakable Rosicrucian symbols. In the top left alcove or window, a child's head is emerging through a cleft in the ground and other symbols of rebirth are significant. In the right alcove, a male figure in front of Jacob's ladder is of particular interest. The left eye is not represented. This symbolises the 'cagastric' earthly or temporal eye that should be kept closed while the right, the 'iliastric' eye, contemplates eternity and the higher sphere of the Creator.

This second portrait depicts a sick man, looking considerably older than his forty-seven years. He grasps his famous sword, from which he was seldom parted, and on the pommel of which is engraved the word AZOTH, the great secret of his world. The word AZOTH contains the first and last letters of the Latin, Greek and Hebrew alphabets. AZOTH was the name of the secret medicine or *Elixir Vitae*, an infallible remedy. According to an old tradition, these miraculous virtues were inherent in a white powder that Paracelsus always carried with him, in a case inside the pommel of his sword. It was for this reason that he never put aside his sword even at night. AZOTH is also the secret name of Mercurius, which was extracted from certain metals, and considered a universal medicine comprising the virtues of all the others. Hence it is also a name for the Philosopher's Stone.

In 1541, the Suffragan Bishop of Salzburg, Ernest of Wittelsbach, invited Paracelsus to come to Salzburg assuring him that, under his protection, he would be welcome. Paracelsus had by now journeyed south from Vienna to St. Veit and Klagenfurt, a town close to Villach where, in 1538, he learned that his father had died some four years previously. Paracelsus arrived in Salzburg early in 1541, but was not destined to enjoy the Bishop's hospitality for very long, for on the 21st September he suffered a stroke while staying at the White Horse Inn near the quay.

There in a little room, sitting on a cot, and on St. Matthew's Day in the presence of a Master Kalbsohr, he dictated his Last Will and Testament, being sound in reason, mind and spirit. In this final document he committed his life, death and his poor soul to the shield and protection of Almighty God, in the steadfast hope that the eternal mercy of God would not allow the bitter sufferings, the martyrdom and death of Our Saviour, Hallowed Lord Jesus Christ, to be fruitless and of no avail to him a miserable man. He then provided for a burial place in the Almshouse of St. Sebastian, and for the services to be held for the repose and salvation of his soul. After his death, his poor belongings were to be sold and a few *gulden* bequeathed to each of his friends and relatives in Einsiedeln and to his universal legatees the poor, the wretched and the needy people who have no stipend or other provision made for them.

Three days later, on 24 September, he died leaving behind a few *gulden*, a few pieces of clothing, some trinkets, a Bible Concordance, a copy of the Holy Scriptures with the commentaries of St. Jerome, an edition of the New Testament, a pharmaceutical compendium and a *Collectiana Theologica*. But where were the tools of his trade and his famous sword?

In 1772, some years after his death, the Church of St. Sebastian in Salzburg was restored and a tablet with the following inscription placed above his burial place. It read, in Latin:

'Here lies buried Phillipus Theophrastus, a distinguished Doctor of Medicine, who with wonderful art cured dire wounds, leprosy, gout, dropsy and other contagious diseases of the body, and who gave to the poor the good which he obtained and accumulated, In the Year of Our Lord 1541, the 24th September, he exchanged life for death. *PAX VIVIS REQUIES AETERNA SEPULTIS* (to the living peace, to the entombed eternal rest).

A prayer of Paracelsus may be appropriate to conclude this section:

'Even if we had in our hands all the Arcana and Elixirs of the great and little world, but not thee O Lord, all this would be nothing! Close to Thee, in Thee, and with Thee alone is the eternal life and the light. In our bodies after the great death, when it is as though renewed by the Divine Fire, the light will be translucent, and only then will it truly shine. God grant that this may be soon! Amen. Amen. Amen. (from *De Limbo Aeterno*)

Nicholas flamel

It is generally believed that Nicholas Flamel was born in France at Pontoise about the year 1330. He received a good education and became a Scrivener and Law Writer in the City of Paris. He had a younger brother Jean who was also a Writer. In 1356 Nicholas married a widow named Perenelle and lived near the Church of St. Jacques la Boucherie, at the corner of the Rue des Escrivains, later building a house for himself and his wife in the Rue de Montmorency. In 1357 he was, unexpectedly, offered a curious old book that he purchased for the sum of two florins. This book was to change his life and to set him on an adventure which brought him success in a search undertaken by so many but achieved by so few – that of the preparation of the Philosopher's Stone. This, apparently, enabled Flamel and his wife to amass great wealth, to build fourteen hostels for the poor, three chapels and to repair seven churches, among their other charitable activities. Perenelle died in 1397 and Nicholas in 1418, although there is some traditional doubt as to the facts of their mortality. In 1712 Paul Lucas, a 17th century traveller, mentions having met a dervish, a Sufi, in Asia Minor who told him his friend Flamel and his wife were still alive and that he had seen them both in India less than three years previously!

The story of Nicholas Flamel has been told many times and is perhaps best related in his own words taken from his introduction to his *Explanation of the Hieroglyphic Figures Placed by me NICHOLAS FLAMEL, Scrivener, in the Church Yard of the Innocents, in the fourth Arch, entering by the great gate of St. Denis Street, and taking the way on the right hand.*[3]

flamel's Background

'Although that I, Nicholas Flamel, Notary, and abiding in Paris, in this year one thousand three hundred fourscore and nineteen, and dwelling in my house in the street of Notaries, near unto the Chapel of St. James of the Bouchery; although I say, that I learned but a little Latin, because of the small means of my parents, which nevertheless were by them that envy me the most, accounted honest people; yet by the grace of God, and the intercession of the blessed Saints in Paradise of both sexes, and principally of St. James of Gallicia, I have not wanted the understanding of the Books of the Philosophers, and in them learned their hidden secrets. And for this cause, there shall never be any moment of my life, when I remember the high good, wherein upon my knees (if the place will give me leave) or otherwise, in my heart with all my affection, I shall not render thanks to this most benign God, which never suffereth the child of the just to beg from door to door, and deceiveth not them which wholly trust in his blessing.'

The Book of Abraham the Jew

'Whilst therefore, I Nicholas Flamel, Notary, after the decease of my parents, got my living by our Art of Writing, by making Inventories, dressing accounts and summing up the expenses of Tutors and Pupils, there fell into my hands for the sum of two florins, a gilded Book, very old and large. It was not of Paper, nor of Parchment, as other Books be, but was only made of delicate rinds (as it seemed unto me) of tender young trees. The cover of it was of brass, well bound, all engraven with letters, or strange figures; and for my part I think they might well be Greek characters, or some such like ancient language. Sure I am, I could not read them, and I know well they were not notes nor letters of the Latin nor of the Gaul, for of them we understand a little. As for that which was within it, the leaves of bark or rind, were engraven, and with admirable diligence written, with a point of iron, in fair and neat Latin letters coloured. It contained thrice seven leaves, for so were they counted in the top of the leaves, and always every seventh leaf was without any writing, but instead thereof, upon the first seventh leaf, there was painted a Rod and Serpents swallowing it up. In the second seventh, a Cross where a Serpent was crucified; and in the last seventh, there were painted Desserts, or Wildernesses, in the midst whereof ran many fair fountains from whence there issued out a number of Serpents, which ran up and down here and there.

Upon the first of the leaves, was written in great Capital Letters of Gold, Abraham the Jew, Prince, Priest, Levite, Astrologer, and Philosopher, to the Nation of the Jews, by the Wrath of God dispersed among the Gauls, sendeth Health. After this it was filled with great execrations and curses (with the word *Maranatha*, which was often repeated there) against every person that should cast his eyes upon it, if he were not Sacrificer or Scribe.

He that sold me this Book knew not what it was worth, no more than I when I bought it; I believe it had been stolen or taken from the miserable Jews; or found hid in some part of the ancient place of their abode. Within the Book, in the second leaf, he comforted his Nation, counselling them to fly vices, and above all, Idolatry, attending with sweet patience the coming of the Messias, who would vanquish all the Kings of the Earth, and should reign with his people in glory eternally. Without doubt this had been some very wise and understanding man. In the third leaf, and in all the other writings that followed, to help his Captive Nation to pay their tributes unto the Roman Emperors, and to do other things, which I will not speak of, he taught them in common words the Transmutation of Metals; he painted the Vessels by the sides, and he advertised them of the colours, and of all the rest, saving the first Agent, of which he spake not a word, but only (as he said) in the fourth and fifth leaves entire he painted it, and figured it with great cunning and workmanship; for although it was well and intelligibly figured and painted, yet no man could ever have been able to understand it, without being well skilled in their Cabala, which goeth by tradition, and without having well studied their books. The fourth and fifth leaves therefore, were without any writing, all full of fair figures enlightened, or as it were enlightened, for the work was very exquisite.

First he painted a young man with wings at his ankles, having in his hand a Caducean rod, writhen about with two Serpents, wherewith he struck upon a helmet which covered his head. He seemed to my small judgement, to be the God Mercury of the Pagans: against him there came running and flying with open wings, a great old man, who upon his head had an hour glass fastened, and in his hand a hook (or scythe) like Death, with the which, in a terrible and furious manner, he would have cut off the feet of Mercury. On the other side of the fourth leaf, he painted a fair flower at the top of a very high mountain which were sore shaken with the North wind; it had the foot blue, the flowers white and red, the leaves shining like fine gold: and round about it the Dragons and Griffins of the North made their nests and abode. On the fifth leaf there was a fair Rose Tree flowered in the midst of a sweet Garden, climbing up against a hollow Oak: at the foot whereof boiled a fountain of most white water, which ran head-long into the depths, notwithstanding it first passed among the hands of infinite people, who digged in the earth seeking for it; but because they were blind, none of them knew it, except here and there one who considered the weight.

On the last side of the fifth leaf there was a King with a great Fauchion, who made to be killed in his presence by some Soldiers a great multitude of little infants, whose Mothers wept at the feet of the unpitiful Soldiers: the Blood of which Infants was afterwards by other Soldiers gathered up, and put into a great vessel, wherein the Sun and Moon came to bathe themselves. And because that this History did represent the more part of that of the Innocents slain by Herod, and that in this Book I learned the greater part of the Art, this was one of the causes why I placed in their Church-yard these Hieroglyphic Symbols of this secret science. And thus you see that which was in the first five leaves.

I will not represent unto you that which was written in good and intelligible Latin in all the other written leaves, for God would punish me, because I should commit a greater wickedness, than he who (as it is said) wished that all the men of the world had but one head that he might cut it off with one blow. Having with me therefore this fair book, I did nothing else day nor night, but study upon it, understanding very well all the operations that it showed, but not knowing with what Matter I should begin, which made me very heavy and solitary, and caused me to fetch many a sigh. My wife Perrenella, whom I loved as myself, and had lately married was much astonished at this, comforting me, and earnestly demanding, if she could by any means deliver me from this trouble. I could not possibly hold my tongue, but told her all, and showed this fair book, whereof at the same instant that she saw it, she became as much enamoured as myself, taking extreme pleasure to behold the fair cover, gravings, images and portraits, whereof notwithstanding she understood as little as I: yet it was a great comfort to me to talk with her, and to entertain myself, what we should do to have the interpretation of them.'

Master Anselme helps Flamel to interpret his Book

'In the end I caused to be painted within my lodging, as naturally as I could, all the figures and portraits of the fourth and fifth leaf, which I showed to the greatest Clerks in Paris, who understood thereof no more than myself: I told them they were found in a book that taught the Philosopher's Stone, but the greatest part of them made a mock both of me, and that blessed Stone, excepting one called Master Anselme, who was a Licentiate in Physic, and studied hard in this Science. He had a great desire to have seen my Book, and there was nothing in the world he would not have done for the sight of it: but I always told him I had it not; only I made him a large description of the Method. He told me that the first portrait represented Time, which devoured all; and according to the number of the six written leaves, there was required the space of six years, to perfect the Stone; and then, he said, we must turn the glass and seeth it no more. And when I told him that this was not painted, but only to show and teach the first Agent, (as was said in the Book) he answered me, that this decoction for six years space, was, as it were, a second Agent: and that certainly the first Agent was there painted, which was the white and heavy water, which without doubt was Argent Vive, which they could not fix, nor cut off his feet, that is to say, take away his volatility, save by that long decoction in the purest blood of young infants: for in that, this Argent Vive being joined with Gold and Silver, was first turned with them into an herb like that which was there painted, and afterwards by corruption, into Serpents; which Serpents being wholly dried, and decocted by fire, were reduced into powder of gold, which should be the Stone.'

Flamel's Early Experiments

'This was the cause that during the space of one and twenty years, I tried a thousand broulleryes, yet never with blood, for that was wicked and villainous: for I found in my Book, that the Philosophers called Blood, the mineral spirit, which is in the metals, principally in the Sun, Moon, and Mercury, to the assembling thereof, I always tended; yet these interpretations for the most part were more subtle than true. Not seeing therefore in my works the signs, at the time written in my Book, I was always to begin again.'

Flamel's Spiritual Pilgrimage to Compostella

'In the end having lost all hope of ever understanding these figures, for my last refuge, I made a vow to God, and St. James of Gallicia, to demand the interpretation of them, at some Jewish Priest, in some synagogue of Spain: whereupon with the consent of Perrenella, carrying with me the Extract of the Pictures, having taken the Pilgrim's habit and staff, in the same fashion as you may see me without this same Arch, in the Church-yard, in the which I put these Hieroglyphical Figures, where I have also set against the wall, on the one and the other side, a Procession, in which are represented by order all the colours of the Stone, so as they come and go, with this writing in French:

> *Moult plaist a Dieu procession,*
> *S'elle est faicte en devotion:*

that is:

> Much pleaseth God procession,
> If it be done in devotion.

Which is as it were the beginning of King Hercules his Book, which entreateth of the colours of the Stone, entitled Iris, or the Rainbow, in these terms, Opera processio multum naturae placet, that is, the procession of the work is very pleasant unto Nature: the which I have put there expressly for the great Clerks, who shall understand the Allusion.'

The Explanation of Master Canches

'In this same fashion, I say, I put myself upon my way; and so much I did, that I arrived at Mountjoy, and afterwards at St. James, where there was a great devotion I accomplished my vow. This done, in Leon at my return I met with a merchant of Bologne, who made me known to a physician, a Jew by nation, and then as a Christian, dwelling in Leon aforesaid, who was very skilful in sublime Sciences, called Master Canches. As soon as I had shown him the figures of my Extract, he being ravished with great astonishment and joy, demanded of me incontinently, if I could tell him any news of the Book, from whence they were drawn? I answered him in Latin, (wherein he asked me the question) that I hoped to have some good news of the Book, if anybody could decipher unto me the Enigmas. All at that instant transported with great Ardour and joy, he began to decipher unto me the beginning. But to be short, he well content to learn news where this Book should be, and I to hear him speak; and certainly he had heard much discourse of the Book, but, (as he said) as of a thing which was believed to be utterly lost, we resolved of our voyage, and from Leon we passed to Oviedo, and from thence to Sansom, where we put ourselves to Sea to come into France.'

The Death of Master Canches

'Our voyage had been fortunate enough, and all ready, since we were entered into this kingdom he had most truly interpreted unto me the greatest part of my figures, where even unto the very points and pricks, he found great mysteries, which seemed unto me wonderful, when arriving at Orleans, this learned man fell extremely sick, being afflicted with excessive vomitings, which remained still with him of those he had suffered at Sea, and he was in such a continual fear of my forsaking him, that he could imagine nothing like unto it. And although I was always at his side, yet would he incessantly call for me, but, in sum, he died at the end of the seventh day of his sickness, by reason whereof I was much grieved, yet as well as I could, I caused him to be buried in the Church of the Holy Cross at Orleans, where he yet resteth: God have his soul, for he died a good Christian. And surely if I be not hindered by death, I will give unto that Church some revenue, to cause some Masses to be said for his soul every day.'

Flamel Discovers the First Principles

'He that would see the manner of my arrival and the joy of Perrenella, let him look upon us two, in this city of Paris, upon the door of the Chapel of St. James of the Bouchery, close by the one side of my house, where we are both painted, myself giving thanks at the feet of St. James of Gallicia, and Perrenella at the feet of St. John, whom she had so often called upon. So it was, that by the grace of God, and the intercession of the happy and holy Virgin, and the blessed Saints James and John, I knew all that I desired, that is to say, The first Principles, yet not their first preparation, which is a thing most difficult, above all the things in the world. But in the end I had that also, after long errors of three years, or thereabouts; during which time I did nothing but study and labour, so as you may see me without this Arch, where I have placed my Processions against the two Pillars of it, under the feet of St. James and St. John, praying always to God, with my beads in my hand, reading attentively within a Book, and poysing the words of the Philosophers; and afterwards trying and proving the divers operations, which I imagined to myself by their only words. Finally I found that which

I desired, which I also soon knew, by the strong scent and odour thereof. Having this, I easily accomplished the Mastery, for knowing the preparation of the first Agents, and after following my Book according to the letter, I could not have missed it, though I would. Then the first time that I made projection, was upon Mercury, whereof I turned half a pound, or thereabouts, into pure Silver, better than that of the Mine, as I myself assayed, and made others assay many times. This was upon a Monday, the 17th January, about noon, in my house, Perrenella only being present, in the year of the restoring of mankind, 1382. And afterwards, following always my Book, from word to word, I made projection of the Red Stone, upon the like quantity of Mercury, in the presence likewise of Perrenella only, in the same house, the five and twentieth day of April following, the same year, about five o'clock in the evening; which I transmuted truly into almost as much pure Gold, better assuredly than common Gold, more soft and more pliable. I may speak it with truth, I have made it three times, with the help of Perrenella, who understood it as well as I, because she helped in my operations, and without doubt, if she would have enterprised to have done it alone, she had attained to the end and perfection thereof. I had indeed enough when I had once done it, but I found exceeding great pleasure and delight, in seeing and contemplating the Admirable works of Nature, within the Vessels. To signify unto thee then, how I have done it three times, thou shalt see in this Arch, if thou have any skill to know them, three furnaces, like unto them which serve for our operations, I was afraid a long time, that Perrenella could not hide the extreme joy of her felicity, which I measured by mine own, and lest she should let fall some word amongst her kindred, of the great treasures which we possessed: for extreme joy takes away the understanding, as well as great heaviness; but the goodness of the most great God, has not only filled me with this blessing, to give me a wife chaste and sage, for she was moreover, not only capable of reason, but also to do all that was reasonable, and more discrete and secret than ordinarily women are. Above all, she was exceeding devout, and therefore seeing herself without hope of children, and now well stricken in years, she began as I did, to think of God, and to give ourselves to the works of mercy.'

The Death of Perrenella

'At that time when I wrote this Commentary, in the year one thousand four hundred and thirteen, in the end of the year, after the decease of my faithful companion, which I shall lament all the days of my life; she and I had already founded and endued with revenues, 14 hospitals in this City of Paris, we had now built from the ground three Chapels, we had enriched with great gifts, and good rents, seven Churches, with many reparations in their Church-yards, beside that which we have done at Bologne, which is not much less than that which we have done here. I will not speak of the good which both of us have done to particular poor folks, principally to widows and poor orphans, whose names if I should tell, and how I did it, besides that my reward should be in this world, I should likewise do displeasure to those good persons, whom I pray God bless, which I would not do for anything in the world.

Building therefore these Churches, Church-yards and Hospitals, in this City, I resolved myself, to cause to be painted on the fourth Arch of the Church-yard of the Innocents, as you enter in by the great gate in St. Denis Street, and taking the way on the right hand, the most true and essential marks of the Art, yet under veils, and Hieroglyphic covertures, in imitation of those which are in the gilded Book of Abraham the Jew, which may represent two things, according to the capacity and understanding of them that behold them: First the mysteries of our future and undoubted Resurrection, at the day of Judgement, and coming of good Jesus (whom it may please to have mercy upon us) a history which is well agreeing to a Church-yard. And secondly, they may signify to them, who are skilled in Natural Philosophy. all the principal and necessary operations of the Mastery. These Hieroglyphic figures shall serve as two ways to lead unto the heavenly life: the first and most open sense, teaching the sacred Mysteries of our Salvation; (as I will show hereafter) the other teaching every man, that hath any small understanding in the Stone, the linearly way of the work; which being perfected by any one, the change of evil into good, takes away from him the root of all sin, (which is covetousness) making him liberal, gentle, pious, religious, and fearing God, how evil so ever he was before, for from thenceforward, he is continually ravished, with the great grace and mercy which he hath obtained from God, and with the profoundness of his Divine and admirable works. These are the reasons which have moved me to set these forms in this fashion, and in this place which is a Church-yard, to the end that if any man obtain this inestimable good, to conquer this rich golden fleece, he may think with himself (as I did) not to keep the talent of God digged in the Earth, buying lands and possessions, which are the vanities of this world: but rather to work charitably towards his brethren, remembering himself that he learned this secret, amongst the bones of the dead, in whose number he shall shortly be found; and that after this life, he must render an account before a just and redoubtable Judge, who will censure even to an idle and vain word. Let him therefore, who having well weighted my words, and well known and understood my figures, hath first gotten elsewhere the knowledge of the first beginnings and Agents (for certainly in these Figures and Commentaries, he shall not find any step or information thereof) perfect to the glory of God the Mastery of Hermes, remembering himself of the Church Catholic, Apostolic and Roman; and of all other Churches, Church-yards, and Hospitals; and above all of the Church of the Innocents in this City, (in the Church-yard whereof he shall have contemplated these true demonstrations) opening bounteously his purse to them that are secretly poor honest people, desolate weak women, widows, and forlorn orphans. So be it.'

Evidence for Flamel's Life and Work

The Book of Abraham the Jew, with Flamel's annotations, apparently passed into the hands of Cardinal Richlieu (1585-1642) the chief minister of Louis XIII. It had been examined at an earlier date by the Count de Cabrines, as is related by Pierre Borel Councillor and Physician-in-Ordinary to Charles VI (Charles the Mad) who ruled from 1380-1422.

According to Langlet du Fresnoy, an historian of alchemy, four items of evidence for the correctness of Flamel's narrative survived until 1742. These were the Arch in the cemetery of the Holy Innocents, a Charnel House in the same cemetery with a Gothic 'N.F.' and the inscription:

Ce charnier fut fait & donne a l'Eglise,
Pour l'amour de Dieu, l'an 1399.

Secondly, there were statues of Nicholas and Perrenelle, one kneeling at the feet of St.James and the other at the feet of St. John by the Marivaux door of the Church of St. Jacques-la-Boucherie, with the Gothic letters 'N' and 'P' on the respective pedestals.

Thirdly there was a statue of Flamel in the rue Notre Dame, at the portal of Genevieve of Arden, said to have been erected in 1402.

The remains of his house, dating from 1407 are still to be seen in the fabric of No. 51 rue de Montmorency and an inscribed marble tablet from his tomb in the old Church of St. Jacques-la-Boucherie, demolished in 1797, is preserved in the Musée de Cluny.

The account of his life and travels given by Flamel and first published in 1612 by Pierre Arnauld in his *Trois Traictez de la Philosophie Naturelle*, is very convincing, albeit slightly prolix, and created a considerable impression at a time when alchemy was approaching its peak in Europe. It was an unsettling period just before the beginning of the Thirty Years War (1618-1648) with the bubonic plague raging across the Continent. The question remains: did Flamel really prepare the Philosopher's Stone and were the transmutations he and his wife Perrenelle carried out responsible for their charitable donations and activities? E.J. Holmyard was of the opinion the extent of these gifts, although large, was not in fact beyond the limit of the fortune that a prosperous notary, publisher and bookseller might have acquired, especially one of a frugal and industrious life and who had perhaps married a wealthy widow.

Nicolas Flamel died on 22 March 1417, and was buried in the Church of St. Jacques-la-Boucherie. Little remains of this Church at the present day. The body of the building was demolished in 1797, although the fine Gothic tower is still in existence. Before his death Flamel had designed a tombstone with an inscription composed by himself. Following the demolition work this disappeared to be discovered many years later in the shop of a fruiterer and herbalist in the rue des Arcis. The polished back surface had been found ideal for chopping up dried herbs! The tombstone ended up in the Musée de Cluny where it remains to this day.

The Inscribed Marble Tablet from the Tomb of Nicholas Flamel

The tablet is a marble stone 58 cm. by 45 cm. and 4 cm. thick and can be divided for the purposes of description into three sections. The top portion shows a representation of Christ between St. Peter and St. Paul. Our Lord is making a gesture of benediction with his right hand and holding a globe surmounted by a cross in his left. Between the three figures are representations of the Sun and the Moon, symbolic of gold and silver.

The inscription on the middle of the stone reads:

'Feu Nicolas Flamel jadis escrivain a laissie par son testament a leuvre de cette eglise certaines rentes et amisons quil avoit acquestees et achetees a son vivant pour faire certain service divin et distribucions dargent chacun an par aumosne touchans les quinze vins lostel dieu et autres eglises et hospitaux de paris Soit prie po les trespasses.'

The bottom section displays a wasted corpse surmounted with the inscription *Domine deus in tua misericordia speravi*, and the inscription below reads *De terre suis venus et en terre retourne Lame rens a toy Ihu qui les pechiez pardon.*

It is said that after his death Flamel's house in the rue des Ecrivans was ransacked until only the cellars were left by the local people in the hope of finding some trace of his alchemical operations. The site was finally searched in 1560 by order of the magistrate of the Chatelet district. Nothing now remains. According to the nineteenth century historian Vallet de Viriville, Flamel's house 'Great Gables' still stands at No. 51 rue Montmorency. This must have been one of the other houses that Flamel owned in Paris.

Perrenelle died either in 1397 or 1404 and was buried in the cemetery of the Holy Innocents. But, according to a Dervish encountered by Paul Lucas, Flamel avoided the accusation of being an alchemist once the stories of his munificence were noised abroad by having the deaths of his wife and himself published in advance and by taking certain small doses of the Stone managed to prolong their lives. According to Lucas' account:

'On his advice she (Perrenelle) feigned an illness and when it had seemingly run its course she was reported as dead, although, in fact, she was in Switzerland, awaiting his joining her (she would have had a long wait for Flamel did not "die" until 1417 at the best some thirteen years later!) Instead of herself, a log of wood was buried, dressed in her clothes; and, so that everything might be done with due ceremony, the interment took place in one of the chapels she had been instrumental in having built. Later on Flamel adopted the same stratagem and, since money can make most things possible, little difficulty was experienced in winning over medical and clerical agreement to the deception. He left a Will in which he requested that he should be buried in the same grave as his wife and that a pyramid should be erected over the sepulchre. So, while the living Adept was on his way to join his wife in Switzerland, a second log of wood was buried in his stead!'

How much reliance can be placed on this narrative, far fetched as it may appear, depends upon belief in the reality of the Philosopher's Stone and its unique properties. A number of alchemists have, apparently, survived their normal life span. Such characters as the Compte de St. Germain, and, more recently, Fulcanelli would seem to have given some evidence for survival to recent times. They appear, however, to have been transformed into rather insubstantial beings, even undergoing a sex change in the process!

Flamel published several works concerned with alchemy, his life and travels. The most significant is his *Exposition of the Hieroglyphic Figures*, reprinted above. This is concerned with both the theory and practice of the Philosopher's Stone. There is also a *Tractatus Brevis* or *Summarium Philosphicum*, both published in Manget's *Bibliotheca Chemicae Curiosa* and in the *Hermetic Museum*. The former gives an interpretation of the basic figures painted on the arch but, surprisingly, does not mention the figures of Abraham the Jew which also appears in the attached illustration.

The *Summarium* draws attention to the basic question of the formation of metals and their ores in the earth and to the changes that are continually occurring in the mineral veins. It proposes that, in accordance with the sulphur-mercury theory, these substances are the components or seeds of all metals, the one representing the male and the other the female principle. The seeds are composed of elementary substances – the former of fire and air and the latter of earth and water – and were represented under the forms of two dragons. Sulphur is shown as a wingless dragon (the female seed) and mercury as a dragon with wings because it is volatile and can evaporate. Mercury, although not the element of that name, is the Mother of Metals and being itself a metal must contain a two-fold metallic substance – the inner substances of the Moon and of the Sun. Nature forming Mercury from these two spirits then strives to transmute them into a perfect bodily form. This is shown in that veins of lead ore also contain small amounts of gold and silver, imparted to them by Nature for the purpose of multiplication and development.

The Hieroglyphical Figures of Nicholas Flamel

As an ore, the metal remains in its 'mercury' and can develop, but if it is extracted all development ceases in a similar manner to a fruit which will not ripen when plucked from the tree.   The alchemists held that if gold and silver be joined together through their proper mercury, they will have power to render all other (imperfect) metals perfect. By this they did not mean common gold and silver which must always remain what they are, can never become anything else and certainly cannot aid the development of other metals. They meant the real *living* gold and silver which must be sought 'on the tree' for only there can it grow and increase in size, according to the possibilities of its nature.

In practice this implies that the 'mercury' should be taken and warmed day and night in an alembic over a gentle fire. Heat and moisture are the food of all earthly things – animal, vegetable and mineral. Coal or wood fires are too violent for this purpose and should be replaced by the heat of the sun. Flamel then gives somewhat mystical instructions for the preparation of the Mercury of the Sages:

'If they would attain it they must go to the seventh mountain, where there is no plain, and from its height they must look down upon the sixth, which they will behold at a great distance. On the summit of that mountain they will find the glorious Regal Herb, which some Sages call a mineral and some a vegetable. The bones they must leave and only extract its pure juice, which will enable them to do the better part of the work. This is the true and subtle mercury of the philosophers which you must take.

This mercury will enable the white and red tinctures to be prepared in sequence. It is a simple task. All that is necessary is that the mercury be placed together with its like (which is fire) into ashes in a glass vessel in a suitable alembic. 'If you do this there will come out a chicken, that will deliver you with its blood from all diseases and feed you with its flesh, and clothe you with its feathers and shelter you from the cold.'

flamel's Exposition of the fiieroglyphical figures

The mountain and the Regal Herb are reminiscent of the back of the fourth leaf in Flamel's Book on which was painted '... a fair flower at the top of a very high mountain which were sore shaken with the North wind.'   The well-known illustration of the Hieroglyphical Figures is discussed in detail in his work of that name. He begins with a general description of 'The Book of Abraham the Jew' and describes the symbolism of the figures that he had painted below the arch in the churchyard of the Church of the Holy Innocents. These are significant in terms of the figures depicted, the colours used and the 'rowls' or Latin texts held by the figures. These last are, apparently, of particular importance.

In the bottom left-hand corner of the illustration is an enclosure with a 'penner' and inkhorn held in a hand. This, and the letters 'N' and 'F' within similar enclosures to the left of St. Paul and to the right of St. Peter, represent athanors, triple furnaces which signify Flamel having prepared the Stone on three occasions. The text indicates the degree of heat to be applied to the reaction vessel in its initial stages. The two Dragons, one with wings and one without, symbolize Sulphur and Mercury (the SOL and LUNA), the two sperms, not to be found on earth and not the vulgar materials of that name commonly available from apothecaries. They are depicted as dragons because, according to Flamel, their stink is exceeding great and their exhalations within the reaction vessel black, blue and yellowish signifying putrefaction and generation. This is the 'head of the crow', the black of the blackest black.

Flamel draws attention to the Serpent of Mars pierced by Cadmus with his lance against a hollow oak. The hollow oak is represented in the rose garden on the fifth page of the Book of Abraham the Jew.

The next figure is that of the Man and Woman clothed in gowns of orange colour upon a field of azure and blue holding their rowls. The man (Nicholas Flamel himself) has written on his rowl *HOMO VENIET AD JUDICIUM DEI* (man will come to the judgement of God) and the woman (Perrenelle) on hers *VERE ILLA DIES TERRIBILIS ERIT* (this will truly be a terrible day). This figure represents the second operation in which the two natures, male and female, are married together. In terms of the elements, the hot and cold, dry and moist, begin amiably to approach together in the presence of 'mediators and peacemakers'. The first operation dissolved the components into solution which was then coagulated to a black earth. As these two natures are married together they form the 'head of the crow' the 'Androgyne' or Hermaphrodite of the Ancients.

In the reaction vessel this embryo can be brought to form a 'most puissant King, invincible and incorruptible' which is the quintessence. The components are divided into two so that the product of the first half may be nourished as it develops. The second half is the *Azoth*, the 'milk of life', which is used to wash and cleanse the first half which is the *Leton*. The *Azoth* is also the serpent Python which must be killed and overcome – again a reference to the split oak of Cadmus! The washings are the teeth of the serpent which when sown in the earth spring up as soldiers, which is the subject of a well-known alchemical engraving. This operation is summarised as 'It dissolves itself, it congeals itself, and it quickens itself.' The Rebis will begin to exhibit its whiteness from its extremities and for this reason Perrenelle is surrounded by the 'rowl' concerning the terrible day of judgement when all will be spiritualised and whitened.

The next hieroglyphic figure to be explained is that of St. Paul holding a sword and supporting Flamel as a young man. His rowl says *DELA MALA QUI FECIT* (Blot out the evil I have done) which as evil equates to blackness and means that the black components should be whitened. The sword symbolises the removal of the head of the crow which, again, is a further term for blackness. To emphasise this the sword itself is wreathed about with a black girdle. This again refers to the blackness and whiteness of the reactants – the five turns on the sword give not only the period of that part of the process but also the number of imbibitions (five – at monthly intervals). This Albification is a critical stage during which:

> '... the Stone before it will wholly forsake his blackness and become white in the fashion of a most shining marble, and of a naked flaming sword, will put on all the colours that thou canst possibly imagine, often it will melt, and often coagulate itself, and amidst these divers and contrary operations (which the vegetable seed that is in it makes it perform at one and the same time), it will grow Citrine, Green, Red (but not of a true red), it will become Yellow, Blue and Orange colour, until that being wholly overcome by dryness and heat all these infinite colours will end in this admirable Citrine Whiteness.'

The next group of figures to be explained are those in the centre of the Arch. These are the three penitents and the figure of Our Lord coming to judge the world. The three represent the unity of Body, Soul and Spirit and their rising to life in the process of Albification. The Saviour 'who shall eternally unite unto him all pure and clean souls, and will drive away all impurity and uncleaness' is the White Elixir, transmuting all impure metals into silver.

The transition of the White Stone to the Red Stone is depicted by two Angels with the rowl *SURGITE MORTUI VENITE AD JUDICIUM DOMINI MEI* (Arise ye dead and come unto the judgement of God, my Lord) which ends in the mouth of the Red Lion and teaches that the operation must not be discontinued until it achieves a true Red Purple colour.

The figures of St. Peter and of Perenelle imply Multiplication, the enhancement of the power of the Stone. She asks to have the rich Accoutrements and colour of St. Peter. Her rowl *CHRE PRECORE ESTO PIUS* (Christ be pitiful unto me) implies particular care in heating during these final stages, for while the Stone is resistant to fire, the vessel containing it 'is always brittle and easy to be broken'.

The final hieroglyph in Flamel's 'Exposition' is described as:

> 'Upon a dark violet field, a man red purple, holding the feet of a Lion red as vermilion, which hath wings, and it seems would ravish and carry away the man'.

These colours symbolise the completion of the Art with its full decoction.

The Stone, alluded to as 'she', is now like a Lion, devouring every pure and metallic nature, and changing it into her pure substance, into true and pure gold, more fine than that of the best mines. In the words of Flamel:

> 'The Stone carrieth this man out of this vale of miseries, that is to say, out of the discommodities of poverty and infirmity, and with her wings gloriously lifts him up, out of the dead and standing waters of Egypt (which are the ordinary thoughts of mortal men) making him despise this life, and the riches thereof, causing him night and day to meditate on God, and his Saints, to dwell in the Imperial Heaven, and to drink the sweet springs of the Fountains of everlasting hope.'

The Work concludes with the following prayer to Almighty God:

O Lord, give us the grace to use it well, to the augmentation of the faith, and the profit of our Souls, and to the increase of the glory of this noble Realm. Amen.

Nicholas Flamel appears to have been one of history's great alchemists. His story covered all that might be expected of an alchemist: the long search, the providential guidance, the participation of a female collaborator, the 'rags to riches' transformation and the evidence for longevity or of some mystical personality change. He flourished at a time when alchemical speculation was at its heyday in Europe, when the West had benefited from the wisdom of the East with the translation of alchemical scripts preserved in Arabic in the libraries of the Islamic world and augmented with concepts from the Silk Road to India and China.

Could Flamel have been a fictional character created by the demand for esoteric knowledge? This is thought to be unlikely. There are those, who defining alchemy as the misguided search for an impossible goal, deny the possibility of transmutation of the elements by an undefined additive as they would the phenomena of 'anomalous water' or 'cold fusion' or, indeed, of anything contrary to the laws of Nature as currently understood. There is, however, something in the narrative of Nicholas Flamel that still rings true and deserves respect and examination!

Denis Zachaire

According to Ferguson, Denis Zachaire or Zeccaire was born at Guienne in France about the year 1510. He was educated privately by an alchemist and went on to study Law at Toulouse. Later he went to Paris and soon made the acquaintance of the alchemical fraternity in that city. It is said after a long study of the works of Arnold of Villanova, Bernard of Treviso, Raymon Lully and a very early work on alchemy the *Turba Philosophorum*, he managed to prepare the Philosopher's Stone himself and transmuted mercury into gold in 1550. He married, and together he and his wife travelled on the continent of Europe. Unfortunately, on reaching Cologne he was murdered, while asleep, by a servant, who escaped with Denis's wife and his store of the Philosopher's Stone.

Denis Zachaire left behind a fascinating autobiographical account of his life and studies which gives much information regarding the way alchemists worked during the second quarter of the 16th century. The present account is based on the abridgement by E.J. Holmyard of the English translation of the 1612 edition, in French, of *Opuscule Tres-Excellent de la vraye Philosophie naturelle des Metaux. Tractant de l'augmentation & perfection d'iceaux. Avec un advertissment d'eviter les folles dispenses qui se font par faute de vraye science. Par Maistre D. Zacaire Gentilhomme Guiennois. MDCXII.*

At the age of twenty Denis was sent by his parents to a college in Bordeaux to attend lectures on arts subjects, but despite making good progress with these mundane studies he soon became attracted to the subject of alchemy and managed to compile a collection of recipes for bringing about transmutation. Following the death of his parents, his relations arranged for him to go to the University at Toulouse with the object of studying Law. Denis was, however, eager to carry out experiments in alchemy.

'Immediately that I was arrived at Toulouse, I set myself to building small furnaces, being confirmed in all things by my master [a private tutor whom he had had at Bordeaux and who went with him to Toulouse]. Then from small I went to large, and soon I had a room entirely equipped with them, some for distilling, others for subliming, others for calcining, others for dissolving in the water bath, others for melting, in such sort that for a beginning I spent in one year the two hundred crowns that had been supplied to us for our support during two years of study. This was spent in setting up the furnaces and in buying charcoal, divers and infinite drugs, divers vessels of glass which I bought for six crowns a time, without counting two ounces of gold which were lost in practicing one of the recipes, and two or three marks of silver for another, in which, if any of it was recovered, which was very little, it was so crude and so blackened by the force of the mixtures which the recipes had ordered to be added that it was almost entirely useless'.

'So at the end of the year my two hundred crowns were gone up in smoke. And my master died of a continued fever which seized him during the summer by reason of the soot that he breathed and swallowed, for he was so desirous of accomplishing something worth while he hardly ever left the room, where he made scarcely less soot than there is in the Arsenal of Venice where cannon are cast. His death was a great grief to me.'

Returning to his home in Guyenne, which now belonged to him, Denis dismissed the resident caretakers and let the property, during the next three years, to bring in the sum of 400 crowns. He then returned to Toulouse to carry on his alchemical work and experiment with a recipe supplied to him by an Italian, who worked with him in his laboratory.

'I kept this man with me to see the outcome of his recipe, to practice which I was obliged to purchase two ounces of gold and a mark of silver. When these were melted together, we dissolved them in aqua fortis, then we calcined them by evaporation. We tried to dissolve them with divers other waters by divers distillations so many times that two months passed before our powder was ready to make the reprojection of it. We used as much of it as the recipe required, but it was in vain. From all the gold and silver which I had used I recovered only half a mark, without counting the other costs, which were not small. So my 400 crowns were reduced to 230, and of this I supplied my Italian with twenty to go to find the author of the recipe, who he said was at Milan, in order that he might write back to us. After this I was at Toulouse all the winter awaiting his return, and I should be there yet if I had decided to wait for him – for I have not seen him since.'

Being forced to leave Toulouse for a time due to an outbreak of the plague, Denis spent a short time in the town of Cahors. Here he was advised as to which of the several approaches to the work might be the most successful. He returned to Toulouse and resumed his experiments, but the results were disappointing.

'But not for that did I cease always to pursue my enterprise. And the better to be able to continue it, I joined with an Abbe near Toulouse who said he had the duplicate of a recipe for making our great work which a friend of his who followed the Cardinal of Armagnac had sent him from Rome and which he took to be perfectly assured. For this I furnished 100 crowns and he the like; and we began to set up new furnaces in a wholly different fashion to work on it.'

Again, although they provided themselves with an adequate supply of charcoal to maintain the heat of the furnaces for a whole year, the powder obtained at the end had no effect when projected upon a quantity of heated mercury. Denis continues:

'I leave you to judge whether we were angry about it, especially Monsieur l'Abbé who had already published to all his monks like a very good public secretary that it remained only to have a beautiful lead fountain melted which they had in their cloister to convert it into gold immediately that our result should be achieved. But he had it melted at another time to provide material in vain for a certain German who stopped at his abbey while I was in Paris.
In spite of all this he did not cease to wish to continue his undertaking, and he advised me to set about to get together three or four hundred crowns, and he would furnish the same amount, in order that I might go and live in Paris, a city which is today more frequented by divers operators in this science than any other in all Europe, and become acquainted with all sorts of people, to work with those in whom I recognised something worthwhile, and to divide it between us like two brothers.'

Raising the necessary sum of money (800 crowns) Denis returned to Paris and set about cultivating contacts within the alchemical fraternity. They would meet regularly either at each others' lodgings or at that great symbolic edifice the cathedral of Notre Dame.

> 'Some would say, "If we had the means to start again, we should do something worth while", while others, "If our vessels had held we should be there", and others "If we had our copper vessel perfectly round and well-closed, we should have fixed Mercury with the Moon", and so on, for there was not one among them who had had any success, and not one who was not ready with an excuse.'

One alchemist did, however, promise to show Denis how to extract silver from cinnabar.

> 'Since he needed filings of fine silver we bought three marks of it and made them into filings. He mixed these with pulverised cinnabar and made little pegs out of the mixture with an artificial paste, and heated them in a well-covered earthen vessel for a certain time. When they were thoroughly dry, he melted them and submitted the material to cupellation, and we found three marks and a little more of fine silver. This he said had come from the cinnabar, the fine silver we had put in having flown away in the smoke. If this was profit, God knows it. And I also knew it, who had spent more than thirty crowns.'

Meanwhile, the Abbé had received a request from the King of Navarre to reveal the secret to him, for which service the King was willing to pay four thousand crowns. It was therefore imperative that Denis should return to Toulouse. This he did.

> 'The word, four thousand crowns, so tickled the ears of the Abbe that making himself believe that he had them already in his purse, he would not rest until I had started on my way to Pau (where the King's palace was). I arrived in the month of May and worked there for about six weeks, and when I had finished I received the recompense I expected.'

Needless to say, the four thousand crowns were not forthcoming although the King promised to reward him later.

> 'This response annoyed me so much that without waiting for his lovely promises (having heretofore been nourished on them at my own expense) I started to return to the Abbé.'

On the way back, Denis heard of a man learned in natural philosophy. On meeting him he was advised he should refrain from any further attempts at transmutation until he had digested as many of the works of the ancient philosophers that he could obtain. Thus fortified, he returned to Paris on the day following the festival of 'All Saints' in the year 1546.

> 'There for ten crowns I bought books on philosophy, ancient as well as modern, part of which were printed while others were written by hand, the 'Crowd of the Philosophers', 'the good Trevisan', 'the Complaint of Nature' and other treatises which have never been printed. And having hired for myself a small room in the Faubourg Saint-Marceau, I lived there for a year, with a small boy who served me, without frequenting anyone, studying day and night on those authors, with the result that at the end of a month I made one conclusion, then another, then I changed it almost entirely. So, while waiting for a conclusion in which there was mot variety nor contradiction between the sentences of the books of the philosophers, I spent an entire year and part of another without being able to gain over my study to the extent of being able to make any entire and perfect conclusion.'

For a time Denis was discouraged and did very little save frequent the haunts of the alchemists.

It was, however, not long before he returned to his studies.

'I began to re-read with very great diligence the works of Raymond Lully, principally his *Testament and Codicil*, which I reconciled so well with a letter which he had written in his time to King Robert [and with a recipe from another source] that I made a conclusion from them wholly contrary to all the operations which I had seen formerly but such that I read nothing in all the books which did not harmonize very well with my opinion, even the conclusion which Arnold of Villanova, who was master of Raymond Lully in this science, has made at the end of his Great Rosary. Thus I lived about one year longer without doing anything except read and think about my conclusion day and night, waiting until the term for which I had let my house should be passed to return to work at my own home.

Determined to practice the above-mentioned conclusion, I arrived there at the beginning of Lent, during which I provided myself with everything that I needed and set up a furnace for working, so that I commenced it the day after Easter. But this was not without having divers impediments [from relatives and friends, who urged the folly of throwing money away and of buying charcoal on such a scale as to arouse suspicion that counterfeiting was going on].

I leave it to you to imagine whether this talk was a bore to me, since at this time I was seeing my work go from better to better, and I was always attentive to the conduct of it in spite of these and other comparable delays which came upon me incessantly, and especially the anger of the plague, which was so great during the summer that no foot-travel or traffic which was not interrupted, in such a manner that a day did not pass that I was not looking with very great diligence for the appearance of the three colours which the philosophers have written ought to appear before the perfection of our divine work. These, thanks to the Lord God, I saw, one after another, for on the very next Easter Day, I saw the true and perfect experience of them on quicksilver heated in a crucible which I converted into fine gold under my own eyes in less than an hour by means of a little of this divine powder.

God knows if I was delighted about it. But I did not boast for all that. But after having rendered thanks to our good God who had shown me such grace and favour, I went away on the next day to find the Abbé at his monastery to fulfil the covenant and promise which we had made together. But I found that he had died six months previously, at which I was greatly grieved.

I therefore went away to a certain place to wait there for a friend of mine and near relative, who had lived with me at my residence and whom I had left there with authority and express instructions to sell all and each of the paternal goods which I had, to pay my creditors with the proceeds, and to distribute the rest secretly to those who were in need of it, in order that my relatives and others might feel some benefit from the great good that God had given me, without anyone being the wiser. But they thought on the contrary that, despairing and ashamed of my foolish expenditures, I had sold my goods in order to retire to another place, as this friend of mine informed me when he came to find me on the first day of July. And we went to Lucerne, I having decided to travel and to pass the rest of my days in a certain very renowned city of Germany, with a very small household, in order that I might not be known even by those who see and read and read this little book of mine during my lifetime, in our country of France.'

Alchemy as Illustrated by the Splendor Solis of Salomon Trismosin

The Harley Manuscript 3469 is one of several versions of the treatise on alchemy entitled *Splendor Solis* which have come down to us. It is ascribed to Salomon Trismosin, an alchemist who is thought to have lived in the latter half of the 15th Century and is said to have had an influence on Paracelsus (1493-1541). The Harley manuscript comprises a text illustrated with twenty-two very beautiful and fascinating miniatures. The series of illustrations is one of the best examples of the attempts of the alchemists to employ art in order to convey the secrets of their processes to those capable of understanding them.

Harley 3469 is dated about 1582 and contained illustrations, it is thought, by the Dutch artist Lucas van Leiden. The work is a compilation of seven treatises with a preface and a conclusion. The text, as is the case with alchemical literature of the period, is fairly enigmatic and cites a number of well – and lesser-known authorities such as Alphidus, Rhases, Aristotle, Hali, Morienus, Sevarius, Hermes, Virgil, Avicenna, Menaldus, Senior, Rosinus, Socrates, Calus, Actor, Miraldus, Ciliator, Lucas, Baldus, Geber, Hortulanus and Albertus Magnus.

Preface: The Entrance of the Aspirants

This scene could well be the '… open entrance to the closed palace of the king …' of Eirenaeus Philalethes, or the temple of Ezekiel, from under the threshold of which a stream flowed out. This goes to emphasise the Art should be approached with reverence, that the guidance of a more experienced operator (or even divine inspiration) is required and that in some ways there is no turning back once the first steps on the path are taken. In this case the stream is a Rubicon. Indeed, the text gives a warning that '… if someone is unable to accomplish something in the Art of the Philosopher's Stone, it were better for him not to throw himself into it at all than to attempt it partially.'

The two Suns in the plate are thought to symbolize the process of alchemical incarnation, the meeting of the ABOVE with the BELOW, for the *Emerald Tablet of Hermes Trismegistus* states:

'It is truth, truth without lies, certain truth that that which is above is liketo that which is below, and that which is below is like to that which is above to accomplish the miracles of one thing. And as all things were made by the contemplation of One, so all things arose from this one thing by a single act of adaption.'

The plate also gives a clue as to the nature of the primal material for SOL is the name for GOLD and as Philalethes says: '… as the end you look for is Gold: so let Gold be the subject on which you work, and none other.'

The First Treatise: The Origin of the Stone of the Philosophers and the Art how to Produce It

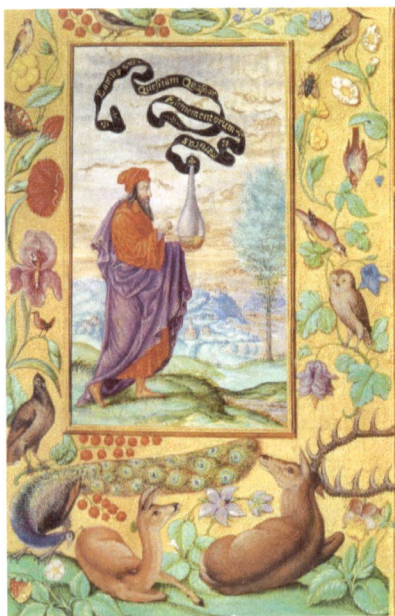

The First Treatise

The legend above the flask of the alchemist reads 'Let us go and seek the Nature of the Four Elements'. The alchemists believed that transmutation of base metals into silver of gold could be achieved by heating with a catalyst, the Philosopher's Stone, when the metal would be broken down into its elements and reconstituted again into a more glorious form. That this form should be gold or silver was not primarily due to the value of these metals in worldly terms, but rather to their reputation as sacred metals which they possessed even before they became a measure of all commercial transactions. As Titus Burckhardt points out, they are the earthly reflections of the Sun and Moon, and thus also of the realities of the Spirit and Soul which are related to the heavenly pair.

The four elements of which all bodies are composed were earth, air, water and fire, possessing the properties of cold and dry, hot and moist, cold and moist and hot and dry respectively. All materials were believed to be built up from these elements in varying proportions. The seven metals of antiquity were named after the known heavenly bodies – Sun (gold), Moon (silver), Mercury (quicksilver), Venus (copper), Mars (iron), Saturn (lead) and Jupiter (tin).

Transmutation of one metal into another could, it was thought, be achieved by altering the proportions of the basic elements to those in silver or gold. It is now known that transmutation is possible on a limited scale by using the techniques of nuclear physics not then available to the alchemists. Might it not have been possible, however, that the alchemist in his transmuted physical state could have used some power, latent in man, perhaps a special form of psychokinesis, to bring about nuclear change? In the plate, the alchemist points to the preparation in the flask which has been raised to a higher form, ready for the incarnation of the spiritual forces.

The Second Treatise: Matter and Nature of the Philosopher's Stone

The legend on the shield of the knight reads: Make one water out of two waters. You who seek to create the Sun and the Moon, give them to drink of the inimical wine and you will obtain vision at their death. Then make earth out of water and you will have multiplied the Stone'. The two waters represent the sulphurous (red) and the mercurial (white) aspects of metals. Salt, or the quintessence, is represented by the sword of the knight which may also be a reference to the 'Secret Fire', the water which does not wet the hand. The legend refers to the *tria prima* hypothesis that all matter is formed from a combination of Salt, Sulphur and Mercury (although not the common forms of these materials) and it could be that the sulphur was initially gold or silver, and Mercury a *regulus* of iron and antimony. The elucidation of the nature of the secret fire is perhaps one of the most difficult problems of the whole work. It has been suggested this might be corrosive sublimate (mercuric chloride).

The Second Treatise

The Third Treatise: the Means whereby the Whole Work of this Mastery is Perfected

The Third Treatise

This plate is of particular importance as it symbolizes the meeting of the polarities in the Lunar Queen and the Solar King. The Queen holds a banner bearing the device *Milk of the Virgin*, while that of the King reads *Coagulate the Masculine*. The illustration enters the realm of the alchemical paradox and of the *mysterium conjunctionis*. The virgin's milk is that liquid substance which arises as an antithesis of the peripheral lunar forces and their solidification, while the coagulation of the masculine is similarly a hardening of the solar forces. The balance of these two components is of particular significance not only on the physical plane but also on the spiritual one.

The King and the Queen are the bridegroom and the bride, and their union in marriage, the supreme union of hostile forces (as Jung has described it) bears a direct relation with the Indian esoteric tradition of Tantra. Sulphur represents the *Pingala Nadi* and Mercury the *Ida Nadi*, the Sun breath and the Moon breath respectively. The Secret Fire corresponds to the

135

Kundalini energy flowing in the *Sushumna*, or central channel, through the seven *Chakras*, the energy centres, and points of contact between the psychic and physical body.

This chymical marriage is the central symbol of alchemy. Again, according to Burckhardt, it is only on the basis of the interpretation of this symbol that a distinction can be made, on the one hand, between alchemy and mysticism and, on the other, between alchemy and psychology. From the point of view of mysticism the soul has become alienated from God, while from that of alchemy, man, as a result of the loss of his original 'Adamic' state, is divided within himself. The King represents Gold or Sulphur (not common sulphur) and the Queen, Mercury. The best symbol for the Sulphur-Mercury couple is the Chinese device of Yin-Yang, with the black pole in a white vortex and the white pole in the black vortex, as an indication that the passive is present in the active, the active in the passive, just as man contains, in part, the nature of woman and woman the nature of man. The legends on the banners can be rendered as the basic command of alchemy: *Solve et Coagula*, dissolve and coagulate – which relates to the multiple processes of distillation and solution involved in the work.

The First Parable

The Third Treatise contains **Seven Parables** which are, in effect, the seven stages of death and rebirth.

The **First Parable** depicts the extraction of the ore forming the *materia prima* being extracted from the earth by two miners or gnomes – not in a mine dug below ground, but in a mountain where the earth meets the heavens. Again, this is a reference to the combination of the material and the spiritual. The scene depicted in the frame of the picture is biblical and represents Esther at the court of King Ahasueras. This may have been included to emphasise the sevenfold nature of the work. For in the book *Esther* there are seven eunuchs, seven administrators and seven special maids for Esther. Not only that, but the action takes place in the seventh year of the king's reign.

The **Other Parable** depicts the Archetypal Tree, relates to Book VI of the *Aeneid*, and shows the three characters Aeneas, Silvius and Anchises.

'Such was the glittering; such the ruddy rind,
And dancing leaves, that wantoned in the wind.
He seized the shining bough with gripping hold,
And rent away with ease, the lingering gold;
Then to Sibyl's palace bore the prize.'

The archetypal nature of the tree is emphasised by the golden crown surrounding its trunk, while the colours of the garments of the recipients of the branches are those of the Red and White Stones. These are the male and female expressions of the active and passive forces, one to raise the earth substance and the other to bring down the cosmic forces.

The Other Parable

The **Third Parable** portrays the subject of death and rebirth for the first time. The King of the Earth sinks in the water which has flowed over the face of the earth and become foul and stinking in the darkness (the Nigredo). Next day, the King is shown with a bright star above his head, rescued and renewed, richly adorned and altogether comely. The three crowns are of iron, silver and gold; a sceptre with seven stars is in the King's right hand and a golden apple and a white dove in his left. The seven stars are interpreted alchemically as the seven planetary spirits or metals. In the text it is emphasised that the death of anything is the generation of something else (Aristotle). The night and the morning emphasise the periodicity of the work, the corruption of the previous night being replaced by the bright sun breaking through the clouds in manifold colours with its stars and lustre and a fragrant scent surpassing all balm arising from the earth. Archibald Cockren refers to an odour 'resembling the dewy earth on a June morning,

The Third Parable

with a hint of growing flowers in the air, the breath of wind over heather and hill and the sweet smell of rain on parched earth'.

The Fourth Parable

This is clearly depicted in the red and white winged being holding an egg in his left hand and another, in section, in his right.

The darkened ringed centre represents the *Nigredo*, the white the *Albedo*, while the yellow and the red represent the *Rubedo*. The egg is in the form of a shield or target and was the symbol of the four elements and the quintessence, the shell the Earth, the space under the shell the Air, the inner membrane the Water, the yolk the Fire and the centre of the yolk the Quintessence. The egg was also the seed and to the alchemist there was a close analogy between the transmutation of a seed into a plant and a base metal into gold. as St. Paul says:

'Whatever you sow in the ground has to die before it is given a new life and the thing that you sow is not what is going to come.'
(I Corinthians 15, v. 36)

The **Fourth Parable: Meeting with the Angel** speaks clearly of making corporeal things spiritual and spiritual things corporeal – one of the main themes of the work. This is in essence the whole purpose of spiritual alchemy. The mire-encrusted man rising naked from the marsh to be clothed by an angelic female figure represents the soul of the alchemist encountering the spiritual ground of his being in the person of his Holy Guardian Angel. This is the zenith of the seven stages of the process. The ore was extracted and killed on the material plane of *Assiah*, the earth substance united with the cosmic forces on the astral plane of *Atziluth*. The next three stages represent a return to the material world. This is in accord with the cabalistic concept of '*Kether* in *Malkuth* and *Malkuth* in *Kether*'.

The **Fifth Parable: The Winged Hermaphrodite with the Egg** – the hermaphrodite results from the union of the King and Queen and is symbolic of the *materia prima* containing the male and female principles, the Philosopher's Sulphur and Mercury.

The Fifth Parable

The **Sixth Parable: The Beheading and Dismemberment of the Body** – in which a dark man dismembers a white body with a golden head. The man has balanced garments of Red and White and the parts of the body represent the four elements and the head, the quintessence. The text emphasises this point:

'I have slain you that you might possess abundant life; but your head I will conceal, that the world shall not see you. I will lay waste your body on the earth, and bury it, that it may become putrid and multiply itself, and bring forth fruits unnumbered.'

The Sixth Parable

The Seventh Parable

The **Seventh Parable: The Bath of Transformation** – again this is a scene of death and rebirth. The text makes reference to a wise old man, written of by the poet Ovid, who wishing to renew his youth, had himself cut up and boiled when his members were reunited and his body rejuvenated. The 'White Dove' is the spiritual essence of the work. The key symbol in the picture is the flask by the side of the bath of transformation. A man with a bellows tends the fire and extracts the essence, the culmination of the seven stages of the work amidst the sure surroundings of a temple or palace of a king.

The sequence of plates now enters its second stage with a further series of seven plates relating to the physical processes of alchemy. It cannot be assumed that the first seven stages were concerned solely with spiritual changes in the alchemist, any more than the second series will be concerned solely with the work in the laboratory.

The fourth Treatise: The means by which Nature attains her Ends

In some versions of the *Splendor Solis* the elaborate borders to the alchemical vessel are omitted. In Harley 3469 however, the border, which depicts a scene of sixteenth century life, is, together with the astrological figure above the flask, an indication of the time of year and the subject of the related pursuits.

The Fourth Treatise, Firstly

The Regimen of Saturn – in this plate, the first in the series, the sign is that of the planet Saturn with the associated astrological signs of Aquarius and Capricorn. This relates to a period from 22 December to 19 February. The border illustrations are very similar to the celebrated series of woodcuts by Hans Sebald Beham (*c*.1530-1540), depicting the planets and their associated occupations. Saturn was usually associated with building, farming, mining and, indeed, with anything to do with the earth. It influenced government, administration and law, philosophers and practical idealists. Some of these aspects are illustrated from the gallows on the top right of the plate to the monastery and the funeral procession on the top left.

In the flask, which is sealed and sitting upon a green wreath, (which should have been a representation of flames to signify heating), a boy with a bellows associated with the man in the previous plate, is nourishing and blowing upon a dragon.

The dragon represents the volatile nature of the matter. If the dragon had been wingless, it would have represented fixed matter. It may, however, also represent the untamed forces of the soul, while the young boy is the new impulse of transformation acting upon these forces or on the Philosopher's Sulphur – the metal gold. This combination forms a threefold process which is carried on to the next regimen.

The Regimen of Jupiter – here the contents of the flask have been completely digested and transformed into three birds, coloured red, white and black. These peck at each other in a never ending antipathy. The Birds represent the Salt, Mercury and Sulphur of the soul and also correspond to the three stages of alchemy: the *Nigredo* (blackness), the *Albedo* (whiteness) and the *Rubedo* (redness). They also symbolise the volatilisation of the fixed by the volatile. The astrological signs of Pisces and Sagittarius are associated with the

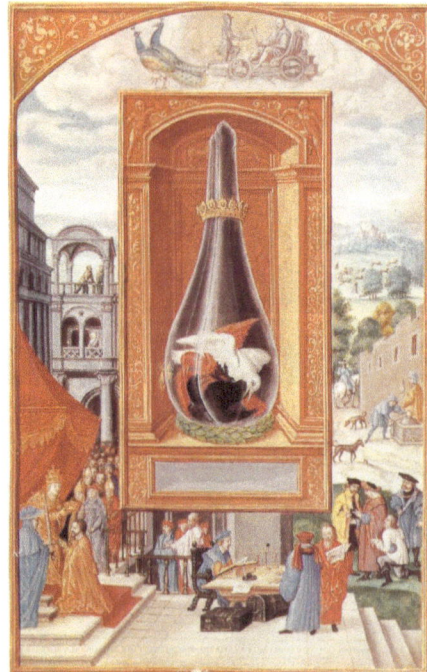

The Fourth Treatise, Secondly

planet Jupiter and, again, the scenes depicted relate to those aspects of everyday life under the influence of that planet. The order of the signs through the seven regimens of the work is determined by the distance of the planet from the Earth in the old geocentric concept for the solar system – Saturn, Jupiter, Mars, Sun, Venus, Mercury and the Moon.

The Regimen of Mars – in this, the third stage, the three birds have fused together to form one bird with three heads. This eagle rises in the flask as a sublimate. The planet is the war-like Mars with its associated signs of Aries and Scorpio. The text states that the heat causes earthly things to be penetrated by a spiritual force, of which it is said in the *Turba* – 'Spiritualises the bodies and make Volatile that which is Fixed.

The Regimen of Sol – in this plate the source of the heat has been removed and the process of integration, that began in the previous stage, has descended lower into a dragon form without wings, indicating solidification. The three colours Red, White and Black are still visible. It is the Sun that controls the work at this stage with its associated sign Leo. The border illustrates the manly activities governed by that planet.

The Fourth Treatise, Thirdly

The Fourth Treatise, Fourthly

The Regimen of Venus – the three-headed dragon has metamorphosed into a Peacock, the symbol of the procession of glorious colours which are displayed once the *Nigredo* stage of the operation is complete. The operation moves under the influence of Venus with the associated signs of Taurus and Libra. The scenes around the flask are those of music and the arts of love, classically associated with the Goddess. The date of the illustration, 1582, is engraved on the seat in the left foreground.

The Regimen of Mercury – this plate signifies the formation of the White Stone or Tincture, here depicted by a Queen holding an Orb in her right hand and a Sceptre in her left. As the previous plate illustrated the Philosophical Mercury, so this plate illustrates the Philosophical Salt. The astrological sign of mercury is Virgo and Gemini presides over the background, the streets of a busy medieval city. In the foreground Sculpture is represented by two men with mallet and compasses, Mathematics by two philosophers with a globe

The Fourth Treatise, Fifthly

and compasses, Literature by an old scholar seated at a desk, Commerce by a man counting money and Music by a company of musicians and singers grouped around a regal or portative organ.

The Regimen of the Moon – the last painting in this series represents the *Rubedo*, the consummation of the Great Work. The Red Stone, or Philosophical Sulphur, is represented by a King holding the Orb in his left hand and the Sceptre in his right, thus complementing the Queen. The Orb and Sceptre are themselves female and male symbols respectively (passive and active) and reflect back to the fourth plate. The astrological sign of the Moon in Cancer is emphasised by the sporting occupations of fishing, shooting, hawking and hunting.

The seven regimens of the operations of alchemy shown by these illustrations are:

1. Solution of the *Materia Prima*
2. Separation of the Three Principles
3. The Unification of the Three Principles
4. The Digestion of the Three Principles
5. The Resolution and Integration of the Three Principles
6. The preparation of the White Stone (Philosophical Salt)
7. The preparation of the Red Stone (Philosophical Sulphur).

The Fourth Treatise, Sixthly

The Fourth Treatise, Eighthly

The Fifth Treatise, Part 1, 1st Chapter

The Fifth Treatise: The Manifold Operations of the Whole Work

The Fifth Treatise, Part 1, 2nd Chapter

The final four plates complete the work and these, as might have been expected, relate back to the four introductory plates which depicted the incarnation of the power from above into the *prima materia*. The whole set of twenty-two plates has suggested to a number of commentators that the plates bear some relationship to the major *arcana* of the Tarot. This is not so, although their interpretation might benefit from association with the paths of the Divine Sephiroth on the Tree of Life.

Chapter 1: Dissolution - The Dark Sun – this plate is particularly strange in that it shows a Dark Sun rising from the earth. The accompanying text is of little help in the interpretation as it claims to expound the manifold Operations of the Whole Work and goes on to say 'The first thing proper to the Art of Alchemy is Solution'. The Law of Nature requires that a body be turned into water; that is into a quicksilver which is so much talked about. The Quicksilver releases the Sulphur which is joined to and present with it.

The Fifth Treatise, Part 1, 3rd Chapter

This separation is nothing less than a mortification of the moist with the dry and is actually a putrefaction; and the same operation will make the matter black. If anything, it is the Sun, the symbol of the Philosopher's Sulphur, that has turned black in its mortification.

Chapter 2: The Play of Children – this is illustrative of one of the most often quoted aphorisms of Alchemy. As stated by Eirenaeus Philalethes:

'... there is one work of ours, which is the Play of Children and Work of Women, and that is decoction by the fire ...'

This statement is said, by some, to be the profound secret of the alchemical process. It may suggest that the alchemist has to listen, as a little child, to what matter is trying to tell him and then, like a woman, listen to the promptings of his inner nature. The text implies that the art is also like the games of children who, when they play, turn everything topsy-turvy. The alchemists always held that to one who really knew the secrets of the process the actual laboratory work involved was extremely simple.

This is, however, to assume a derogatory interpretation of the play of children and the work of women. Rather it is the trusting faith of, in Our Lord's words, 'Except ye become as little children ye shall in no wise enter the kingdom of Heaven'. It is this that implies a purity as well as a simplicity and, above all, a steadfast faith as a precursor of success.

Chapter 3: The Work of Women – this is a complementary plate to the former depicting the work of women, here shown as washing and bleaching strips of linen cloth. The philosophical sublimation by which perfect whiteness is achieved with women's work – to washing since it renders the garment clean and white and to cooking and roasting until it is done.

Chapter 4: The Splendour of the Sun – the final illustration depicts the setting of the purified Sun. That which is above has finally united with that which is below, and all is accomplished. Thus is the Splendour of the Sun revealed to the alchemist at the conclusion of his work.

The operations summarised in the four chapters of the Fifth Treatise are: solution, coagulation, sublimation, separation and reunification. The remainder of the text of the **Splendor Solis** is without illustration.

The Fifth Treatise, Part 1, 4th Chapter

On the Government of the Fires

The first part of the Fifth Treatise concludes with a section on the government of the fires. Heat is necessary for the process, but must not be excessive during the initial stages - in fact, not very much higher than fermentation temperatures. As the work continues the heating should be maintained and increased gradually to that of bright sunlight on a hot summer's day.

The Fifth Treatise:
Part 2: The Colours which Appear in the Preparation of the Stone

The principal colours are black, white and red, and while other colours may appear during the course of the process they are only transient. Overheating during the early stages may hasten the development of the colours, but this should be avoided.

The Sixth Treatise:
The Properties of the Whole Work

The preparation of the Philosopher's Stone is carried out by the application of external heat to the *materia prima*. This calcinations separates it into its three principles – Mercury, Sulphur and Salt. Thereafter the material is subjected to a series of operations – sublimation, fixation, putrefaction, separation of the elements, trituration, assation (or roasting), distillation (or clarification) and finally, coagulation.

The Seventh Treatise: Of the Whole Work's Manifold Effects – the Seventh treatise repeats the process of the work described in the previous Treatise with additional details culled from the early alchemists and concludes with a section setting out the uses of the tincture. These are:

1. It gives heath and cures men of various diseases
2. It makes perfect the metallic bodies
3. It changes all base stones into precious ones
4. It softens and colours every kind of glass.

Discussion and Conclusions

After this brief encounter with the sixteenth century, though through the subject of the Philosopher's Stone, and albeit seen through twentieth century eyes, the question arises as to whether anyone ever prepared the Stone or whether it was all an illusion based on an imperfect knowledge of the nature of matter.

Carl Jung was very decided in his opinion on the matter. He wrote 'It is certain beyond all doubt that no real tincture or artificial gold was ever produced during the many centuries of earnest endeavour'. He thought that in carrying out the alchemical processes the alchemist experienced his projection as a property of matter, but what he was really experiencing was his own unconscious. Indeed, Jung's extensive studies on alchemy and its literature certainly indicated that there was a close connection between the concepts of the alchemists and the dreams and mental processes common to the human mind and, perhaps, to the collective unconscious.

It is known that modern chemistry, having benefited extensively from alchemy during its history, would not consider the type of operation carried out by the alchemist as capable of leading to the nuclear transformation of metals such as lead or mercury into gold. While it is now possible to transmute elements, on a minute scale, by bombardment with high energy particles in a nuclear reactor or cyclotron, the technique involving the Philosopher's Stone in which a small quantity of the Stone, wrapped in wax or paper, was added to a crucible of molten metal and with further heating for a short time, about fifteen minutes, to give a mass of virtually pure gold on cooling, appears very unlikely. But as St. Augustine said 'Miracles are not contrary to Nature but only contrary to what we know about Nature.' Perhaps the alchemists had discovered some psycho-chemical process by which nuclear change might be brought about. Many of their accounts are very convincing!

Chapter Thirteen

The English Alchemists

Roger Bacon (1214 - 1294)

I t has been inferred that Roger Bacon was born about the year 1214 for in 1267 he wrote:

'I have laboured much at sciences and languages, and it is now forty years since I first
learned the alphabet: I have always been studious, and except for two of those forty years
I have always been in studio'.

As at that time young boys were generally sent to the University at the age of about twelve or
thirteen, it has been deduced that 1214 was the year of Roger's birth. He was born near Ilchester
in Somerset and appears to have been the scion of a wealthy family which, it is said, sacrificed
their wealth in the cause of the king, Henry III, in his struggle with the barons from 1258 to 1265.

At the University of Oxford Roger came under the influence of Robert Grosseteste (1175 -
1253), Bishop of Lincoln, and with his encouragement undertook a study of the Greek language.
It may well have been Grosseteste's encouragement that encouraged him to join the Franciscan
Order about 1247.

Roger Bacon was a contemporary of those early European alchemists and scholars Albertus
Magnus (1193 - 1280), Thomas Aquinas (1225 - 1274) and Vincent of Beauvais (dates uncertain),
all three of whom were Dominicans.

Roger Bacon died in England in 1292. John Rous in his *Hist. Regum Angliae* wrote:

*'Anno Christi 1292 obiit Rogerus Bacon professor theologie et quasi eruditus ut
magister in octo scienciis ubi alii clerici non potuerunt preter vii scire'.*

Bacon was buried in Grey Friars in Oxford on the Feast of St. Barnabus the Apostle (11th June,
1292).

During the years 1234 to 1250 Roger Bacon studied and lectured at the University of Paris.
It was here that he came to know and follow the teachings of Peter Peregrinus, who he referred to
as 'the Master of Experiments', and devoted his studies to the subjects of chemistry, mathematics
and astronomy. Returning to Oxford in 1250 he continued his studies until, in 1257, he was sent
back to Paris by his superiors in the Franciscan Order. At one time it was believed during the
next decade Bacon was kept 'in close confinement', but it would appear from his own account,
discovered in 1897, he was at that time in poor health and for this reason took no part in the
outward affairs of the University.[1] It is believed that Bacon could often be free in his criticism
of authority, capable of a considerable independence of thought and, when the occasion arose,
could be extremely quarrelsome.

Plans for the *Compendium Philosophae*

For a long time Bacon had intended to compile a magnum opus which summed up the extent
of his knowledge of science and philosophy. It would appear this period of confinement could
have offered an excellent opportunity for at least making a start on this project. It now seems,
however, that at this time he actually put very little on paper, or parchment, except perhaps a few

chapters 'now about one science and now about another' at the insistence of friends. We know this from his report to Pope Clement IV in 1267 when explaining the absence of any completed work for submission to the Holy Father Roger speaks of being in exile:

> '*Exhil enim aliquando factus fui, longe vadens ad loca metallica ut experiri possem naturas metallorum*' (*Mineralium*, III, I, 1).

It would appear Clement was given to understand that the work was already in existence even before he was elected to the Chair of St. Peter on 5th February, 1265 and was still Guy de Foulques, Archbishop of Narbonne, and Cardinal Bishop of St. Sabina. The Cardinal had heard of Roger's work, which was expected to have been a comprehensive dissertation on philosophy, through a clerk in his service and had commissioned him, one Raymond de Laon, to obtain a copy. So insistent was the Cardinal that on becoming Pope he repeated his request in the form of a Papal mandate in 1266. The text of the mandate has survived and runs as follows:

> 'To our beloved son Friar Roger called Bacon of the Order of Friars Minor. We have gratefully received the letters expressing your devotion. We have also noted carefully the word by which our beloved son G. called Bonecar, Miles, both faithfully and wisely gave verbal explanation of your letter. We wish to be clearer as to what you intend, and we order you to send us the work, which we have already asked you to communicate to our beloved son Raymund de Lauduno, a minor official, notwithstanding the order of any prelate to the contrary or any rule of your Order as quickly as you can; that you may declare to us through your writing what remedies seem to you fitting for dealing with those matters which you recently intimated to be of such moment; and do this secretly as far as you are able and with as little delay as possible. Given at Viterbi *X Cal. Julii, Anno II*.' [2]

Bacon was apparently delighted rather than apprehensive on the receipt of this mandate and excited at the interest shown by the Holy Father in his work. He was to write:

> '... the Head of the Church has sought out me, the unworthy sole of its foot [...] I feel myself elevated above my ordinary strength; I conceive a new fervour of spirit. I ought to be more than grateful since your Beatitude has asked me for that which I have most ardently desired to communicate, which I have worked on with immense toil and brought into light after manifold expenses.'

It appears that Roger had intended to compile a systematic and scientific treatment of the various branches of knowledge, entitled the *Compendium Philosophae*. But there were problems including lack of money and assistants, obstacles placed in his path by his superiors in the Order, the impossibility of finding competent and trustworthy copyists and, above all, his own methods of composition which involved re-writing the text three or four times to achieve the desired result. In the end Bacon started again on a more modest project which resulted in his *Opus Maius* written in 1266 to 1268.

This work was subsequently supplemented by two further works, the *Opus Minor* and the *Opus Tertium*. The *Compendium*, had it been completed, would have been a revised and enlarged edition of these three works. It had been planned in four volumes and was intended to cover the six main branches of knowledge at that time. Volume I covered Grammar and Logic, Volume II Mathematics, Volume III Physics (Naturalia) and Volume IV Metaphysics and Morals. It was in Volume III that '*Alkimia*' (*Scientia de omnibus rebus in animatis que fiunt primo ex elementis*) was to be treated.

The Opus Maius

The Opus Maius was composed in seven parts:

I.	On the four general causes of human error
II.	On the connection of philosophy with theology
III.	On the study of language
IV.	On mathematical science
V.	Optics
VI.	Experimental science
VII.	Moral philosophy.

It begins with the following consideration:

'A thorough consideration of knowledge, consists of two things, perception of what is necessary to obtain it and then of the method of applying it to all matters, that they may be directed by its means in the proper way. For by the light of knowledge the Church of God is governed, the commonwealth of the faithful is regulated, the conversion of unbelievers is secured, and those who persist in their malice can be held in check, by the excellence of knowledge, so that they may be driven off from the borders of the Church in a better way than by shedding of Christian blood.'

Throughout the work Roger was acutely aware theology or 'divine wisdom' is, or should be, the objective of all human thought and that morality is the supreme science. In all his studies it is their utility to the Church of God that is paramount and the opening paragraph of the Opus Maius is typical of this guiding principle. Lynn Thorndike is of the opinion that:

'there is no other book quite like the Opus Maius in the Middle Ages, nor has there been one like it since; yet it is true to its age and is still readable today. It will therefore always remain one of the most remarkable books of the remarkable thirteenth century.' [3]

Roger Bacon the Alchemist

Roger's views on alchemy are summarised in the Opus Maius :

' The dignity of this science can be exemplified in alchemy. For that whole art is scarcely so perfected that the greater metals may be produced from the lighter ones, as gold from lead and silver from copper. But that art never suffices to show the natural and artificial grades of gold and the modes of its grades. For experimental science has brought both to light, since it has discovered both the four natural grades and their seventeen modes and the artificial ones. By experiment it can be produced beyond twenty-four. Thus the vessel in which the liquid was contained, by means of which the ploughman became the messenger of the king, had a purity of gold far beyond the twenty-four...'

' The art of alchemy not only omits these modes, but this gold of twenty-four degrees is very rarely found, and with the greatest difficulty. There have always been a few who during their life have known this secret of alchemy; and this science does not go beyond that. But experimental science by means of Aristotle's Secret of Secrets knows how to produce gold not only of twenty-four degrees but of thirty or forty degrees and of as many degrees as we desire. For this reason Aristotle said to Alexander, "I wish to disclose the greatest secret"; and it really is the greatest secret, for not only would it procure an advantage for the state and for everyone his desire because of the sufficiency of gold, but what is infinitely more, it would prolong life. For that medicine would remove all the impurities and corruptions of a baser metal, so that it should become silver or the purest gold, is thought by scientists to be able to remove the corruptions of the human body to such an extent that it would prolong life for many ages. This is the tempered body of elements, of which I spoke above.' (*Opus Maius*)

Now alchemy has both an exoteric and an esoteric side. There is the practical laboratory work, seeking by the repetition of various processes such as calcinations, solution, separation, conjunction, putrefaction, congelation, cibation, sublimation, fermentation, exaltation, multiplication and projection to prepare the White and Red Stones which will transmute base metals into silver or gold. On the spiritual side, which is also reflected in these operations, changes are brought about in the persona of the alchemist himself. Lynn Sadler is of the opinion that the historical Roger Bacon was a believer in higher alchemical theories, although he probably was not an actual practitioner of laboratory alchemy.

In his *Opus Tertium*, however, Bacon writes:

'I have laboured from my youth in the sciences and languages, and for the furtherance of study, getting together much that is useful. I sought the friendship of all wise men among the Latins, and caused youth to be instructed in languages and geometric figures, in numbers and tables and instruments, and many needful matters. I examined everything useful to the purpose, and I know how to proceed, and with what means, and what are the impediments: but I cannot go on for the lack of the necessary funds. Through the twenty years I laboured specially in the study of wisdom, careless of the crowd's opinion, I spent more than 2,000 livres [about £100,000] in these pursuits on occult books (*libros secretos*) and various experiments, and languages and instruments and tables and other things.'

From this account it would appear Roger Bacon was not without some kind of laboratory experience. Throughout the *Opus Maier* he repeatedly states that without experience nothing can be known with certainty.

'… there are two modes in which we acquire knowledge, argument and experiment. Argument shuts up the question, and makes us shut it up too; but it gives no proof, nor does it remove doubt, and cause the mind to rest in the conscious possession of truth, unless the truth is discovered by way of experience, e.g. if any man who had never seen fire were to prove by satisfactory argument that fire burns and destroys things, the hearer's mind would not rest satisfied, nor would he avoid fire; until by putting his hand or some combustible thing into it, he proved by actual experiment what the argument laid down; but after experiment had been made, his mind receives certainty and rests in the possession of truth, which could not be given by argument but only by experience.'

The science of the 13th century advanced entirely *per argumentum* and verification by experiment was the exception rather than the rule. Bacon appears to have based his knowledge not only on traditional authors, such as Aristotle, but also on practical experiment in the laboratory. Indeed it has been said his work on practical alchemy was carried out in private places, sometimes in the suburbs of Oxford. A tower, traditionally known as 'Friar Bacon's Study', stood on Folly Bridge on the South side of Oxford until it was demolished in 1779. This could well have been the site of his early laboratory work.

The Opus Minus

This work was written in the year 1267 and appears to have been intended both as an introduction and as a supplement to the *Opus Maier*. This work also had seven parts:

1. Introduction and dedicatory letter

2.   Mathematics - *de notitia caelestium*
3.   Practical alchemy '*in enigmatibus*'
4.   Notes on the chief points in the *Opus Maier*
5.   Treatise on the seven sins in theological study
6.   Speculative alchemy
7.   *Remedia studii.*

The *Opus Tertium*

This work, written in 1267 and 1268, was like the *Opus Minus* both as an introduction and a supplement to the Opus Maior. In the comparatively modern edition of the *Opus Tertium* in *Fr. Rogerii Bacon, Opera quaedam hactenus inedita*, edited by J.S. Brewer (London, 1859) there are seventy-five chapters. Chapters 1 to 21 deal with Bacon's gratitude to the Pope, his difficulties, the relative importance of the sciences and their relationship to each other with special reference to the *Opus Minus*. The commentary on the *Opus Maier* begins at chapter 22 and includes a long digression on vacuum, motion and space. The remainder of the work emphasises the utility of mathematics in relation to sacred and secular subjects, as treated in Part IV of the *Opus Maier*.

Theoretical and Practical Alchemy

Roger Bacon distinguished two kinds of alchemy – speculative (or theoretical) and practical. Speculative alchemy, he writes:

> 'Treats of the generation of things from the elements and of all inanimate things and of simple and composite humours, of common stones, gems, marbles, of gold and other metals, of sulphurs and salts and pigments, of lapis lazuli and minium (red lead) and other colours, of oils and burning bitumens and other things without limit, concerning which we have nothing in the books of Aristotle. Nor do the natural philosophers know of these, nor the whole assembly of Latin writers. And because this science is not known to the generality of students it necessarily follows that they are ignorant of all that depends upon it concerning natural things, namely of the generation of animate things, of plants and animals and men, for being ignorant of what comes before they are necessarily ignorant of what follows.'

Operative and practical alchemy, on the other hand:

> 'teaches how to make the noble metals, and colours, and many other things better and more abundantly by art than they are made in nature. And the science of this kind is greater than all those preceding because it produces greater utilities. For not only can it yield wealth and very many other things for the public good, but it also teaches how to discover such things as are capable of prolonging human life for much longer periods than can be accomplished by nature… It confirms theoretical alchemy through its works and therefore confirms natural philosophy and medicine, and this is plain from the books of the physicians. For these authors teach how to sublime, distil and resolve their medicines, and by many other methods according to the operation of that science, as is clear in health-giving waters, oils and many other things.' (*Opus Tertium*)

De Secretis Operibus Artis et Naturae

A tractate generally attributed to Roger Bacon reproduced as an appendix to the *Opera Inedita* is 'The Secret Works of Art and Nature' which has a chapter on 'the method of making the philospher's egg' and another on the Philosopher's Stone. The 'Egg' may be prepared as follows:

'Make a diligent purification of the calx with the water of alkali, and other acute waters, grind it by several contritions with the salts, and burn it with many assations, that the earth may be perfectly separated from other elements... understand if you can... Then take oyl of the form of saffron cheese, and so viscous as not to be smitten asunder by a stroak, divide the whole fiery virtue, and separate it by dissolution, and let it be dissolved in acute water, of a temperate acuteness, with a slight fire, and let it be boyled till his fatness, as the fatness of flesh be separated by distillation, that nothing of the unctuousness mat issue forth; and let this fiery virtue be distilled in the water of urine; Mind and search what I say, for the speech is difficult... You must observe whether I speak enigmatically or according to the truth... Farewell: whoever unlocks these, hath a key which opens and no man shuts; and when he hath shut, no man opens.' [7]

This passage is typical of the usual obscure style adopted by alchemists and would appear unworthy of Bacon's more scientific approach. It could well be that the work is not actually attributable to Bacon but only ascribed to him in accordance with established alchemical tradition! Roger Bacon, however, regarded gold as the most perfect metal and in his *Opus Minus* taught that it was formed by interaction of the purest forms of mercury and sulphur. Other metals suffered from infirmities and sickness and while they were also formed from these two basic components, they were also either too humid, too dry, too hot or too cold. In a further treatise *De Arte Chymiae* he writes :

'Let all workers in alchemy know that true species cannot be transmuted. For they say, and it is said also here, that lead is always lead, even granted that its impurities are purged away so that silver may be seen; thus they deceive men not understanding the words of Aristotle on solution. To whom it may be answered that it is not for us to be labouring about the transmutation of bodies in order to make some sort of being from non-being (*ut faciamus aliquod ens de non ente*) and to make something from this mineral. But that we may reduce a corrupted mineral to an incorrupted mineral. Since lead is a species of silver, the sickness of which mineral – to wit, softness, blackness and foulness – have invaded it, when these are put aside, there is silver true and good; and so it is reduced to its true mineral, and in accordance with its first primary origin (*secundum primam radicem*) it is not translated nor transmuted from its own mineral. Similarly iron is silver, but it is corrupted by the power of its impure sulphur and of its impure quicksilver. From which source hurtful things enter it, to wit, blackness, hardness and dryness; which being removed there is good silver. In like manner copper is the soul and sister of silver, in all its dispositions, to wit, in softness, hardness, in fusibility and malleability, but it is red; we take away from it its redness, and then there is genuine silver. Similarly tin has softness, the grating sound of teeth, and blackness, which being removed is reduced to silver. In like manner silver is gold save in its colour, because the colour of gold was taken away from it, in its mineral state, by the power of its quicksilver. But we give colour to it, and then there is good gold. This is in keeping with what is said by Aristotle. If the quicksilver is pure and bright, and the sulphur is clean and red, and the temperate heat be used, mineral gold is made from these, in accordance with the inward disposition of that body. What nature does in a thousand years, we are able to accomplish artificially in a short time, perhaps in one day, or in some hours, with the proper medicine, that takes a long time to prepare, by use of which bodies suffering from mineral corruptions are reduced to uncorrupted mineral substances.' [8]

Pattison Muir in his study *Roger Bacon: His Relations to Alchemy and Chemistry*[9] makes a number of observations which may be summarised as follows:

1. Roger Bacon laid stress on the usefulness of alchemical knowledge without apparently having studied the subject much at first-hand.

2. Despite the times in which he lived and the fact that he was a Franciscan monk, his writings were singularly bold and outspoken. He adopted what would now be called 'a scientific approach'.

3. Bacon used the traditional concept of the four elements and their properties, but in a manner different from and more suggestively than had been the practice of other alchemists.

4. He protested against and completely discarded the central concept of alchemy that there was a universal primary matter.

5. He recognised the existence of distinct material things, each characterised and distinguished from all others by its own particular properties.

6. Bacon realised that effort, activity and change are the characteristics of physical things – a concept not unlike the modern notion of energy.

7. The substance of Bacon's ideas and knowledge is contained in the three works he produced for the Pope - the *Opus Maius*, the *Opus Minus* and the *Opus Tertia*. The last two works were secondary and supplementary to the *Opus Maius*.

8. Bacon was interested in the scholasticism of his time as well as in natural science. Two of the characteristic traits of the former are found in the *Opus Maius*, namely the continual use of authorities and the highest regard for Aristotle, whose works he regarded as the foundation of all wisdom.

9. Bacon was critical rather than constructive in his approach and this resulted in a false reputation for the depth of his learning. His criticisms apply to but two countries, France and England (Paris and Oxford). He apparently knew little of the state of learning in other countries such as Spain, Italy and Germany, and knew little of the work of his great contemporary Albertus Magnus in Cologne.

10. Bacon asserted that credulity was a necessary component of experimental science. He stated 'First one should be credulous until experience follows second and reason comes third… At first one should believe those who have made the experiments or who have faithful testimony from others who have done so, nor should one reject the truth because he is ignorant of it and because he has no argument for it.' [*Unde oportet primo credulitatem fieri, donec secundo sequitur experientia ut tertio ratio comitetur… Et ideo in principio dedet credere hic qui experti sunt, vel qui ab expertis fidelitur habuerunt nec debet reprobare veritatem propter hoc, quod eam ignorat, et quis ad eam non habet argumentum.*]

11. Bacon was interested in speculating upon mechanical devices such as machines for navigation without rowers, for flying in which artificial wings are made to beat the air like a bird, and a machine, small in size, for raising or lowering enormous weights. (*Epistola de secretis operibus caput IV*)

12. He expressed the view that through alchemy life can be greatly prolonged and that Artephius was enabled by this means to achieve a life-span of a thousand and twenty-five years!

13. As J.H. Bridges concluded in his *Life and Work of Roger Bacon* 'the *Opus Maier* remains the one work in which the central thought of Bacon is dominant from first top last; the unity of science, and its subordination to the highest ethical purpose conceivable by man'.

The *Speculum Alchemiae*

The *Speculum Alchemiae* and the *Radix mundi*, or *Tractatus de lapido philosophico*, are generally regarded as doubtful or indeed spurious works of Roger Bacon. The former was included in Zetzner's *Theatrum Chemicum* (Nurnberg, 1541) and the latter in an English translation in Salomon's *Medicina Practica* of 1707. There are some thirty alchemical works ascribed to Bacon due to his reputation in this field and, as with Raymon Lull, many minor works on alchemy were ascribed to these authors to gain recognition.

Roger Bacon or the anonymous author of the Speculum, stated:

'It is not species that are transmuted, but rather their subject matter. Therefore the first work is to reduce the body into water, that is into mercury, and this is called solution, which is the foundation of the whole art.'

The *Speculum* is a brief work of seven chapters. In the introduction (chapter 1) alchemy is defined as:

'a science teaching how to make and compound a certain medicine, which is called 'Elixir', the which when it is cast upon metals or imperfect bodies, does fully perfect them in the very projection.'

Chapter 2, 'Of the Natural Principles and Procreation of Metals' identifies the natural principles as mercury and sulphur and comments on the purity of the metals gold, silver, steel, lead, copper and iron. The nature of gold, for example is:

'Gold is a perfect body, engendered of mercury pure, fixed, clear, red and sulphur clean, fixed, red, not burning and it wants for nothing.'

While lead

'is an unclean and imperfect body, engendered of mercury impure, not fixed, earthy and dross, being somewhat white outwardly, but red inwardly, and of such a sulphur in part burning. It has not purity, fixation, colour and firing.'

Chapter 3, 'Out of What Things the Matter of the Elixir Must Be More Nearly Extracted', emphasises that all metals are engendered by mercury and sulphur and therefore these components must form the basis of any purifying agent such as 'our Stone'. The author then introduces the concept of male and female, gold being the perfect masculine body and silver a body most perfect and feminine. These metals, however, are not suitable for the matter of the Stone as 'we should hard and scantily find fire working in them'. This should contain mercury, clean, pure, white and red, not fully complete, but equally and proportionally mixed after a due manner with a like sulphur and congealed into a solid mass and capable of being treated with 'our artificial fire' to give the Stone.

Chapter 4, 'Of the Manner of Working and of Moderating and Continuing the Fire', draws attention to the fact that while we endeavour to perfect the imperfect Nature has delivered us the imperfect only with the perfect by continual heating over long periods of time. Heat perfects all things! All the elements need is fire and *Azoth*. The fire must be gentle and easy, the whole mastery being performed in one vessel patiently and continually, the fire must always by little and little be increased and augmented to the end.

Chapter 5, 'Of the Qualities of the Vessel and Furnace', points out that as minerals are generated in the earth of a mountain by continual application of heat so the furnace must be capable of prolonged heating of a glass vessel, round with a small neck and sealed with a glass stopper or lute, which contains the mercury-sulphur primary matter and is enclosed within a similar vessel 'so that the temperate heat may touch the matter above and beneath.'

Chapter 6, 'Of the Accidental and Essential Colours Appearing in the Work', refers to the many stages marked by colour changes from the black of putrefaction to the true red of the crown of perfection. During the change from black to red the work passes through the green of the soul, citrine, the colours of the peacock's tail, the eyes of fishes and then an ash colour before the final red. Of the ash colour it is said 'Do not set light by the ashes, for God shall give it to you molten and then at the last, the King is invested with a red crown by the will of God.'

The final Chapter 7, 'How to make the Projection of the Medicine upon any Imperfect Body', mentions some metals are further off from perfection than others and that these require longer treatment with the Stone than those nearer perfection. Since one part of the Stone will transmute several thousand parts of the metal it is necessary that:

> 'One part is to be mixed with a thousand of the next body and let all this be surely be put into a fit vessel and set it in a furnace of fixation, first with a lent fire, and afterwards increasing the fire for three days, till they be inseparably joined together, and this a Work of three days. Finally, every part thereof by itself must be projected upon another thousand parts of any near body and this is the work of one day, or one hour, or a moment, for which our wonderful God is eternally to be praised.'

The Legend of the Talking Head

In common with many of his contemporaries Roger Bacon was believed to have been a sorcerer or necromancer and was said to have constructed a 'talking head' which like those of the poet Virgil, Pope Silvester II, Albertus Magnus and even Bacon's friend, Robert Grosseteste, could be induced to speak and foretell the future! This legend was brought to the delight of Elizabethan audiences in the play *The Honorable Historie of Friar Bacon and Friar Bongay* by Robert Greene (1560 -1592). The two friars having constructed a head of brass

> '... that by the inchaunting forces of the Devil, Shall tell out strange and uncouth Aphorisms, And girt faire England with a wall of Brasse.'

were awaiting the time when it should speak and after three weeks Bacon, who had watched day and night, fell asleep leaving the head to the attention of his servant, Miles. Very soon the head spoke saying 'Time is' but Miles refrained from waking his master for so trivial a remark and let him sleep on. The head then spoke again 'Time was' and relapsed into silence. Finally the head spoke for the third and last time 'Time is past' and a hand then appeared that 'breaketh down the Head with a hammer' with a loud noise which awoke Friar Bacon who was naturally somewhat displeased with the unthinking Miles.

> 'Tis past indeed. Ah, villaine! time is past: My life, my fame, my glorie, all are past, Bacon the turrets of thy hope are ruin'd downe. Thy seven years of study lieth in the dust: Thy Brazen Head lies broken!'

Sir George Ripley (1415 - 1490)

One of the best known of the 15th century English alchemists was Sir George Ripley, who was born at Ripley, near Harrogate in Yorkshire, about the year 1415. William Camden (1551 - 1623), the antiquary and historian, was of the opinion that his birthplace was Ripley in Surrey. Little is known of his early life save that he studied alchemy and other subjects in Rome and to enhance his knowledge travelled widely on the Continent, particularly in France, Germany and Italy. He lived in Rome for some years, becoming Chamberlain to Pope Innocent VIII in 1477, and returning to England the following year when he became a Canon-Regular of St. Augustine at Bridlington. Part of his continental tour had been spent on the Island of Rhodes where he had been a guest of the Knights of St. John of Jerusalem. There is a story, quoted by Holmyard, that through his ability to transmute base metals into gold he was able to donate to the Knights no less a sum than £100,000 yearly to support their military operations against the Turks.[10]

Back in England he returned to the Priory at Bridlington where the fumes and unpleasant odours issuing from his alchemical laboratory soon became irksome to the Abbot and other Canons. He was released from the Order and joined the Carmelites as an anchorite at the monastery of St. Botolph at Boston, Lincolnshire, where he died and was buried in 1490.

George Ripley was one of the first scholars to popularise the alchemical teachings ascribed to Raymon Lully and is said to have composed some twenty-five different works, most of which are still in manuscript. In 1470 he dedicated The Compound of Alchemy to Edward IV and in 1476 his *Medulla alchimiae* to George Neville, Archbishop of York. Most of his published works can be found, in English translation, in Elias Ashmole's *Theatrum Chemicum Britannicum* of 1652.

The one work of Ripley that is universally acknowledged as genuine is *The Compound of Alchymy*. It was extremely popular from the time of its publication in 1591. It was first printed in London and according to Ferguson:

> '*The Compound of Alchymy Or the ancient hidden Art of Archemie*: Conteining the right & perfect meanes to make the Philosophers Stone, *Aurum potabile* with other excellent experiments. Diuided into twelve Gates. First written by the learned and rare Philosopher of our Nation George Ripley... Whereunto is adioned his Epistle to the King, his Vision, his Wheele, and other his Workes, neuer before published.... Set forth by Ralph Rabbards Gentleman... London Imprinted by Thomas Orwin, 1591 small 4, A, *, B to M, in fours. The title has a woodcut border; there is an ornamental capital E containing a portrait of Queen Elizabeth, to whom the book is dedicated, and M3 recto is taken up with the diagram called Ripley's Wheel.'

Elias Ashmole reprinted *The Compound of Alchymy* in his *Theatrum Chemicum Britannicum*, adding a note on the author. Ashmole also reprinted several other works by Ripley; verses belonging to his *Scrowle*, *The Mystery of Alchymists* and the Preface to his *Medulla* written in 1476 and dedicated to George Nevell, then Archbishop of York, and another 'Shorte Worke'. All these works by Ripley were written in verse.

Ripley's Wheel

The Wheel is a complex figure based on eleven concentric circles containing legends in English and Latin. It may give an indication of the influence on Ripley of Raymund Lully. The wheel is drawn by one John Goddard and headed 'Here followeth the Figure conteyning all the secrets of the Treatise both great & small'. At the foot of the Wheel is the legend *Caelum Philosophorum* (heaven of the philosophers):

> Our heaven this Figure called is
> Our table also of the lower Astronomy
> Which understood thou may not misse
> To make our Medicen parfetly
> On it therefore set thy study
> And unto God both night and day
> For grace and for ye Author pray

The Wheel comprises an outer rim oriented like a compass, the top being south west and the bottom north east. Four small circles mark the cardinal points, each containing three signs of the zodiac:

> South - Attractive - Summer - Dry - Hot (Aries - Leo - Sagittarius)
> West - Retentive - Autumn - Dry - Cold (Taurus - Virgo - Capricorn)

North - Expulsive - Winter - Moist - Cold (Cancer - Scorpio - Pisces)
East - Digestive - Spring - Moist - Hot (Gemini - Libra - Aquarius)

Before the South circle is the legend *Jupiter tenet aerum*, before the west *Sol tenet ignum*, before the north *Saturnus tenet terram* and before the east *Mercurius tenet aquam*.

The outer rim contains four symmetrical circles spaced between the cardinal circles. That at the top has the legend 'The altitude of the stone in the South fiery in quality having more than perfect quintessence the end of practice, temperate South. Then follows a seven line stanza:

> As holy scripture maketh mention
> Into the womb of a virgin immaculate
> Christ descended for our redemption
> From his high throne to be incarnate
> So the sonne descendeth here from his estate
> Forth from the south passing into the west
> Through the ocean labouring withouten rest

That on the right has the legend 'The first side or west latitude of the stone and the entry into the practice. Earthy in qualitie. *Occasionate occidental*'. This is followed by a further seven line stanza:

> As Christ his Godhead hidd from our sight
> When he our kinde on him did take
> Right so the sonne his beames of light
> As for a time here him forsake
> For under the wynges of his make
> The moone he hideth his glory
> And dieth in kind that he may multiplie

The circle at the bottom has the legend:

The profundity of the stone in the north and the sphear of purgatory watry in quality, variable in colour darke and imperfectly eclips of the sonne, in north. The east stanza follows:

> As Christ in tombe was tumulate
> After his passion and death on tree
> And after in body glorificate
> Uprose indewed with claritie
> Right so by corporall penaltie
> The sonne eclipsed with colloures variable
> In the east upriseth in whitnes incomparable

The circle to the left reads:

The second side or east latitude of the stone and entering into the specula true fiery in qualitie the light spheare of paradise the full moone, oriental. Finally the south stanza:-

> As Christ from earth to heaven did ascend
> In Cloudes of glory up to his high throne
> And reigneth there shining without end
> To whose clearness comparison may be none
> Right soe the sonne now made our stone
> To his first glory renewed with youth
> Ascendeth to be intronizate in the south.

There follows nine spheres or inner circles leading to the centre of the wheel which contains cusps of the four elements fire, air, water and earth and the legend '*centrum lapi vis*'.

The first circle has the legend:

'Heere is the red man to his white wife. Beespoused with the spirit of lyfe - Heere to purgatory must they goe - There to be purged with paine and woe - Heere they have passed their paines all - And made resplendent as is crystal - Heere to paradise they goe to wonne - Brighter made then is the sonne.'

In circle two the alchemical regimens follow with legends in Latin:

'*Sphaera Solis I - melior lapis omnium est ille qui magis yt decoctus a sole et igni yt proximus qui plus caeteris sustinet ignem et tardius frangitur ab io et qui maioris fuerit ponderis unde inter metalla magis valet aurum quia nec igne nec aqua corrungitur.*' (The best stone, which is ripened by the Sun, and is close in nature to the fire, and sustains the fire of the other stones. Gold, the Sun metal, is among metals the greatest and neither fire nor water corrupts it.)

Circle III again starts in the north west:

'*Sphaera Lunae III quae est tinctura albedinis fulgidi splendoris mater lapides sartens ipsum in ventre sua adiuua solutionem cum ea sicut congelationem cum sole qui continet in se virtutes omnium ductibilium et omne metallum tinget.*' (which is the White Tincture of flashing splendour the mother giving birth to these stones in her womb, helps solution as the Sun brings congelation, as she contains in herself all the virtues of softening and also tinges all metals.)

Circle IV reads:

'*Sphaera veneris VIII dea amoris, qua est medium coniugendi tinctures inter solem et lunam, et est corpus de facile convertibile ad virumque et ideo panitur in opere pro imperfecto corpoe et dicitur Leo viridis.*' (which forms the middle of the tinctures of the Sun and Moon, is called the Green Lion.)

Circle V reads:

'*Sphaera Mercury XII qui est spiritus fuminic aureus non differens ab auro nisi in hoc quod aurum est fixum ipse autem infixus, eujus manifestum et humidum occultum vero igneum, quod tinget quia aurum non tinget donec.*' (which is in the highest degree the Golden Spirit, does not differ from Gold, except that Gold is fixed, whereas this is unfixed and manifests coldness and moistness, and its fire is hidden.)

Circle VI is that of the colour sequence of the operation:

'*Sphaera colorum principalium*' - pale - blacke - white - redd.

Circle VII is linked with Circle VIII the former being that of the primary qualities and the latter of the secondary qualities:

'X terra sit aer X aqua sit ignis X aer sit terra X ignis sit aqua X quinta ex quidus oritur essentia'

Circle IX around the cusps of the four elements reads:

'When thou hast made the quadrangle round, then is all the secret found.'

Here followeth the Figure conteyning all
the secrets of the Treatise both great & small

Sol tenet ignem

Cælum Philosophorum.

Our heaven this Figure called is
Our table also of the lower Astronomy
Which vnderstood thou may not misse
To make our Medicen parfetly
On it therefore set thy study
And vnto God both night and day
For grace and for y Author pray

John Goddard sculpsit

Ripley's Wheel

It is thought Ripley's Wheel was an alchemical mandala intended for meditation. It conveyed its meaning directly through its geometrical arrangement of ideas. It synthesised the fourfold division of the world into one, leading from a quadrangular conception of the world structure to a circular one (squaring the circle!). The Philosopher's Stone at the centre unites the four elements in the circle while they are contained with their usual correspondence and the astrological triplicities in the four outer circles arranged in a quadrangle. The unification of these four outer realms and their descent through the concentric circles makes 'the quadrangle round'.[11]

The practice of the work begins in the West and proceeds clockwise to purification in the North both of soul and substance. Proceeding round to the east, it becomes more speculative as the practical work exerts its influence on the alchemist. The uppermost circle or globe in the South concludes the practice of the Stone.

Between the four stages of the circles or globes the seven-line stanzas parallel the alchemical work with the spiritual stages in the life of Our Lord: the Incarnation, the Passion, the Resurrection and the Ascension. Then follow the four spheres of the Sun, the Moon, Venus and Mercury, the principal colours of the work and the sphere of the first and second qualities which give rise to the quintessence. Thus is achieved the unification of the fourfold into the One - central Stone - the squaring of the circle or circling of the square!

The Compound of Alchymie

This was a major work of Sir George Ripley, composed in 1471, and was made more generally available in Ashmole's *Theatrum Chemicum Britannicum* of 1652. It was there described as ... 'a most excellent, learned and worthy worke, written by Sir George Ripley, Chanon of Bridlington in Yorkeshire, conteining twelve Gates.'

The work is prefaced by an Epistle to King Edward IV to whom the work is dedicated. An exposition on the Epistle was written by Eirenaeus Philalethes and published as part of Ripley Revivd in 1677. As the Epistle is reputed to comprise the whole secret of Alchemy, Philalethe's summary may be of some interest here.

1.   That as all things are multiplied in their kind, so may be metals, which have in themselves a capacity of being transmuted, the imperfect into the perfect.
2.   That the main ground for the possibility of reduction of all metals, and such minerals as are of metallic principles, into their first Mercurial matter.
3.   That among so many Metalline and Mineral Sulphurs, and so many Mercuries there are but two Sulphurs that are related to our work, which Sulphurs have their Mercuries essentially united to them.
4.   He who understands these two Sulphurs and Mercuries aright, shall find that one is the most pure Red Sulphur of Gold, which is Sulphur in manifesto, and Mercurius in occulto, and the other is most pure white Mercury, which is indeed true Quicksilver in manifesto, and Sulphur in occulto, these are our two Principles.
5.   If a man's principles be true, and his operations regular, his event will be certain, which event is no other than the true Mystery.

In Ripley's work the Epistle is followed by a Prologue and a Preface, and then by twelve 'Gates' each representing a different stage in the operations of the work: calcination, solution, separation, conjunction, putrefaction, congelation, cibation, sublimation, fermentation, exaltation,, multiplication and projection. According to Lynn Thorndike, Ripley was 'a popularizer of the alchemical doctrines ascribed to Raymond Lull, and his poems are in general a rehash of previous alchemical commonplace.' Be this as it may, his reputation was such that Ashmole included most of his works in his *Theatrum Chemicum Britannicum*. In conclusion to his major work Ripley added a final verse:

> Now to God Almyghty I thee Recommend,
> Whych graunte the by Grace to knowe thys one thing,

For now ys thys Treatys brought to an end:
And God of hys Mercy to hys blysse us bryng,
Sanctus, Sanctus, Sanctus where Angells do syng:
Praysing without ceasynge hys gloriose Magestye,
Whych he in hys Kyngdome graunte us for to see.

The Ripley Scrolls

There are a number of these Scrolls or 'Scrowles' in existence, all of which are thought to derive from the 16th and early 17th centuries. They vary in size. That in the Huntington Library in California is 10 ft. 8 in. long, while one in the Fitzwilliam Museum, Cambridge is twenty-two feet. long. The British Museum has four examples, of various lengths, one of which is inscribed 'This long roll was drawn for me in Cullers at Lubeck in Germany *anno* 1588.'

It is thought that the Scrowles are studied from the bottom to the top as they are intended to convey the exaltation of matter, in caballistic terms from *Malkuth* to *Kether*. At the foot of the Huntington Scrowle two figures are examining a scroll. The one on the right is a crowned king with a long sceptre, while the other on the left may well be the alchemist himself with what appears to be an unlit torch under his arm and a red sash from his right shoulder to his left hip. The text of the scroll held by these initial characters is difficult to decipher in its entirety but it begins In the name of the Trinitie and refers the reader in a series of rhyming couplets to the Booke of the *Turba Philosophorum* and to such well-known alchemists as Aristotle, Geber, Hermes, Lully, Morien, Albertus Magnus, the Blacke Monke, and Mary the Sister of Moses. It indicates that the wisdom of these forerunners has been included in this one work.

Then follows a version of the verses published by Elias Ashmole as 'Belonging to an Emblematicall Scrowle Supposed to be invented by Geo: Ripley:

I shall tell with plaine declaration,
Where how and what is my generacion:
Omogeni is my Father,
And Magnesia is my Mother:
And Azot truly is my Sister,
And Kibrick forsooth is my Brother:
The Serpent of Arabia is my name,
The which is leader of all this game:
That Somertyme was both wood and wild,
And now I am both meeke and mild;
The Sun and the Moone with their might,
Have chastised me that was so light;
My wings that me brought,
Hither and thither where I thought
Now with their might they downe me pull,
And bring me where they woll,
The blood of myne heart I wiss,
Now causeth both Joy and Blisse:
And dissolveth the very Stone,
And knitteth him ere he have done;
Now maketh hard that was lix,
And causeth him to be fix.
Of my blood and water I wis,

> Plenty in all the world there is.
> It runneth in every place;
> Who it findeth he hath grace:
> In the world it runneth over all,
> And goeth round as a ball:
> But thou understand well this,
> Of the worke hou shalt miss.
> Therefore know ere thou begin,
> What he is and all his kin,
> Many a name he hath full sure,
> And all is but one Nature:
> Thou must part him in three,
> And then knit him as the Trinity:
> And make them all but one,
> Loe here is the Philosophers Stone.

Then follows the first set of emblematical pictures in which a bleeding dragon stands upon a winged globe containing three black spheres which it nourishes with its blood while supporting in its mouth, in addition to its tail, a crescent moon and a formalised Sun with in its centre three spheres now black, red and grey, linked together to form a Trinity. Above is the legend 'THE BIRD OF HERMES IS MY NAME, EATING MY WINGS TO MAKE ME TAME'.

The scroll next depicts the crowned Bird of Hermes eating its wings and standing upon a feathered sphere or globe, with the following verse:

> In the sea withouten liffe,
> Standeth the Bird of Hermes:
> Eating his wings variable,
> And thereby making himself more stable;
> When all his Fethers be agon,
> He standeth still there as a stone;
> Here is now both White and Red,
> And also the Stone to quicken the dead,
> All and sume withouten fable,
> Both hard and nesh and malliable
> Understand now well aright,
> And thanke God of this sight.

Below the verses is the legend 'THE RED SEA THE RED SOLL THE RED ELIXIR VITA'.

Above the Bird of Hermes is a Sun with a face issuing White and Red drops. This is flanked by a further verse:

> Take thou Phoebus that is so bright,
> That sitteth so high in Majesty;
> With his beames that shineth soe light,
> In all places where ever that he be,
> For he is Father to all living things,
> Maynteyner of Lyfe to Crop and Roote,
> And causeth Nature forth to spring;
> With his Wife being Soote,
> For he is salve to every sore,
> To bring about this precious worke;
> Take good heede unto his lore,

I say to the learned and to Clerk,
And Omogeny is my Name:
Which God shaped with his own hand,
And Magnesia is my Dame;
Thou shalt verily understand,
Now heere I shall begin,
For to teach thee a ready way:
Or else little shalt thou wyn,
Take good heed what I say;
Devide thou Phoebus in many a parte
With his beames that byn so bright,
And thus with Nature him Coarte,
The which is mirrour of all light:
This Phoebus hath full many a Name,
Which that is full hard for to know;
And but thou take the very same,
The Philosophers Stone thou shalt not know,
Therefore I councell ere thou begin:
Knoe him well what it be,
And that is thick make it thin;
For then it shall full well like the.
Now understand well what I meane,
And take good heed thereunto,
The worke shall else little be seene:
And tourne thee unto mikle woe,
As I have said in this our Lore,
Many a Name I wiss it have,
Some behinde, and some before;
As Philosophers of yore him gave.

Above the Sun and his scalloped sky is the legend 'HERE IS YE LAST OF YE RED &
YE BEGINNING TO PUT AWAYE YE DEAD YE ELIXIR VITA' and on either side of
a flaming archway 'The Red Lyon' and 'The Greene Lyon'. Above the Lyons are four further
sections of verse:

On the Ground there is a Hill,
Also a Serpent within a Well:
His Tayle is long his Wings wide,
All ready to fly on every side,
Repaire the Well round about,
That the Serpent pas not out;
For if that he be there agone,
Thou loosest the virtue of the Stone,
What is the Ground thou mayst know heere,
And also the Well that is so cleare,
And eke the Serpent wth his Tayle
Or else the worke shall little availe,
The Well must brenne in Water cleare,
Take good heede for this thy Fyre,
The Fire with Water brent shalbe,
And Water with Fire wash shall he;
Then Earth on Fire shalbe put,

And Water with Air shalbe knit,
Thus ye shall go to Putrefaccion,
And bring the Serpent to reduction.
First he shalbe Black as any Crow,
And downe in his Den shall lye full lowe:
I swel'd as a Toade that lyeth on ground,
Burst with bladders sitting so round,
They shall to brast and lye full plaine,
And thus with craft the Serpent is slaine:
He shall shew Collours there many a one,
And tourne as White as wilbe the bone,
With the Water that he was in,
Wash him cleane from his sin:
And let him drinke a little and a lite,
And that shall make him faire and white,
The which Whitnes is ever abiding,
Lo here is the very full finishing:
Of the White Stone and the Red,
Loe here is the true deed.

Ascending the scroll, the next stage is the rectangular Bath of the Elements above a Green Dragon with a Black Toad issuing from its mouth. The Dragon is the active redeeming alchemical agent which, in the initial stage of the scroll, may be a pre-view of the entire process. The Christ nature of the Dragon is shown by the outpouring of his blood to nourish the three spheres of the *Prima Materia*. The Toad may be symbolical of this primal substance in its triune state and is preserved by the Dragon to ascend the scroll and take part in the neck of the double pelican vessel held by the divine Alchemist at the top of the scroll.

The next section, usually the bath of the Elements, has four corner towers, one for each of the elements (earth, water, air and fire) and a central pillar supporting an heptagonal bath. In the lower bath are three figures – a red bearded man by the central column, a winged female touching the column and a male figure surrounded with an aura of clouds and flames (a glory). The essence of the element is contained in a sealed flask surmounting each tower.

The heptagonal bath, labelled The White Sea, has turrets at each angle with figures perched upon them holding flasks, some of which are being poured into the bath, and are thought to be the essences of the planetary metals. In the bath stand the Sun King and the Luna Queen holding branches of a vine which climbs up the central tree. Halfway up this tree is seated a male figure above whom is a Lilith-like half woman, half serpent, crawling down from the upper foliage. The man is surrounded by a golden glory.

Above the tree is a furnace or Athenor with the inscription: 'HERE IS YE LAST OF YE WHITE STONE AND YE BEGINNING OF YE RED STONE'. On the base of the furnace are a further four sections of verse:

Of the sonne tak the lighte
The redd goume that is so brighte,
And of the moone doe allso
The white goume there keep to,
The philosophers sulphur wiffe
This I-called withouten stryfe,
Kyberte and Kybtyte I-called allsoe
And other names many moe;

Of him draw out a tincture,
And mak then a marriage pure
Between the husband and the wife
Spoused with the water of lyfe;
But of this worke you must beware,
Or els thy worke wil be ful bare.
He must be made of his one kinde,
Marke thou well now in thy minde;
Acetum of philosophers men call this,
And water abidinge, soe it is
The maydens milke of the dewe,
That other workes doe renew,
The spirit of life men call allsoe,
And other names many moe.
The which causeth our regeneration
Between the man and the woman.
Soe looke, that noe devision
be there in the conjunction
Of the moone and of the sonne;
After the marriage is begonne,
And all while they ben a weddinge
And him to her drinkinge
Acetum it is very fyne
Better to them than any wine;
Nowe when this marriage is done
Philosophers call this a stone,
The which hath great nature
To bring a stone that is pure,
So he have kindly norrishinge
Be perfitt heate and decoction;
But in the matrices where they be put
Looke never thy vessel be mishutte
Till they have engendered a stone
I   n the world there is not such one.

At the top of the scroll above the depiction of the Divine Alchymist is the legend: EST LAPIS OCCULTUS SECRETO FONTE SEPULTUS FERMENTUM VARIAT LAPIDEM QUI CUNCTA COLORAT (The burial place where the fountain is hidden from which is born the ferment or stone coloured in many ways.). This last section of the scroll is of considerable interest as it shows the Divine Alchemist holding a double pelican upon a water bath. On the pelican is the legend 'Ye must make water of ye earth & earth of ye ayre & and ayre of ye fyre & fyer of ye earth'. The vessel contains feathers representing vapours ascending and drops of liquid descending. In its neck is the Blacke Toade. The body of the vessel is taken up with a mandala of eight circles linked to each other and surrounding a central larger circle containing the King and the Alchemist holding a book with seven seals, all being linked with chains to the outer circles. The rim of the central circle bears the legend '*Spiritus – Anima – Corpus – Spiritus – Anima – Spiritus*'.

Each of the smaller circles, with the exception of that on the top right which is bound to the inner circle with a band labelled Prima Materia, depicts a vessel on a furnace with a variable number of figures, each surrounded by an individual legend. Clockwise the legends on the circles are:

1. *Spiritus – Anima – Corpus – Leo Rubens – Viridis*
The naked King and Queen are being killed, the former with a spear and the latter with an axe. The Red and the Green Lions await their prey.
2. The Soule forsooth in his Sulphur not brenninge
The first stage of the work, the vessel containing the dissolved bodies of the royal pair is attended by four operatives.
3. *Accrido & Humido – Primo ex illis pasce quoniam Debilis sum*
(Because of the former weakness I am fed with fire and the first moisture.) A Stillhead has been placed on the reaction vessel, there are now three receivers and three operatives.
4. *& Leniter digisius animalus sum exalta me crasionibus*
(and I am gently divided by my friends commencing in the second exhaltation).
There are now two operatives attending the vessel. The still head and one of the receivers have been removed while birds (? crows) perch on the other two.
5. *Exulto severa subtilia me vi possum reducere ad simplex*
(I am able to be exhalted to the highest subtlety according as I am reduced to simplicity).
The vessel with four attendants discloses a figure giving birth to a child.
6. *Sitio desicio posa me albifica*
(I am dried up, I am deserted, I drink and I am made white). The child in the vessel now has five attendants adding the contents of the receivers.
7. I am bereaved and am far from my characteristic home (the Latin
    inscription is unreadable),
The reaction vessel now has six attendants holding further additives.
8. *Leniter cum igne amicitium fac vi asiqua violatum nos separare*
(Thus gently with friendly fire according as some other violence is not able to overcome).
The previous scene has increased its number of attendants to six.

Adam McLean has dated the Scroll as being produced within the last few decades of the 16th century and ascribed to Sir George Ripley in a similar manner that alchemical documents were attributed to well known adepts to impart an air of verisimilitude and importance.[12] The verses on the scroll are in the style of Ripley and may well have been composed by him, but the designs are rather later. McLean's valuable interpretation is based on a version of the scrolls in the library of the Royal College of Physicians in Edinburgh, originally in possession of the Earl of Cromarty. The late Betty Jo Teeter Dobbs in a valuable paper entitled 'Alchemical Death and Resurrection: The Significance of Alchemy in the Age of Newton', published in 1990 a version of the Ripley Scroll which dates from about 1570 and which is now held in the Huntington Library.

The Mistery of the Alchymists

This long poem of some 300 lines is included in Ashmole's *Theatrum*. It is in the form of question and answer between a father and his son. In the very first line reference is made to Sol being in Aries and Phoebus shining bright, indicating that Spring is the time for the work to begin. The answers to the questions asked as to the nature of the starting materials and where they may be found are, as might be expected, very diffuse and imprecise. The father admits that the First Principle is Sulphur comprising Water and Fire and that the Philosopher's Stone may be likened to the Most Holy Trinity, with Sulphur acting as the Holy Ghost. This may well be a reference to the *Tria Prima* of sulphur, mercury and salt, although the salt component is never mentioned as such. The 'sulphur' may be found almost everywhere and can easily be identified by its taste and colour. From this our 'mercury' may be obtained having three 'vestures' – white, red and purple. It is later stated that the white and red sulphurs are the same kind.

The preparation of the white and red stones given the proper starting materials is then discussed in more detail. Gentle, controlled, heating is required in a well-sealed vessel. A 'moist fire' (water bath) is used for the initial work and during the first forty days the reactants undergo a series of colour changes – first to blackness and then, through a series of colours, to whiteness, at which stage a 'dry fire' (sand bath) is applied to convert the white stone to the red. It is reckoned that the whole process would be completed in one year as the Sun passes through the Zodiac. As might be expected, the Mystery of the Alchemists remains a mystery!

The Preface to the '*Medulla*'

This work comprises nineteen verses of seven lines, each with a clearly defined rhyme scheme as is shown in the following example:

> 2. This Stone divine of which I write,
>   Is knowne as One, and it is Three;
>   Which though it have his force and might,
>   Of Triple nature for to be,
>   Yet doe they Metalls judge and try.
>   And called of Wise men all,
>   The mighty Stone that Conquer shall.

The general impression produced by the work is that these verses were written much later than the *Compound of Alchymy* and are more in the style of later alchemists such as Eirenaeus Philalethes. As the Preface was included in Ashmole's collection it must have been in manuscript prior to 1652, but it is unlikely it was the work of George Ripley himself. The alchemical tradition of ascribing works to well-known alchemists of the past might well have been followed in this case. It is interesting to note Ashmole did not include the *Medulla* itself but only the preface and the title suggesting that it was composed in 1476 and dedicated to Geo: Nevell, then Archbishop of York. The poem tells of the three natures of the Stone – animal, mineral and vegetable. It is a taste of 'the wine that is to follow' in the *Medulla* and sings the praises of the Stone with no details of its preparation.

The '*Short Worke*'

This is also reprinted in the *Theatrum* in the more characteristic verse style of Ripley, extending to ninety-six lines. The work is described in four stages. In the first the *materia prima* (heavy, soft, cold and dry) is cleansed, reduced to a calx and dissolved in 'water of the wood' (? acetic acid). The second stage involved congealing the water of earth and earth of water when the calx is repeatedly dissolved and calcined until the third stage of albification is achieved, the colour of the material progressing from black to white. It is then dissolved in the 'water of life', referred to as 'mercury', and enclosed in a vessel. This, then, becomes 'our sulphur' from which is made 'our antimony', white and red. The fourth stage is distillation. The man and wife (king and queen) enclosed in the vessel produce a child which is nourished until he become to full age . A marriage is then made between the daughter and the son and the work is complete. The ashes at the bottom of the vessel become the Diadem of our Craft.

The Bosom Book of Sir George Ripley

The 'Bosom Book' was published in the *Collectanea Chemica* in 1893 based on a manuscript in the collection of the late Mr. Frederick Hockley.[13] It is a unique prose composition ascribed to Ripley and claims to disclose the whole process for the preparation of the Philosopher's Stone, of the great elixir and the first solution of the gross body. In thirty brief paragraphs is set out in apparent clear detail the whole alchemical process from the *prima materia*, in this case called *Sericon* or Antimony, to the production of the perfect Red Stone.

1. The *materia prima*, *Sericon* or Antimony, 30lb. of which is dissolved in twice distilled vinegar (30 gal.) and the insoluble residue, called the *terra damnata*, removed.
2. The liquid is evaporated on a water bath to give a green gum (Our Green Lion).
3. The gum is placed in a sealed retort to give a distillate and a sublimate in the neck of the retort.
4. The sublimate (our Dragon) is removed and calcined to a white calx, the *Ferrum Philosophae*.
5. The remainder of the Black Dragon is spread out on a clean marble top and ignited with a hot coal. The mass takes fire and produces a citrine coloured product 'glorious to behold'.
6. This product is dissolved in vinegar, filtered and evaporated to a gum from which is drawn out the Dragon's Blood. This operation is repeated until all the residue is transformed into 'our natural and blessed liquor', which is then added to the Green Lion's Blood and allowed to putrefy for fourteen days to separate the elements.
7. The purified *menstruum* is placed in a still (of fine Venice glass) with the receiver sealed on with linen cloth soaked in egg white and immersed in a water bath to cool it.
8. The distillate is re-distilled seven times, until it will burn a linen cloth when touched with a flame. This is our 'Ardent Water' which by frequent rectifying will produce white and yellow oils. The sublimate produced, when powdered and placed on an iron plate, will dissolve in water and draw all the mercury in the form of a light Green Oil, the Oil of Mercury.
9. The solution is separated and placed on a water bath until there remains a liquid pitch-like residue at the bottom of the still.
10. Ardent Water is mixed with the pitch for three hours, decanted and filtered. Fresh Ardent Water is added and the operation repeated three times. The product is called 'Man's Blood rectified'.
11. Water is added to the black residue and the whole distilled to give a distillate which separates into an oil and an aqueous layer. The residue is called 'the Earth of the Stone'.
12. This earth is ground up and mixed with the rectified 'Man's Blood' for three hours and then distilled a further three times. The product, the 'Fiery Water', is said to be exalted in three of the elements into the virtue of the quintessence (water, air and fire).
13. The black earthy residue is calcined into a very fine white calx.
14. The calx is mixed with the fiery water and distilled, the whole operation being repeated seven times to give the 'Water of Life' (our Mercury and our Lunary).
15. Cornish tin is molten and reduced to a fine white calx which is placed in a glass still and treated with 'our Mercury' by a repeated process of imbibition and distillation to give Lunary perfect. This will flow on a copper plate like wax, will not evaporate and will then convert the copper into fine silver.
16. The distillate from (11) is distilled at a higher temperature to leave a Red Oil at the bottom of the still. This is the 'element of fire'.
17. The original White Residue (Mars) is then treated with Ardent Water in a Chymia kept cool for eight days until all of the liquid has been absorbed. A further like quantity of the Water is added and stood for a further eight days. This is repeated until the residue will take up no further liquid. The vessel is then sealed and set on a water bath in a temperate heat to putrefaction.
18. The White Stone is digested by leaving the vessel on the water bath for a further 150 days. During this time the contents become first russet coloured, then whitish green and, finally, very white (like unto the eyes of fishes). This is the White Stone ready for fermentation.

19. This product, Sulphur of Nature, will silver hot glass.

20. The White Stone was divided into two parts and one part heated in a hermetically sealed vessel in a hotter fire (cineriton) until the colour changed to red and then to purple. This was the Red Stone ready for fermentation.

21. Dissolved pure silver in 'our mercury', one part silver to one part mercury and set it on warm ashes. When dissolved the liquid will be green in colour. Rectify the mercury several times to give the Oil of Silver, the white Ferment of Ferments.

22. Place the other half of the White Stone into a vessel with the White Ferment and leave under the fire for two or three days.

23. When the White Stone and the White Ferment become a fine powder it is imbibed with the white oil (our Lunary, **n**.14 *supra*) drop-wise. The product will flow in the fire like wax and is the Perfect White Stone capable of transmuting copper or iron into pure silver.

24. Dissolve pure gold with 10 parts of antimony in our mercury and treat as for the Oil of Silver. The contents of the sealed vessel becomes the black oil of gold. On putrefying on a water bath the colour changes through white to citrine, the Ferment of Ferments for the Red Work.

25. To the other half of the Stone (20 *supra*) add a fourth part of its weight of the Gold Ferment and fix them in the fire for two or three days.

26. When the Red Stone and the Red Ferment become a fine powder it is imbibed with the red oil (as in 23 *supra*) to give the Red Stone capable of transmuting mercury, lead, and silver to pure gold.

27. The stone may be exalted by heating in a sealed vessel in a water vapour bath.

28. The way of Projection is described in which 100 oz. of mercury in a crucible is treated with 1 oz. of the Red Stone and 1 oz. of the product used to treat a further 100 oz. of mercury and this operation is repeated up to four times to produce gold in quantity. Praise be to God!

29. These operations may be shortened considerably if the white sublimate in the neck of the retort (**n**.4 *supra* is heated in a sealed vessel with the Lunar Ferment (21 *supra*) for three days and then imbibed with the White Oil of our Stone (our mercury).

30. A similar shortening may be adopted for the preparation of the Red Stone by heating the sublimate after solution in the Red Mercury to achieve a purple colour and then subjecting to heating in a sealed vessel with the Gold Ferment for three days when a fine red powder will result which is imbibed with the Red Oil of our Stone to give the perfect Red Stone for which thank God.

Thomas Norton of Bristol

The *Ordinall of Alchimy* written by Thomas Norton is the first poem in the great alchemical collection put together by Elias Ashmole. It was first published, in a Latin translation, in the *Tripus Aureus* of Michael Maier in 1618 which included, in addition, the *Twelve Keys* of Basil Valentine and the *Testament* of Abbot Cremer.

While the Ordinall was originally designed to be anonymous as Ashmole himself commented:

'From the first word of this Proeme, and the Initiall letters of the six following chapters (discovered by Acromono syllabiques and Sillabique Acrostiques) we may collect the Author's name and Place of Residence: for those letters, (together with the first line of the Seventh Chapter) speak thus,

Thomas Norton of Briseto,
A parfet Master ye maie him call trowe.'

Ferguson, in his Bibliotheca Chemica, states the author of the Ordinall was the son of another Thomas Norton, and was born in Bristol towards the end of the 14th century. He was the member of Parliament for the Borough of Bristol in 1436, a member of the Privy Chamber of Edward IV, acted as a contact with the Embassies and accompanied the King and the King's brother, Richard, when he fled to Burgundy in 1470. Norton is said to have studied alchemy under George Ripley for a short period of forty days when he was in his late twenties. He was reputed to have achieved, and lost, the Elixir of Life on two occasions. On the first, it was at the hands of a servant and on the second by the designs of a woman, the wife of William Canynges, a master mason, who, it is said, re-built the Church of St. Mary, Radcliffe, with the proceeds of the sale of the stolen Elixir.

Doubt has been cast by Nierenstein and Chapaman[14] on the identification of Norton with the member of Parliament of that name by Nierenstein and Chapman. Norton, the M.P., died in 1449, whereas the poem refers to the alchemical work having been started in 1477. J. Reidy is of the opinion that Nierenstein and Chapman's evidence is faulty and there is no cause to doubt that the Ordinall was in fact written by Thomas Norton of Bristol, a member of Edward IV's Privy Chamber, who was called Thomas Norton III. At least one Bristol historian has correctly identified this Norton as the alchemist.[15]

The first publication of the work was in a Latin translation by Michael Maier in 1618, reprinted in Manget's *Bibliotheca Chemica Curiosa*,[16] and again in the original English by Ashmole in his *Theatrum* (1652). Ashmole notes that:

'In the search I have made after Authentique manuscripts to compleate this worke, a private Gentleman lent me a very faire one of Norton's Ordinall, which I chiefly followed; yet not omitting to compare it with fourteen other copies. It was written on Velame and in an ancient sett hand, very exact and exceeding neate.'

This copy, according to Ashmole, may have been the dedication copy to George Nevell, Archbishop of York in the reign of Edward IV (1442 - 1483).

In Ashmole's version the poem commences with a preface in Latin, followed by its English translation which contains the four 'prophetic lines' ascribed to Sir John Abbot of Bridlington in his '*Prophecies, Ubi de Tauro &*:

Heaven doth all things gratis give,
For Sin protracts the gifts of Heaven,
Shall change the old for better things.
Most goodly Graces shall descend,
(when least looked for: to Crowne his End.)'

The section of the poem following the Preface is entitled 'The Proheme'. This sets the tone of the whole work by first dedicating it 'To the honor of God, One in Persons Three' and pointing out the profundity of 'the subtill science of holy Alkimy'. Norton goes on to discuss the obscurity of the work of earlier alchemists such as: Hermes, Rhases, Geber, Avicenna, Merlin, Hortulanus, Democritus, Morien, Bacon, Anaxagoras and Raymond Lully. Norton's work will, however, shew the trewth in fewe words and plaine, and he emphasises it is important to remember that 'Nothing is wrought but by his proper Cause'. The book is called the *Ordinall of Alchimy* in allusion to the order imparted by the service book used by priests in the Western Church, particularly for Ordination when special powers are imparted to the Ordinands. There then follows a section which sets out the purpose of and analyses the structure of the work:

The first Chapter shall all men teache
What manner People may this Science reache,
And whie the trew Science of Alkimy,
Is of old Fathers called Blessed and Holy.
In the second Chapter may be sayne,
The nice Joys thereof, with great paine.

> The third Chapter for the love of One,
> Shall trewly disclose the Matters of our Stone;
> Which the Arabies doon Elixir call,
> Whereof it is, there understonde you shall.
> The fowerth Chapter teacheth the grosse Werke,
> A foule laboure not kindly for a Clerke.
> In which is found full great travaile,
> With many perills, and many a faule.
> The fift Chapter is of the subtil Werk,
> Which God ordayned only for a Clerke;
> Full few Clerks can it comprehend,
> Therefore to few Men is the Science send.
> The sixt Chapter is of Concord and Love,
> Between low natures, and heavenly sphaeres above:
> Whereof trew knowledge advanceth greatly Clerks,
> And causeth furtherance in our wonderfull werks.
> The seaventh Chapter trewly teach you shall,
> The doubtfull Regiments of your Fires all.'

Chapter 1.   Those who may practice this Science

The tincture of Holy Alchemy must be regarded as a gift and grace from God. It cannot be bought or achieved by labour alone but may be imparted by teaching or by divine revelation. It is essential that the alchemist is worthy of the secret and then God will send him a Master to impart the secret knowledge to him. The secret can, apparently, only be imparted from mouth to mouth accompanied by 'a most sacred dreadfull Oath'. Norton would appear to suggest that it is incumbent upon the alchemist in due time to pass on the secret to a suitable candidate, and to one candidate only, 'soe that noe man shall leave this Arte behind'.

Norton goes on to stress the importance of preserving the secrets of alchemy from the profane 'for this Science must ever secret be'. As these secrets may only be obtained by the grace of God so due reverence must be always paid to Holy Alkimy. It is because so many have sought for the secret and so few actually found it that the general disbelief in alchemy has arisen. This has, however, not stopped men from pretending to a knowledge of the subject. Norton gives a warning against such pretenders.

Chapter 2 - The Joys of Alchemy

Norton describes this chapter as one of joy and pain. He first tells the story of a monk from Normandy who pretended that he was an adept, well grounded in alchemy, who had the wish to immortalise his name by founding some fifteen abbeys across Salisbury Plain. He promised Norton that he had 'in fire a thing that should fulfill his desire'. After forty days the Stone would be prepared to produce assayable amounts of transmuted gold. However, forty days came and went but there was no sign of either Stone or gold! Norton used this as an example that no deceiver should study this science but promised never to name the man responsible. He made the promise before a statue of St. James, the patron saint of alchemy, whose relics at Compostella in Northern Spain became an object of pilgrimage. The shrine in the cathedral there became the greatest of all such Christian shrines. St. James (the Greater) was the brother of St. John the evangelist and a son of Zebedee. Legend has it that St. James made a missionary journey to Spain, although this has been disputed by scholars. Nicolas Flamel gives an account of one such pilgrimage to Compostella.

Norton then tells of a parson in a little town not far from the City of London who, as an alchemist, compared himself with such great practitioners of the art as Raymond Lully and Roger Bacon. This alchemist was so sure of his art that he wished to build a bridge across the Thames with pinnacles 'shining as goulde', the bridge being lit with lamps or carbuncle stones. Worries over the provenance of the means of illumination was so distracting to the alchemist that he lost the product of his preparation – after a year the reaction vessel contained nothing and 'all was gone!'

These two examples were cited by Norton to emphasise that 'ignorant hope is Fooles Paradise!' He then goes on to contrast the joys of alchemy with the pain it can produce. Firstly, many seek but few actually achieve anything and those who do often fail before they achieve the Stone. Secondly, even embarking on the study of Alchemy is a painful process involving, as it does, decisions as to the correct path to follow. Thirdly, even when a Master is found there is the pain of uncertainty as to whether or not he is true and, fourthly, the work may be subject to haste, despair and deceit, for all haste is the devil's work *'omnis festinatio ex parte diaboli est'*.

The criteria for a 'true' Master would appear to have been like those of Norton's own Master, one who loves justice and abhors fraud, secret, unwilling to advertise his abilities or to impart the secret until he was fully assured that his pupil was really worthy of that grace. The secret must not be written for that would break the master's fealty but must be communicated 'mouth to mouth'. Norton then tells how, after receiving a letter from his master, he rode 'an hundred miles and more' and stayed with him for forty days during which time he learned all the secrets of Alchemy.

The choice of suitable assistants to help with the work is also fraught with problems:

> Some be negligent, some sleeping by the fire,
> Some be ill-willed, such shall let your desire;
> Some be foolish, and some overbold,
> Some keep no Counsell of Doctrine to them told;
> Some be filthy of hands and of sleeves,
> Some meddle strange Matter, that greatly greeves;
> Some be drunken, and some use much to jape,
> Beware of thes if you will hurt escape,.....

Norton goes on to relate the story of Thomas Daulton:

'..... an alchemist with a great store of the Red Stone was leading a peaceful life in an abbey in Gloucestershire when one of Edward IV's courtiers, one Thomas Harbert, brought Daulton against his will to the Court where another courtier, Sir John Delves, revealed to the King that Daulton had once made for him 'a thousand pound of as good Goulde as the Royall was, within halfe a day'. In telling the King this Delves broke his oath, but excused himself by saying it was for the good of both the King and the realm. Daulton then told the King he no longer had any of the Red Stone, having thrown it into a lake near his Abbey to ease his conscience. He added, remarkably, that the Stone would have produced enough gold to send twenty thousand men to the Holy Land on a crusade! Moreover, Daulton denied making the Stone himself, so it was uncertain whether he could make any more. He had obtained it from a Canon of Lichfield, now dead. The King then let him go with a present of four marks but Harbert decoyed Daulton to Stepney, where his servants stole the money before putting him in prison at Gloucester Castle in an attempt to force him to prepare some more of the Stone. Eventually Daulton was given the option of revealing the secret or being executed. True to his obligation Daulton chose death but when Harbert saw that he was willing to die rather than disclose the secret he had pity and released him. All three protagonists, however, had little time left. For Daulton 'It was not his will to live one year'. Harbert died soon after in his bed, and Sir John Delves was beheaded after the Battle of Tewkesbury in 1471.'

Chapter 3. – The Matter of the Stone

Norton then goes on to mention a number of 'labourers in the fire' who had worked for many years with no success. One in particular, called 'Tonsile', had experimented with a variety of vegetable and mineral substances without achieving the Stone. He had approached Norton with a request for help:

> For Christ his love then said he teache me'
> Whereof the substance of our Stone should be:
> Tonsile (said I) what should it you avayle
> Such thing to know: your lims doth you faile
> For very Age, therefore cease your lay,
> And love your Beades, it is high time to Praye;
> For if you knew the Materialls of our Stone,
> Ere you could make it your dayes would be gone

Despite this Tonsile pressed Norton to tell him the nature of the Stone and its preparation who recommended the study of many books, while Tonsile asked many questions as to the part played by such materials as gold, silver, sulphur and almoniack. These substances all had a part to play and, although the Stone was one, two materials were involved in its preparation, both different from each other as are a mother and child or male and female. One substance is a subtill Earth, browne, roddy and not bright, called 'Lithage', which, suitably treated, becomes Markasite. The other material, glorious, faier and bright and costing some twenty shillings an ounce, is Magnetia. These two materials when mixed with sal ammoniac and sulphur and heated will yield the White Elixir!

Chapter 4. – The Gross Work

As promised in the Proheme, the fourth chapter sets out 'the grosse Werke', starting with the two *materia prima*, lithage and magnesia whose cleansing should proceed without undue haste. Arnoldus is then cited with regard to these two components, the identity of which is the central secret of the Art. Reference is also made to Roger Bacon, who is said to have dwelt more fully on this point when he said:

> 'Divide all parts into their cognate elements. For the unlearned do not proceed in this way; but they continue pertinaciously and senselessly to add more and more to a divisible substance – and while they fancy they are on the point of bringing to perfection the flower of our Art, all they really effect is the multiplication of error.'

He then quotes Avicenna:

> 'You must go forward to perfection by true teaching in accordance with the facts of Nature: you must eat to drink, and drink to eat, and in the main season to be covered with perspiration.'

Norton goes on to cite Rhases to the same effect, although he warns workers in the field to avoid allowing the matter to consume its food too quickly – 'Let it assimilate its aliment little by little'. It appears that it is necessary to feed or nourish the materia at times, when the body craves sleep. This led Arnold (of Villanova) to observe:

> 'Let all poor men eschew this experiment, as this Art is for the rich of this world....let the enquirer be patient on of even temper, for those who are in a hurry will never reach the goal.'

Norton emphasises the gross work will be found to be generically impure; and it is a matter of great difficulty requiring the utmost wisdom of the wise, and confounding the folly of the ignorant, to purge 'our Substance' from all foreign matter. 'All men' says Anaxagoras, 'need to be taught discretion by bitter experience'.

The substance must be prepared with gentle heat, and so long as there is no violent effervescence heating may be continued over the fire and the substance consumed by gentle coction. Any haste in this operation will be indicated by excessive bubbling.

As far as the servants or laboratory assistants are concerned, Norton has some strict recommendations:

'If your servants are faithful and true, you will be able to carry out your experiments without constant vexation. Therefore, never engage married men but hire single men on a daily basis paying them over and above the standard rate, being prompt and ready in your payments. This will result in a spirit of zeal in the conduct of the work committed to them for they know that they are liable to be discharged if they are negligent in your service'.

Chapter 5. – The Subtil Work

The gross work of purifying and preparing the basic materials having been discussed in Chapter 4, Chapter 5 discusses the subtle work in which the materials must be separated into their four elements. Reference is made to a treatise by Hortulanus on the subject, and emphasis is placed on knowledge of the effects of the four qualities – heat, cold, moisture and dryness – the first two being active qualities and the second pair passive. The alchemist is also recommended to know how colours are generated. The substance should be both fixed and fluxible and have an abundance of colour. To conjoin these three contraries in one substance is one of the great secrets of the Art. The true foundation of Alchemy consists in the proper graduation of the work, in the correct adjustment of heat and cold, moisture and dryness and in the knowledge that through these qualities others are generated. Qualities such as hardness and softness, heaviness and lightness, roughness and smoothness are formed according to the addition of these primary qualities in certain proportions of weight, number and measure.

Norton then proceeds to lay down rules for the conjunction of the elements:

1. Combine your elements grammatically, in accordance with their own proper rules. These rules are the principal instruments for aiding the learned in this work: for the two greatest contraries upon earth are fixedness and volatility.
2. Join them together also after the manner of the rhetorician, with purified and ornate essences. Inasmuch as your tincture must be pure and fair, take pure earth, water, fire and air.
3. In accordance with logical methods, combine such things as admit of a true and natural union.
4. Combine them also arithmetically, in accordance with those subtle natural proportions, of which little was known when Boethius wrote 'Bind together the elements by numbers'.
5. Combine your elements musically, for two reasons: first on account of the melody, which is based on its own proper harmonies and secondly for these musical proportions closely resemble the true proportions of Alchemy.
6. Combine your elements also by means of Astrology, that all their operations may prosper and the simple rude and unformed substance may, in due course of time, and in the proper order of its development, be brought to perfection through the blessed influence of the stars.
7. The science of perspective (optics) also affords much help to those who labour in the noble Art.
8. As does the science of plenum and the vacuum.
9. The mistress of all sciences in the help of this Art is the science of Natural Magic.

These rules are taken from the English translation of the Latin version of *The Golden Tripod* in the *Hermetic Museum*. (Vol. II, p. 41)

The principal agent may be discerned by four signs or symptoms: colour, taste, smell and by fluxibility. The colour is determined by the principal agent at that time. Norton then digresses on the theory of colour based on the effect of the elements on the initial transparency and clearness. Whiteness results from transparency and blackness from the obscuring of the clearness of a dense body by the thickness of its parts and may be produced from an earthy substance by combustion. The range of intermediate colours as shown by the ruby, amethyst, chalcedony, beryl, emerald, lead, sapphire, lazulite, quicksilver and gold are described by Anaxagoras in his work Natural Changes:

'Let there be all the colours in their proper order and then let Nature bring about the process of generation in her own way till among the great variety of colours one is found to predominate which resembles the colour you are seeking to discover.'

These colours are due to the properties of Magnesia, the nature of which is capable of being changed into any proportion or degree, just as crystal, for instance, exhibits the colour of any substance which is placed under it. Hence it is well and generously said by Hermes that 'for performing the miracles of one thing, God has so ordained it that out of one thing all these marvels should spring forth'. For this reason common philosophers cannot find this virtuous Stone, because it transcends their comprehension.

The sense of smell will also enable the recognition of the predominant element and enable a distinction to be made between a subtle and a gross substance and when the stage of putrefaction has set in. As black and white are the two extremes of colour, so stench and fragrance are the extremes of odour.

The third sign and test is that of taste. Norton expresses the view the test of the palate would be more certain than that of the eye or the nose, if it were not dangerous to taste the Stone, seeing it is destructive of both health and life, so penetrating is its quality; hence it is inexpedient and even dangerous to taste it too often. Although the palate is the best judge of the progress which has been made in the Art, yet it is of little practical use because the taste of the substance (the Stone) is both horrible and hurtful. He goes on the describe the nine varieties of taste as distinguished by the ancient writers: acrid, oily and vinegary (indicative of a subtle substance), biting, salt and watery (characteristic of intermediate substances) and bitter, acid and sweet (inherent in substances of great thickness and density).

The fourth method is the fluxability (or viscosity) of the liquid. The liquid, according to Norton, is the strength of 'our' substance, and its condition affords the most striking evidence of the progress of the work; moreover, by its means the elements are both combined and dissolved. The liquid joins together the male and the female and causes the dead to be restored to life. It purges by ablution and is the principal nutriment of 'our Stone'. It would appear that many liquids may be needed for this nutriment depending upon its requirements. The views of the philosophers are then cited on this range of liquids. In *The Turba*, Aristeus, together with Pythagoras and Plato, in their mention of 'the gentle dropping of the dew' draw attention to the month of May as being the beginning of the alchemical year. There is a mystical reference 'No liquid is sufficient for the great work but the water of Lithage, which together with the water of Azoth produces Virgin's Milk'. Democritus refers to 'permanent water' in the preparation of the Stone while Rupescissa says that *Aqua Vitae* (brandy!) is the best of all liquids because it renders thick and dense substances spiritual. Even Pythagoras calls *Aqua Vitae* the vivifying principle as it volatilizes that which is fixed and fixes that which is volatile. Hermes refers to the 'Water of Mercury' as being of paramount importance. Liquids further cause the generation of the Stone by the conjunction of many things into one, assisting their fluxability and motion.

Then follows a description of the three spirits by which the soul is joined to the body. The first of these, the vital spirit, is located in the heart, the second, the natural spirit abides in the liver, while the third, the animal spirit, dwells in the body and sustains life. When these spirits are unable to abide in man, the soul must leave the body. In the work of alchemy it is necessary to distinguish between body, soul and spirit and the intermediate substances which join body and soul together by partaking of the nature of both.

Norton also makes reference to the seven circulations of each element – agreeing with the seven 'planets' and their metallic counterparts. One circulation begins with fire, the most exalted of all the elements, and ends with water, which of all the elements is the most unlike fire. Another begins with air and ends with earth.

'From earth to fire, thence to pure water, thence again to fire, and after this to a mean passing to earth, finally once again recurring to fire – by such circulations, the Red Tincture is perfected. Other circulations are more suitable for the production of the White Tincture'.

Each circulation has its proper time according to the planet under the influence of which it takes place and after all the gross and crude operations have been performed the work may still require a further twenty-six weeks, not the forty days imagined by the inexperienced.

Attention is drawn to the effects of the 'seven waters' although for this Norton recommends that the student of Alchemy seeks further instruction in the books of others. The transmutation of metals implies bringing about a change, not only of the colour of the metal, but also of its substance. The elements of the substance which is to undergo change must, therefore, become the elements of the final product. The ancient writers are said to have called the Philosopher's Stone 'a microcosm' as its composition greatly resembles that of the sublunary world.

According to Norton, when the Stone is first prepared it has abundant power in imparting its colour to other substances. This power increases with time. The Stone may go on growing in quantity and becomes more excellent in quality over the years. It is, however, necessary to augment it at the earliest possible moment by dividing it accurately into two equal parts. One half is for the Red Tincture and the other for the White. Both tinctures are composed from the same substance, in the same vessel and by the same methods, until the living matters have been mortified. Norton confesses that he had never actually prepared the Red Tincture but he understood the method perfectly and would be able to explain it to others! The redness is contained in the whiteness and may be brought out by the gentle and compelling heat of the fire. As Hermes observed:

'there lies the snowy wife wedded to her red spouse.'

Chapter 6. – Concord and Love

Again, according to Norton, the alchemist should be guided by five rules or concords.

1. The alchemist's mind should be in perfect harmony with his work. Only men of constant and persevering minds are fitted to be students of this art. They should learn the truth of the saying; 'Let us do everything from beginning to end strenuously, and yet softly and gently'.
2. He should know the difference between this art and those who profess it. The succouring of fit and suitable assistants is imperative. No assistant should be chosen who is not sober, discreet and diligent, faithful, vigilant, a keeper of secrets and a pure liver; a man of clean hands and of a delicate touch, obedient and humbly content to carry out orders.
3. There should be harmony between the work and the instruments. The different parts of the experiment require their own proper utensils. Divisions and separations should be carried out in

small vessels, with a broad vessel for humectation and a vessel of larger capacity for the process of circulation. Those for precipitation should be long while for sublimation both short and long. Narrow vessels are more appropriate for the process of correction. Vessels may be made of clay, lead, glass or stone. Other apparatus includes suitable furnaces. Without the right instruments there can be no certainty of success.

4. The work should be carried out in a place most suited to its execution. The experiment cannot succeed unless it is performed in a suitable place. Different places are differently influenced by the celestial bodies. 'The very worst of all possible places are those which have been defiled by lechery'.

5. There should be sympathy between the work and the Celestial Sphere. During the preparation of the Stone its elements follow their own proper constellations and all adverse and evil influences should be excluded. The White Tincture requires a lunar aspect, together with the ruler of the fourth house (the treasure of hidden things). Provided that things on earth correspond to things in heaven the Elixir will be obtained.

Chapter 7 – The Regimen of the Fire

The most formidable impediment to success in alchemy is ignorance of the regimen of heat and fire. There are, according to Norton, thirteen different degrees of heat for different purposes during the work:

1. For decoction of intermediate minerals and for covering lithage with sweat (scalding of pigs and geese).
2. For Air operations (drying of linen) and for divisions (roasting of meat).
3. For separation of dividents (similar heat to 2 *supra*).
4. For circulation of the elements a uniform white heat is required throughout the operation.
5. A 'moist' fire should be used to remove substances from the sides of the vessel.
6. Moist substances should be dried with a moderate heat.
7. A different heat is required for conservation.
8. For the preparation of Magnesia an effusion of fire is used, which is dangerous due to the poisonous nature of the smoke.
9. For the separation of related elements a corrosive fire is required.
10. For smelting very hard minerals a fire of consuming fierceness is required.
11. For the purging of impure metals, the essential qualities of which would be impaired by smelting, a calcining fire is necessary.
12. For sublimation a more gentle heat is required.
13. For the Projection of the Stone a most important degree of heat is required. If a mistake is made at this stage the Work must be started all over again!

Norton concludes the chapter and the Ordinall with the following lines:

> All that have pleasure in this Boke to reade,
> Pray for my Soule, and all both Quick and deade.
> In this yeare of Christ One thousand foure Hundred seaventy and seaven,
> This Warke was begun, Honour to God in Heaven.

Thomas Charnock (1526 - 1581)

In Elias Ashmole's compendium of alchemical poetry the *Theatrum Chemicum Britannicum*, published in 1652, there is a long poem by one Thomas Charnock entitled 'The Breviary of Natural Philosophy'. The title page states the poem was compiled by the unlettered scholar Thomas Charnock, student of the most worthy science of astronomy and philosophy. It is dated the 1st January, 1557 and is in six 'chapters', with a preface by the author in verse and an introduction entitled 'The Booke Speaketh'. According to its author, the poem was completed on 20th July in the year 1557. A copy of a seventh and last chapter, in the handwriting of Sir George Wharton (1617-1681), is contained in the manuscript, Ashmole 1445, and has been reprinted by Sherwood Taylor.[17]

Thomas Charnock was born in Faversham in Kent in 1526 or, as stated in the poem, in the year 1524. The reference to his being an 'unlettered scholar' probably only means that he was not particularly proficient in Latin and Greek. He could certainly write well in English and was a fluent versifier. In his early twenties he travelled through the length and breadth of England seeking alchemical knowledge and settled for a time in the City of Oxford. It was here that he became acquainted with 'James S., a spiritual man' who lived at Salisbury, and who was to take on Charnock as his laboratory assistant. As Charnock writes:

> Master I. S. his name is truly:
> Nighe the Citty of Salisbury his dwelling is,
> A spirituall man for sooth he is;
> For whose prosperity I am bound to pray,
> For that he was my tutor many a day,
> And understood as much of Philosophie,
> As ever did Arnold or Raymund Lullie:

Master I. S., whose name may have been Sauler, died in 1554 (the first and second year of King Phillip and Queen Mary) having 'gave me his worke and made me his Heire.' Charnock continued the work 'intending to my fire both Midday Eve and Morne' until 'a greate mishap unto my Worke befell'. The wooden Tabernacle in which the Philosopher's Egg or reaction vessel was being kept warm caught fire and all was lost. Nothing else could be done but to start all over again, presumably with some of the original *prima materia*. Charnock hired a 'good stoute Groome' to look after the fire and thought that all would be well. Unfortunately the groom was far from helpful. He would oversleep and let the fire go out and then increase the temperature too much by adding tallow to the fire, ruining the work and gaining the description from his master of 'Knave and whorson Lout'.

Charnock now had problems for on continuing the work himself he suffered many sleepless nights and continual anxiety the fire might get out of control. Not only that but the work was expensive to carry out, the fire alone cost £3 a week to maintain! As he wrote:

> Above a hundred pounds truly did I spend,
> Only in fire ere 9 months came to an end;
> But indeede I begun when all things were deare,
> Both Tallow, Candle, Wood, Coale and Fire:
> Which charges to beare sometimes I have sold,
> Now a Jewell, and then a ring of Gold:

There was also the need to seek further advice from a genuine adept so that Charnock could obtain further supplies of the necessary materials. During his earlier travels in search of knowledge he had met the last Prior of Bath Abbey, one William Holloway, who was known to have been successful in preparing the Philosopher's Stone (the White Stone), the Medicine and the Elixir. When Bath Abbey had been surrendered to the Crown in 1525, Holloway had received a pension of £80 a year. He later told Charnock he had in fact prepared the Red Elixir but had hidden it in a wall at the time of the dissolution of the Abbey. When he went back to recover it some years later, he was unable to find it and was so overcome with grief that he became temporarily deranged and ended up physically blind, requiring the assistance of a boy to lead him about the country. However, according to Elias Ashmole:

> 'Shortly after the dissolucion of Bath Abbey, upon pulling down some of the walls, there was a Glasse found in a wall full of Red Tincture, which being flung away to a dunghill, forthwith it coloured it, exceeding red. This dunghill (or Rubbish) was after fetched away by boate by Bathwicke men, and layed in Bathwicke field, and in the places where it was spread, for a long tyme after, the Corne grew wonderfully ranke, thick and high: insomuch as it was there look'd upon as a wonder.'

According to Charnock, neither he nor John Sauler had achieved more than the White Stone – 'But yet to the Red Worke they never came neere'. Charnock eventually managed to locate the old Prior and Holloway, having established Charnock's credentials and swearing him to further secrecy, accompanied him to a Church where they confessed their sins to the priest and received the Blessed Sacrament. After dinner they departed for a neighbouring field when, having dismissed the boy, and then, according to Charnock:

> 'Now Master quoth I the coast from hearers is clear,
> Then quoth he 'my Sonn hearken in thyne ear;'
> And within three or four words he revealed unto me,
> Of Minerals prudence the greate Mysterie.'

Charnock stayed with this master for a further nine days, during which time he learnt many of the secrets of Nature. Shortly after leaving Holloway he returned to ask him from whom the Prior had learned the Great Art. Holloway told him.

> 'Marry quoth he and speake it with a harty joy,
> Forsooth it was Ripley the Canon his Boy:'

Armed with this information Charnock started again on the work and with some success for after a time the Crowe had begun to appear black indicating that the Nigredo was setting in. All looked well, but even in the 16th century 'Murphy's Law' apparently operated and when he was within a month of completion war was declared by England against France and 'a Gentleman that ought me great mallice' caused him to be press-ganged into service in defence of Calais. Being poor and without influence to avoid this military service Charnock was desolated and:

> In my fury I tooke a hatchet in my hand,
> And brake all my Worke whereas it did stand;
> And as for my Potts I knocked them together,
> And also my Glasses into many a sliver;

It is recorded that in the July of 1557 a force of 7,000 men under command of the Earl of Pembroke crossed to France to support the Spanish King who besieged St. Quentin and on 10 August utterly routed a relieving army under Montmorency. This left the way to Paris open. Philip hesitated, however, and in so doing gave the French the opportunity to re-group. The Duc de Guise returned from Italy and strong forces were concentrated at Abbeville. The siege of the English stronghold at Calais was even contemplated. While the English commanders at Calais were by no means unaware of their danger they did realise its imminence. Their defences were in need of repair and there was discontent among the garrison due to lack of pay. The English Council, confident there would be no campaign during the winter dismissed what troops it had and there was also no fleet available to act in any emergency. On 1 January 1558, Calais was attacked and, despite the bravery shown by the remaining defenders, Lord Wentworth had to yield the city six days later. England lost the bridgehead that it had held on the continent for over 200 years. It is uncertain whether Charnock was one of the 7,000 'conscripts' but, if so, he may well have been back in England by the end of the year.

The next information concerning Charnock was that he married Agnes Norden of Stockland, Bristow, in Somerset in 1562. They had a son, Absolon, who died in infancy the following year, and a daughter, Bridget, who was later to keep house for him and assist him in his alchemical work. From Stockland he moved to Combewich near Bridgewater, where he set up a laboratory and continued his experiments until his death, which occurred in April 1581. He was buried in the neighbouring village of Otterhampton.

In 1681, a certain Mr. Andrew Pascal, *B.D.* sent a letter to John Aubrey, the antiquarian. Aubrey (1626-1697) was an associate of Elias Ashmole and the letter is one from the series amongst those manuscripts bequeathed by him to the University of Oxford. In 1946 this letter was published by Sherwood Taylor in his article on Thomas Charnock in the journal *Ambix*[17] It is reproduced here as it contains important information relating to the discoveries made at the house in Combwich where Charnock lived for many years and where he carried out his experiments which led to the preparation of the White Stone. The letter is dated 26th May, 1681 and reads:

Sr:
'This day a parchment roll about 6 ft. long and 9 in. wide was communicated to me by a friend, of which I think it will not displease you if I give some account. In the inside tis fild' from one end to the other with Scheames most circular, giving the Chimicall process of making the Elixir, and one Table, the same with what I have seen in Ram: Lully his *Ars Magna*. At the bottom are written these words. Thomas Charnock of Stockland Bristow who, travelled all the Realme of England over, for to attain unto the Secrets of this Science, which as God would he did attaine unto A.D. 1555 as it appeareth more planely in the Booke which I dedicated unto Q. Eliz: of England 1566. By this is written Borne at Faversham in Kent A.D. 1526. On the backside of the Roll about a foot from the top is written thus 'At Stockland Bristow 4 miles from Bridgewater 1566. The principall Rules of Natural Philosophy figuratively set forth to the obteyning of the Philosopher's Stone, collected out of 40 authors by the unlettered Scholar Thomas Charnock Student in the Science of Astronomy physick and Natural Philosophy. the same year that he dedicated a Book of this Science to Q: Eliz: of England wch was A.D.1566 and the 8 yeare of her reign'

'Then follow the Scheames of Sol and Luna with others of Chemicall Furnaces and Vessells. Then his Posy on the white and red Rose in 6 verses, with Thomas Charnock, subscribed in red letters 1572. [The six verses or lines are as follows]:

> XL or black appeare sartayne
> and XX or it wax bright
> And LX after to black againe
> And XX or it be white
> And it or all quick things be dedd
> — or this Rose be redd.
>   Thomas Charnock 1572

Then the Philosophers dragon described in 6 verses and a Scheame and the dragon afterward speakes in 52 verses too long to transcribe.

The dragon speaketh

> This is the philosophers dragon which eateth up his own Taile
> Being famished in a doungell of glas and all for my prevail
> Many years I kept this dragon in prison Strounge
> Before I could mortiffy him, I thought it lounge
> Yet at the length of Gods grace yff ye believe my worde
> I vanquished him wythe a fyrie sword.]

'This was lately found by accident in an old wall at the place named wch I know, and brought to a friend of mine who is seene in Judiciary Astrology, as supposed to conteine misteries belonging to Conjuration. I feare I shall not make myself Master of this Curiosity, but shall have it in my hands 10 or 12 daies.

Two or three of the inside Scheames at the Top are decaied but all the rest very faire, to the number of 30 or 40. With this was found the ruins of a paper booke in which he had written many old prophesies. such as Englishmen have been always too apt to dote upon, which I suppose he collected in his Travells.'

This discovery of a scroll and the association of Holway with Canon Sir George Ripley might indicate that it had something in common with the well-known Ripley Scrolls of which detailed descriptions are available. Sherwood Taylor examined a parchment roll (MS 2640) in the British Museum which is written in a late 17th century hand and which corresponds in all ascertainable aspects to Charnock's Roll as described by Pascal in the above letter. Some of the schemes have been identified with those contained in so-called Lullian manuscripts but their interpretation is made difficult, if not impossible, by Lull's complicated code of symbols for the different operations and materials.

On 11 June Mr. Pascal again wrote:

'The Roll was some tyme since remanded from me, and I heare that the finder of it (said to be a crafty — [*sic*]) having received informacion that it was of value, sold it, and that to a person that may deeme it an inestimable treasure.

There were 9 quarto leaves of the Prophesies in the Booke I had, one might see 50 had been eaten out by tyme.

I was Tuesday last at Stockland to make enquiry after Mr. Charnock. The oldest men were strangers to the name, which had gone out of rememberance, but in the Registers I found the tyme of his Marriage the Birth of his Son, who soone died, and the marriage of one, that I suppose must be his daughter, but no mention of his burial there. There are none of the name of the Families into wch he and his daughter married, but they are remembered vizst. Norden and Thatcher. They were of meane condition, having noe estates of their owne, but lived by renting little Tenements there.

While abroad to enquire after this Master of the Great Secret, I met a plaine Miller, who had laboured about 20 yeares, he lent me the Theatrum Chemicum and also gave me hopes to have from him an ingenious invention of his, to doe the business of a Lamp Furnace by a Candle that shall last 24 hours.'

A third letter from Mr.Pascal to Mr. Aubrey reads:

'Sr,

'I recd: and retourne thanks for yours. Since my last I got leave to transcribe what Mr. Charnock wrote on the back side of this Roll which I here send you. I keepe as neere as I could to the very errors of his pen, by which it may in part be seene, that he was as he professeth an unlettered Schollar. The inside of the Roll which is all in Latin, and perhaps is the same with the Scrowle mentioned in Theat. Chem: p. 375, was composed by a great Master in the Hermetic Philosophy, and written by a Master of his pen. Some notes written in void spaces of it by Mr. Charnock's hand shew he did not (at least thoroughly) understand it. But it seems to me, that this Roll was a kinde of vade mecum or Manuall, that the Students of that wisdome, carried about with them. I presume they are drawn out of Raym: Lully, of wch I shall be able to give fuller satisfacion, when I have his workes come down. I was also since my last at mr. Charnock's house in Comage (Combwich), where the Roll was found, and saw the place where it was hid. I saw a little roome, and contrivance he had for keeping his worke, and found it ingeniously ordered, so as to prevent a like accident to that which befell him New Yeares day 1555, and this pretty place joining as a closet to his Chamber was to make a servant needles, and the work of giving attendance more easy to himself. I have also a little Iron Instrument found there which he made use of about his Fire. I saw on the dore of his little Athanor-room (if I may so call it) drawne by his owne hand, with couse colours and worke, but ingeniously, an Embleme of the Worke, at which I gave some guesses, and soe about the walls of his Chamber, I think there was in all 5 panes of his worke, all somewhat differing from each other, some very obscure and almost worne out. They told me that people had been unwilling to dwell in that house, because reputed troublesome, I presume from some traditionall stories, of this person, who was looked on by his |Neighbours as no better than a Conjuror. As I was taking horse to come home from this pleasant entertainment, I see a pretty ancient man come forth of the next door. I asked him how long he had lived there, finding that it was the place of his birth, I enquired of him, if he had ever heard anything of that Mr. Charnock. He told me he had heard his mother (who died about 12 or 14 years since and was 80 years of age at her decease) often speake of him. That he kept a fire in, divers years; that his daughter lived with him, that once he was gon forth, and by her neglect (whome he trusted it with in his absence) the fire went out and so all his worke was lost. The Brasen head was very neere coming to speake, but soe was he disappointed. I suppose the pleasant humoured man (for that he was so appears by his Breviary) alluding to the Friar Bacon's story did so put off the inquisitions of his simple neighbours, and thence it is come down by tradition till now. Indeede it appears by the enclosed papers, that when he wrote the Roll, he had attained but to the White Stone, which is perhaps not halfe the way to the red (put me to my sister mercury, I will congeale into silver). And if the old woman's tale were true, he might afterwards be going on, and become neare to the red, and then the vexing incident might befall him. And this might be notwithstanding what is said in the fragment referred to the year 1574, for being so neare to the red as the traditional story saies he was, he might see in that 50th year of his age that the white was ferment to the red. You may observe my Calculation differs in one thing from Mr. Ashmole's in his Notes upon Theatr. Chem. p. 296. I mention this to give a reason of my dissenting from yr worthy friend, to whom I must entreat you to communicate these Informations that I have had opportunity to gather and also to present my humble service. Sr I thought when I set pen to paper to have given you an account of some conversation with a Person who is a zealous friend and admirer of this sort of knowledge, but I see I have already gone beyond my bounds. I shall only say he hath almost convinced me that it is not so hidden and obscure, so difficult and unaccountable, as men commonly seeme to believe. I have not yet seen my Miller and his invention, though he promised to bring it to me, I presume it is not yet ready, I expect him daily'.

Sr: your very affect: and faithful Servt. Th. July 19, 1681 A.P.

A further letter, dated 18th July, 1684 reads:

'Sr: Mr. Wells performed his promise. He writes that the house was lately puld downe, and is now built from the ground; all except the wall at the East End. Hee could make nothing of what only left over the Chimney. But he found the little Dore, that led out of the lodging chamber into the little Athanor roome, of that you have an account in the inclosed draught [Figure 1]. The two roses I take to be the white and the red, termes common with Charnock

Wells's Figure I

for the two Magistries, and the two Animals over them I suppose are wolves, denoting the abounding with volatile and used to the preparing and purifying one of the principall ingredients into the worke; out of it growing what beares, if those authors may be credited, most precious fruits. obliged a Painter to goe over soone after I had ben there, and take all he could find exactly. He was there, but I could never get anything from him, an ingenious man, but egregiously careless. Looking back I finde this note by me June 22, 1681. The place in the Ath: roome in wch he kept his Lampe was stone worke, about 15 inches deepe, and soe much square, but by the Collar, contrived probably after the accident of burning his Tabernacle mencioned in his printed pieces: I finde this added, 'twas painted about the Chimney thus. On the left side of the Chimney proceeded from a red stalke streaked with white, first a paire of red branches, then a paire of white, then of red, then one white to the top. Something like a Rabits head painted looking from the Chimney, to the foote of this stalke. The next picture separated as by a piller on the Chimney, from one stalke 2 white branches, of either side one, then 2 red above then 2 white then at the top this [here is inserted in the text a figure showing an object like a headstone bearing seven balls arranged six in an elongated hexagon and one in the centre] the Balls of a dusky yellow. The next picture is also distinguished by a piller on the Chimney to the right side, this quite obscured by smoak. In the last corner of the roome an other Picture described with double branches, first white and then red then white. The one at the top red. This is all that I can say of that place, of which I wish I were capable of sending a better account. The other side of Mr. Well's paper gives you one of the Scheames in the middle of this Roll, which is now by me. The transcription of the thing, said to be Ripley's should cost Mr. Ashmole nothing, were I not under an obligation not to impart it to any. It may be greatly to his loss who did communicate it to me, if the owner should know I have it, if I can contrive a way to send it with leave, I shal be ambitious to gratify that worthy person.'

Charnock and the Great Work

It may be asked what do the writings of Thomas Charnock convey on the progress he had made in the preparation of the Philosopher's Stone? Clearly by July of 1557 he had only achieved the nigredo stage of the process despite being in possession of the secret from Prior Holloway. Disgusted with the way that events had turned against him, and being unsure whether he would even return from his forced visit to the Continent, he wrote his Breviary and closed it with the words:

> 'Now for my good Master and Me I desire you to pray,
> And if God spare me lyfe I will mend this another day.'

A note at the foot of the Charnock scroll reads as follows:

> 'Thomas Charnock of Stokeland, Bristow, who travelled all the realm of England over for to
> obtain unto the secrets of this science, which, as God would, he did attain unto, *Anno Domini*

1555, as it appeareth more plainly in the Book which I dedicated unto Queen Elizabeth of England, Born at Feversham in Kent, 1526.'

A holograph copy of a Booke Dedicated to the Queenes majestie in Charnock's hand can be found in the British Museum (Lansdowne MS 703 ff. 1-53). This is dated from his house in Stockland, Bristol as November 1565 and has Lord Burleigh's autograph on f. 1.

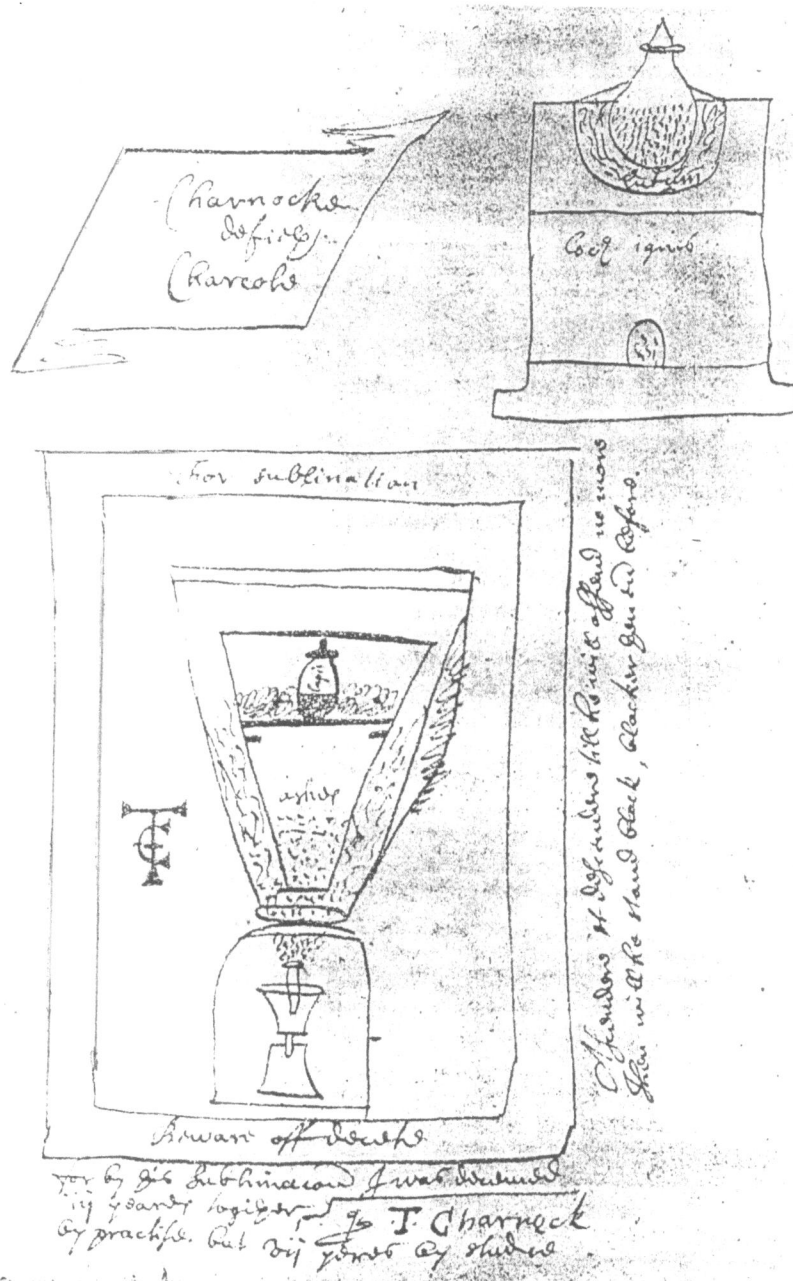

Charnock's Apparatus

In 1572 Charnock produced two short poems entitled *Aenigma de Alchimiae*, both of which are included by Ashmole in his *Theatrum*. The first of these reads:

'AENIGMA AD ALCHIMIAM

When vii tymes xxvi had run their race,
When Nature discovered his blacke face.
But when an C. and L. had overcome him in a fight,
He made him wash his face white and bright:
Then came xxxvi wythe great rialitie,
And made Blacke and White away to fle:
Me thought he was a Prince off honoure,
For he was all in Golden armoure;
And one his head a Crowne of Golde
That for no riches it might be solde:
Which tyll I saw my hartte was colde
To think at length who should wyne the filde
Tyll Blacke and White to Red dyd yelde;
Then hartely to God did I pray
That ever I saw that joyful day.

1572. T. Charnocke.

It is known from Charnock's work that the Nigredo is reached after some nine months or 36 weeks. Could this be the greate rialitie?'

The only other information that Charnock left was collected by Elias Ashmole and printed in his *Theatrum* or noted in the various manuscripts [Ashmolean MSS 972, 1420, 1441, 1445, 1452, 1478, 1492 and MS Sloane 2640]. The Booke Dedicated unto the Queenes maiestie, written in 1565, is to be found in the Lansdowne MS 703. Nowhere, however, does Charnock give any indication of the type of starting materials, although he would appear to have adopted the pseudo-Lullian system based on such devices as figures, diagrams and alphabets. The two major works by Charnock, the Breviary and the Booke may be summarised as follows:

The Breviary of Natural Philosophy

The title continues 'Compiled by the unlettered Scholar THOMAS CHARNOCK. Student in the most worthy Scyence of Astronomy and Philosophy. the first of Ianuary *Anno. Dom.* 1557. The first day of the new yeare / This Treatise was begun as after may appeare'. The Breviary commences with a section entitled 'The Booke Speaketh' and Charnock concludes:

'Wherefore in good order, I will anon declare,
What instruments for our Arte you neede to prepare.'

The next section 'The Preface of the Author' concludes with the promise:

'That of Alchimy Scyence the dore hath let open;
Sufficient for thee if thou have any Braine,
Now sharpen thy wits that thou maist it attaine.'

In the first chapter the implements of the work are discussed, starting with the potter who must make a double vessel with a third 'to guide up the heate' ostensibly to distil water for a 'Father which is somewhat blinde'! Then reference was made to a joiner for the preparation of a wooden 'Tabernacle' with keys and a lock to enclose the vessel and, finally, an egg-shaped glass vessel 'To open and close as close as a haire'. Charnock specifically recommends a glassmaker living at Chiddingsfold in Sussex for this purpose.

The second chapter begins with a prayer to God and an exhortation to keep his commandments. Then follows a parable:

> 'And we are now ready to the Sea prest,
> Where we must abide three months at the least;
> All which tyme to Land we shall nor passé,
> No although our Ship be made but of Glasse,
> But all tempest of the Aire we must abide,
> And in dangerous roades many times to ride,
> Bread we shall have none, nor yet other foode,
> But only faire water descending from a cloude:
> The Moone shall us burne so in process of tyme,
> We shal be as black as men of Inde:
> But shortly we shall passé into another Clymate
> Where we shall receive a more purer estate;
> For this our Sinns we make our Purgatory,
> For the which we shall receive a Spiritual body:
> A body I say which if it should be sould,
> Truly I say it is worth his weight in gold:'

The third chapter is called 'The Chapter of Fire' and relates to the disaster of 1557:

> 'It was upon a Newyeares day at Noone,
> My Tabernacle caught fire, it was soone done:'

and also to his Master who had given him his work and made him his alchemical heir:

> 'I obteyned his grace the date herefro not to varie
> In the first and second yeare of King Phillip and Queene Mary.'

He goes on to relate the problems with 'a good stoute Groome', who overslept and let the fire go out and then overheated the vessel by the excessive use of tallow to start it again.

The fourth chapter is again autobiographical in nature and tells how Charnock restarted the work, but when he was 'within a Moneths reckoning, Warrs were proclaimed against the French King' he was conscripted or pressed into military service to serve at Calais. In his fury and disappointment he smashed all his apparatus. The fifth chapter, however, tells of his 'greate seacret' and of his meeting with the Prior of Bath Abbey who, before imparting the secret to him, insisted that they both went to Confession and received the Holy Sacrament. Then sending his boy away, the Prior 'within three or foure words he revealed unto me,/ Of Mineralls prudence the greate Misterie.'

The sixth chapter recapitulates on the burning of the Tabernacle and the destruction of his apparatus on being pressed into military service. In Ashmole's collection the *Breviary* ends here, but Sherwood Taylor[17] has disclosed that in MS Ashmole 1445 there are a number of items of interest concerning Charnock, one of which is a copy of a seventh and last chapter of the *Breviary* that, he thinks, must have been added after the MS used by Ashmole had been compiled. It consists of seven pages, believed to be in the hand of Sir George Wharton (1617-1681).

The chapter tells how while in Calais Charnock encountered two 'yeomen come from Ingland', an apothecary and a leech, calling themselves 'philosophers', who endeavoured to obtain the secret from him by much talk among themselves of the various stages of the alchemical art. They even showed him a sample of the Philosopher's White Stone which they claimed to have prepared. Charnock appears to have set them right for before leaving their company he 'proved to their faces ere that I went; / and then they began themselves to repent.'

The Booke Dedicated unto the Queenes maiestie

In 1566 Charnock dedicated a 'booke of philosophie' to Queen Elizabeth, delivering it to her chief secretary William Cecil, later Lord Burghley, and according to him the book was to be included in the Queen's library. The book cannot now be found in the library which was transferred to the British Museum in due course. Allan Pritchard of the University of Toronto, located a manuscript, written in Charnock's own hand and bound in gold-tooled leather, in the Lansdowne collection of manuscripts in the British Museum.[18] It is thought that the work remained among the papers of Lord Burghley and is now known as Lansdowne MS 703. The book, which has not yet been published, consists of fifty-three folio sheets written in a regular, generally quite legible hand, on both sides of the parchment. The title page has the following text within an ornamental frame:

> A Booke Dedicated vnto the
> Queenes maiestie: / by master Thomas
> Charnocke, studient in the most
> Worthie sciences, off Astronomie,
> physicke, and philosophie: //
> Contayneing the worke, off
> natural philosophie. /
> *Nihil est opertu quod non reveletur*
> *et occvltu, quod non sciatur, mathaei* x
> *Anno a virgineo partu*
> .1565

After a dedicatory epistle to the Queen, Charnock claims to have had revealed to him the secret of making the Philosopher's Stone and that great time, patience and care was necessary to perfect it. He then makes specific proposals in the form of a dialogue between 'master charnocke and an oxefoordeman'. These proposals are to set up four 'woorkes' to be carried out in the Tower of London over a period of seven years. An initial sum of £100 would be required to obtain the necessary materials and a further £100 each year to keep the works going and maintain himself and his servants. In all some six people would be involved, tending the fires – himself, his wife and four servants, two men and two maidens.

The first of the four 'woorkes' will be the production of the perfect cordial or elixir and the exploitation of its medicinal use. The second will result in the production of gold intended for trial and assay to demonstrate its equality with conventional fine gold. If this gold is not found to be up to standard, Charnock declares that 'his head should be struck off on the scaffold at Tower Hill'. The third work will be to ensure a supply of the elixir for such medicinal purposes as may be 'commanded by the Queen and her council' and to preserve the health of himself and his friends, while the fourth will be for the steady production of gold. Charnock estimated that at the end of eight years he will have produced some fifty ounces of gold, and he will then increase the rate of production by fifty ounces a year until at the end of fourteen years and every year thereafter he will provide 350 ounces valued at £1,050. However, he would be unable to produce both gold and the elixir at the same time.

Following these proposals Charnock embarked upon a history of alchemy or 'natural philosophy' in England which included reference to Roger Bacon, Sir George Ripley, two monks of Bath, a Canon of Lichfield, Thomas Dalton, Thomas Norton, Henry VII and his own uncle Thomas Charnock, D.D. Charnock also mentions such European alchemists as Raymond Lully, Arnold of Villanova, and Cornelius Agrippa of Nettesheym.

The Miscellaneous Writings of Thomas Charnock

Charnock appears to have been a great note maker, both in the margins of his books and on the back of his great scroll. These jottings are usually in verse and relate to stages in the work, the number of operations or repetitions to that point. For example, at the end of his copy of Ripley's poem Cantalena he wrote:

> Abowt 653. I dare be bold,
> This Chyld shall put on a Crowne of Gold;
> Or at 656, at the moste,
> This Chyld shall rule the roste.

and again:

> Fro the tyme that he be Black and Ded,
> Wash him 7. tymes, or he be perfect Red.

and also:

> Looke you conceive my words aright,
> And marke well this which I have sede;
> For Black is Ferment unto the Whyte,
> And Whyte shalbe Ferement unto the Rede:
> Which I never saw till I had white heres upon my head.
> T.C. 1574 the 50 yeare of my age.

Charnocke his trewe similitude
(MS. Ashmole 1420)

'Knowe ye well that when fire, ayre and water are departed from the earth, then the body remaineth ded, black and stinking, and is by reason so called: *terra nigra, fetida et spongiosa* for that as it be is nothing worth: Knowe ye therefore the pueriste and the heist element that our god made is the fyre. And therefore it is I set in the heist sphere next to the hevens by which ellament every earthly creature shalbe purified before the day off doume or he shall receive his spirit agane then shall that coruptable body be changed into a glorious bodie bright and shining and shall remayne dere and pretious in the lords sight ever world without end.

Even so shall that earth wch was so black through corupption, be puryffied by lavigation off the ayre and purging off the fire and so receve an immaculate bodie bright and shining as the Sson.'

Charnock's Parchment Roll

In a letter from the local parson, a Mr. Andrew Paschal, written to John Aubrey Esq. Fellow of the Royal Society, and dated 26th May, 1681, an account is given of finding a parchment roll about 6 ft. long and 9 in. wide in Charnock's house at Comage near Stockland. It is uncertain what became of this roll, although it was for a short time in the possession of Mr. Paschal when he wrote a description of it and sent this to John Aubrey in a series of letters. The inside of the roll was:

'fild' from one end to the other with Scheames most circular, giving the Chimicall process of making the Elixir, and one Table, the same with what I have seen in Ram: Lully his Ars Magna. At the bottom were written these words. Thomas Charnock of Stockland Bristow who travelled all the Realme of England over, for to attain unto the Secrets of this Science, which as God would he did attaine unto A.D. 1555 as it appeareth more planely in the Booke which I dedicated unto Q: Eliz: of England 1566. By this is written Borne at Faversham in Kent. A.D. 1526'

'On the backside of the Roll about a foot from the top is written thus 'At Stockland Bristow 4 miles from Bridgewater 1566. The principall Rules of Natural Philosophy figuratively set forth to the obteyning of the Philosopher's Stone, collected out of 40 authors by the unlettered Scholar Thomas Charnock Studient in the Science of Astronomy phisick and Natural Philosophy, the same yeare that he dedicated a Book of this Science to Q: Eliz: of England wch was A.D. 1566 and the 8 yeare of her reign.'
'Then follows the Scheames of Sol and Luna with others of Chemicall Furnaces and Vessells. Then his Posy on the white and red Rose in 6 verses, with Thomas Charnock, subscribed in red letters 1572'
'Then the Philosophers dragon described first in 6 verses and a Scheame, and the dragon afterward speakes in 52 verses too long to transcribe.'

Apparently Paschal was only allowed to keep the roll for transcribing purposes for 10 or 12 days, although he borrowed it again to transcribe Charnock's writings on the back of the roll, the text on the front being all in Latin, and originally thought to be similar to the 'emblematicall scrowle described in the *Theatrum Chemicum Britannicum*.[19]

Sherwood Taylor describes the parchment roll (in MS Sloane 2640) which was written in a late 17th or early 18th century hand, and the contents of which correspond 'in all ascertainable respects to those of the latter part of Charnock's roll as described by Pascal. It contains only nineteen circular schemes and, in addition, a number of illustrations of alchemical vessels, a drawing of an '*arbor philosophorum*' and a triangular table of combinations of the letters of the alphabet, taken two at a time. It was concluded that the roll was a transcription of schemes and drawings of apparatus to be found in manuscripts of Lull, but not in his printed works. Sherwood Taylor concludes that Charnock's operations were chemical, involving distillation and circulation in the pelican or other hermetic vessel, and not simply psychological, and that they were conducted by continuous repetitions of a process which completed itself in periods of a week.

These writings and a number of other jottings by Charnock are reproduced by Sherwood Taylor, but despite the indication of the years which the experiments occupied him Charnock neither divulged any practical process used in the Great Art nor did he use any of the Philosopher's Stone, on the assumption that he achieved success around 1574 to enrich himself. On the contrary, he warned people from seeking gold for their own profit. Burland is of the opinion that Charnock too had achieved a peace of mind and soul that transcended the worth of material gold.[19]

William Backhouse – Alchemical Father of Elias Ashmole

On 3 April 1651, Elias Ashmole recorded in his diary: 'oH. 30' P.M. Mr. Will Backhouse of Swallowfield in Com. Berks, caused me to call him Father thence forward.'

In the *Theatrum Chemicum Britannicum* which was published in London in 1652, Ashmole comments:

'There has ever been a continued succession of Philosophers in all Ages, although the heedlesse world hath seldome taken notice of them; For the Ancients usually (before they dyed) Adopted one or other of their Sonns, whom they knew well fitted with such qualities, as are sett downe in a letter that Norton's Master wrote to him when he sent to make him his Heire unto this Science. And otherwise then for pure virtues sake, let no man expect to attaine it... Rewards nor terrors (be they never so Magnificent or Dreadfull) can wrest this secret out of the bosome of a Philosopher: amongst others, witnesse Thomas Dalten. Now under what Tyes and Ingagements this Secret is usually delivered, (when bestowed by word of mouth) may appear in the weighty obligations of that Oath which Charnock tooke before he obtained it, for thus spake his Master to him:

> Will you with me to Morrow be content
> Faithfully to receive the blessed Sacrament
> Upon this Oath that I shall here you give,
> For ne Gold ne Silver as long as you live,
> Neither for love you beare towards your kinne,
> Nor yet to no great man preferment to winne,
> That you disclose this Secret that I shall you teach,
> Neither by writing nor by no swyft Speeche;
> But onely to him which you be sure,
> Hath ever searched after the Secrets of Nature,
> To him you may reveale the Secrets of this Arte,
> Under the covering of Philosophie before this worlde you depart.

'And this Oath he charged him to keep Faithfully and without Violation 'As he thought to be saved from the pitt of Hell', And if it so fell out, that they met not with any, whome they conceived in all repects worthy of their Adoption, they then reseigned it into the hands of God, who best knew where to bestow it. However, they seldome left the World before they left some written Legacy behind them, which (being the Issue of their Braine)stood in roome and place of Children, and becomes to us both Parent and Scholmaster, throughout which they were so universally kinde, as to call all Students by the deare and affectionate Tytle of Sons (Hermes being the first President) wishing all were such, that take the paines to tread their Fathers stepps, and industrially follow the Rules and Dictates they made over to posterity, and wherein they faithfully discovered the whole Mystery:

> As lawfully as by their fealty thei may,
> By licence of the dreadful Judge at domes day.'

'In these Legitimate Children they lived longer than in their Adopted Sons, for though these certainly perished in an Age, yet their Writings (as if when they dyed their Souls had been Transmigrated into them) seemed as Immortal, enough at least to perpetuate their Memories, till Time should be no more. And to be the Father of such Sons, is (in my opinion) a most noble happinesse.'

> Let Clownes get Heires, and Wealth; when I am gone,
> And the great Bugbeare grisly death
> Shall snatch this Idle breath,
> If I a Poem leave, that Poem is my Son.

So overjoyed was Elias Ashmole by the event of his adoption as a landmark in his lifelong interest in alchemy that he expressed his gratitude and joy in an ode (which is preserved in MS Ashmole 36-37, ff. 241-241).

To my worthily honour'd
William Backhouse Esquire
Upon his adopting of me to be his Son.

From this blest Minute I'le begin to date
My Yeares & Happines; (since you create
What wise Philosophers call Lyfe;) & vow
I ne're perceived what being was till now.

See how the power of your Adoption can
Transmute imperfect Nature to be Man:
Nay, with one word may yet refine it more,
Then all ye best digested Indian Oare.
Your Son! 'Tis soe! for I begin to finde,
Your Auncestors large Thoughts grow in my Minde:
I feele that noble Blood spring in my Heart,
Which doth intytle me to some small parte
Of grand sire Hermes wealth; & hope to haue
Interest in all the Legacies he gave,
To his Successiue children; from whome too,
I must derive what is confer'd by you.

To prove each mie Descent, I neede not see,
A byast Herald for my Pedigree;
That I'me true bred, question it he that dare,
If these my Aeglete Eyes on th' Sun can stare.
Or cause a in Crest I hold
Since my crude Mercury's transmute to Gold.
Ile vouch my fate for Honor, Witt Descent,
And all, which to th' Hermetick Tribe is lent.
Then be you blest my Starrs, who gaue to me
So blest a tyme for this nativity.
That plac'd the golden Lyon in the East
When Sol within the ram, the Nynth possest,
As if their Influence meant to ope the way,
To make Night Mysteries shine cleere as Day.
Hast yee some good direction that shall lead
My Fathers hand with's Blessing to my Head
And leave it there.   His leaves of Hermes Tree
To deck the naked Ash bequeath to me;
His legacy of Eyes to'th blinde Mole spare
And (though a younger Son) make me his Heire.

On 13th May, 1653 Ashmole notes:

> 'My father Backhouse lying sick in Fleete Street over against St. Dunstans Church, & not knowing whether he should live or dye, about eleven a clock, told me in Silables the true Matter of the Philosophers Stone: which he bequeathed to me as a Legacy.'

In his edition of Elias Ashmole's autobiographical and historical notes C.H. Josten remarks: 'The obvious explanation for William Backhouse's speaking in syllables would seem that his physical condition prevented him from speaking coherently, yet this would hardly have been worth noting. Gerard Heym suggested that the secret of the true matter, which William Backhouse conveyed, might have been a magic formula consisting of syllables. This would appear to tally with an idea Ashmole expressed in the course of the same year, namely, that by joining 'Formall with Vocall' the philosopher can reach 'th' unfathom'd depth of greatest Misteries'. There is an anonymous alchemical manuscript in the *Bibliotheque Nationale*, Paris (MS *Français* 12335) which lends further support to this theory. It dates from the late 17th or early 18th century and contains at ff. 89-90 a chapter entitled '*Sillabes Chimiques*'. The author of the manuscript explains that certain syllables to be derived from seven hieroglyphic signs, which he placed at the beginning of the chapter, will form '*un mot significatif ou un charactere universel*', revealing '*le veritable nom et charactere de la matiere premiere*'. As far as is known, this information does not occur in other alchemical texts. It indicates that there was a secret tradition among certain alchemists, according to which the substance of the prime matter could be expressed in syllables.'

Elias Ashmole noted down an alchemical riddle (MS Ashmole 1417, f. 9):

> Of one part of mans Frame,
> Six letters make ye name,
> One P: add unto them.
> Then change S: into M:
> This done you do uncage,
> The Subject of ye Sage.

It is thought that the solution of this riddle would indicate the true matter of the Stone. It is uncertain whether Ashmole made this riddle up himself or merely copied it from another author.

William Backhouse of Swallowfield

From a collection of horoscopes cast by Elias Ashmole himself it appears William Backhouse was born on 17 January 1593 at Swallowfield, near Reading. His father, Samuel Backhouse was High Sheriff of Berkshire in 1598 and again in 1601. It is recorded that, at the age of seventeen William matriculated at Christ Church College at Oxford, although there is no evidence he ever acquired a degree. In an account of Swallowfield and its owners, written by Lady Russell and published in 1901, it is mentioned:

> 'there is a curious MS written to William Backhouse about this date by John Blagrave, the celebrated astrologer and mathematician, who lived at Southcot near Reading and also had a house at Swallowfield and land at Eversley. It is probable that it was this and similar communications that induced William Backhouse to enter deeply into the study of Rosicrucian philosophy.'

It has been suggested by Josten, Backhouse may have met Robert Fludd (1574-1637) who was also a member of Christ Church and was later to become an eminent defender of the Rosicrucian cause in England. Their mutual interest in alchemy may have led them into the same circle. Anthony a Wood calls Backhouse 'a great Rosy Crucian' and it was in the period from 1612 to 1616 that the appearance of the Rosicrucian pamphlets in Germany was raising considerable interest and controversy. Very little is known of Backhouse's movements after he left Oxford, although from his translations of French alchemical works it would seem he may have travelled for some time on the Continent.

In Ashmole's *Theatrum Chemicum Britannicum* is a poem entitled 'The magistry' by one W.B. In Ashmole's own copy of the *Theatrum* the author's name is spelt out in full and the whole poem runs as follows:

THE MAGISTERY

> Through want of Skill and Reasons Light
> Men stumble at Noone day;
> Whilst buisily our Stone they seeke,
> That lyeth in the way.

Who thus do seeke they know not what
　　　　Is't likely they should finde?
Or hitt the Marke whereat they ayme
　　　　Better than can the Blinde?

No, Hermes Sonnes for Wisdome aske
　　　　Your footsteps shee'le direct:
Shee'le Natures way and secret Cave
　　　　And Tree of lyfe detect.

Son and Moone in Hermes vessel
　　　　Learne how the Collours shew,
The nature of the Elements,
　　　　And how the daisies grow.

Greate Python how Apollo slew,
　　　　Cadmus his hollow-Oake:
His new rais'd army, and Iason how
　　　　The fiery Steers did yoke.

The Eagle which aloft doth fly
　　　　See that thou bring to ground;
And give unto the Snake some wings,
　　　　Which in the Earth is found.

Then in one Roome sure bind them both,
　　　　To fight till they be dead;
And that a Prince of Kingdomes three
　　　　Of both them shalbe bred.

Which from the Cradle to his Crowne,
　　　　Is fed with his own blood;
And though to some it seemeth strange,
　　　　He hath no other Foode.

Into his Virgin-Mothers wombe,
　　　　Againe he enter must;
Soe shall the King by his new-byrth,
　　　　Be ten times stronger just.

And able is his foes to foile,
　　　　The dead he will revive:
Oh happy man that understands
　　　　This Medicen to atchive!

Hoc opus exigium nobis fert ire per altum
DECEMBER, 1633　　　W.B(ackhouse).

This poem is full of alchemical imagery and brings to mind many of the illustrations by Johann Theodor de Bry, in Michael Maier's *Atalanta Fugiens* which was published in 1618. In the first verse the 'Stone that lyeth in the way' might refer to *Emblema* XXXVI – 'The Stone is projected upon the Earth, and exalted upon the mountains, and dwells in the air, and feeds in the river: that is Mercury.'

The second verse relates perhaps to the eighth key of Basil Valentine in which the Marke or Target, the key of the work, is dissolution leading to putrefaction which precedes the glorious rebirth and stems from the saying of Jesus in St. John's Gospel:

> 'In truth, in very truth I tell you, a grain of wheat remains a solitary grain unless it falls into the ground and dies; but if it dies, it bears a rich harvest.'

The third verse returns to Michael Maier and *Emblema* XLII. 'To him who concerns himself with Alchemy, may Nature, Reason, Experience and Reading be guide, staff, spectacles and lantern.' The 'Secret cave' suggests an illustration from the *Museum Hermeticum* of 1625 which illustrates the hermetic maxim 'As Above, so Below' and the conjunction of the four elements, earth, air, water and fire with the Seal of Solomon or Star of David, which is the hieroglyph of the Philosopher's Stone wherein all the elements are reconciled in perfect balance. Apollo and the Six Muses represent the seven metals of antiquity: gold, silver, mercury, iron, copper, tin and lead and their celestial counterparts.

Verse four starts with 'Sun and Moone in Hermes Vessel', clearly the setting for the conjunction of King and Queen. Their royal marriage is the supreme union of hostile opposites which must be followed by death and rebirth, giving rise to the phenomenon of the alchemical nigredo and the formation of the hermetic androgyne. Historically the relationship of the bride to the bridegroom is incestuous as in the story of Oedipus. As the reaction vessel is heated in the Athanor, the black of the nigredo stage gives rise to a progressive display of colours often referred to as 'The Peacock's Tail'. The reference to the growth of daisies may indicate the gentle nature of the process, with moisture circulating in the vessel like rain upon the fresh earth.

The reference to Cadmus in verse five could well be to the fixation of the volatile for Cadmus or Sulphur transfixed the Great Serpent on the hollow oak. Cadmus killing the dragon or python represents the Philosopher's Gold vanquishing Nature or the Philosopher's Sulphur, represented by the four corruptible elements. The hollow tree is the reaction vessel of the philosophers. In Greek mythology Cadmus was then able to sow the teeth of the serpent which came up as armed men and attacked the sower. Cadmus, however, on the advice of Athene survived this attack by throwing a precious stone (the Philosopher's Stone) among them, for the possession of which they fought amongst themselves, killing each other. Jason was associated with Cadmus as in the legend he sowed the remaining dragon's teeth. Jason was, of course, involved in the quest of the Golden Fleece which had been hung on the sacred oak by Phryxus at Colchis. The quest for the Stone has often been likened to that for the Golden Fleece. Jason also was a dragon slayer.

In the sixth verse there is reference to an eagle which is symbolic of the volatile as that to a serpent is to the fixed. Here then is the essence of alchemy – volatilising the fixed and fixing the volatile in a continuous reflux process with in the reaction vessel, the Philosopher's Egg. As Solomon Trismosin puts it:

> 'Si fixum solvas faciasque volatile
> Et volucrum figas, faciet te vivere tutum'

translated by Backhouse as:

> If thou dissolue ye fixt & make it fly
> And fix ye bird thou shalt lieu happily.

It is interesting to note that the armorial crest of the Backhouse family included 'an eagle displayed vert, holding by each claw a serpent on its back, the tail knotted.'

In the seventh verse there is reference to the conjugation of the King and Queen within the hermetic vessel. The union of the opposites proceeds through death to a glorious resurrection with all the colours of the peacock's tail. The three kingdoms may well be the animal, mineral and vegetable, which emphasises the universality of the Stone.

Verse eight is difficult to understand, although as the pelican feeds its young with its own blood so the King must be nourished to enhance the transmutatory power of the Stone. This is reflected in the Parabola of Hinricus Madathanus from the *Geheime Figuren*. The familiar alchemical symbol of a dragon or serpent eating its own tail is an alternative to the pelican.

The enhancement of the power of the Stone 'by entering again into his Virgin-Mothers wombe' is continued in verse nine. This suggests the importance of returning the Stone to the reaction vessel for its final treatment, after which its full properties become manifest. Re-birth is also necessary on the spiritual plane to achieve perfection.

The tenth verse continues and concludes by hinting at the strength and power of the Stone. Indeed happy is the man that understands and achieves the goal or hits the marke!

Le Vraye Livre de la Pierre Philosophale du Docte Synesius, Abbé Grec

This treatise by Synesius was translated by William Backhouse and can be found in Ashmole MS 58, (pp. 79 et seq.). Writing of 'Sublimacion', the first operation of the art, Synesius must have influenced Backhouse, and throws further light upon the symbolism of the Eagle and the Serpent in the Arms of the Backhouse family. Synesius writes:

> 'What is noways vulgar but physical sublimacion with wch wee remove ye superfluousness of ye stone, wch is indeede ye elevacion of ye not fixed parte wch doth remaine in ye bottom, but we would not have them separated ye one from ye other but by yt abidinge together they fixe each other. And know he that shall sublime the philosophers O (as he ought) which hath all ye virtue of our stone, he shall perfecte ye maistery. Geber thearfore saieth all ye perfeccion consisteth in sublimacion, and in his sublimacion are all other operacions, that is distilacion, assacion, destruccion, coagulacion, putrefacion, fixacion, reducion of ye white and red tinctures procreated & engendered in one furnasse and on vessel and this is ye righte way ouerto ye finall conclucion of wth ye philosophers have diverse chapters to deceive ye ignorante.'

As C.H. Josten has pointed out 'Eagle and Serpent, therefore, represent the Philosopher's Mercury as well as its sublimation and the Stone itself – the matter, the method and the result' – an interpretation which is confirmed in the Second Key of Basil Valentine where Mercury, wearing a crown, stands between duelling knights. A serpent winds itself round the drawn sword of one knight, while an eagle perches on the sword of the other.

William Backhouse and Elias Ashmole

There are a number of references to William Backhouse in Ashmole's diary, the first of which, as has been mentioned, was recorded on 3 April 1651 when Elias was to call Backhouse 'father'. Some days later, on 26 April there is a further entry:

> '5H 30' p.m. my Father Backhouse brought me acquainted with the Lord Ruthin, who was a most ingenious person'

Lord Ruthin (or Ruthven) is thought to have been the Right Honorable and Learned Chymist, who was the author of *The Ladies Cabinet, Enlarged and Opened*, published in London in 1654.

On 10th June, 1651 there is a further reference to Backhouse:

'Mr: Backhouse told me I must now needes be his Son, because he had communicated soe many Secrets to me.'

Such records of these 'Secrets' as have been traced from Ashmole's shorthand notes indicate that they were of a general chemical nature and not, apparently, concerned with the 'first matter' of the alchemical work.

Again on 9th October, 1651:

'My Father Backhouse & I went to see Mr: Goodier (the great botanist) at Petersfield.'

On 10th March, 1652:

'This morning my Father Backhouse opened himselfe very freely, touching the Great Secret.'

and on 12th April, 1652:

'This morning I received more satisfaction from my Father Backhouse, to the Questions I proposed.'

On 13 May 1653, Ashmole received 'the True matter of the Philosopher's Stone', but he is discreet and records nothing concerning the nature of the information which he received. Again there is no record of Ashmole being involved in any practical work in the laboratory. He spent most of his time on his antiquarian studies and in the collection and publishing of alchemical poems in *Theatrum Chemicum Britannicum*, the *Fasciculus Chemicus* (London 1650) and an edition of an Elizabethan manuscript entitled 'The Way to Blisse' (published in 1658).

On 30th May, 1662 Ashmole writes:

'My Father Bachus died this Evening at Swallowfield.'

and on 17th June, 1662:

'This afternoon my Father Backhouse was buried in Swallowfield Church.'

The substance of the great Secret passed by Backhouse to Ashmole may never be known. Nor is there any indication that Ashmole passed it on to any suitable successor, although Sherwood Taylor is of the opinion that when getting on in years and despairing of achieving the alchemical arcana himself, he might have passed it to Dr. Robert Plot (1640-1696). Plot and Ashmole were introduced to each other by John Evelyn in 1677. Ashmole subsequently appointed Plot as Custodian of the Ashmolean Museum in Oxford and also probably procured his appointment as Professor of Chemistry there.

Chapter Fourteen

Contemporary Alchemists

Fulcanelli – Master Alchemist

In 1926 there was a book published in Paris by Jean Schemit of 45, Rue Lafitte in the Opera district, entitled *Le Mystere des Cathedrales* and written by an author with the pseudonym 'Fulcanelli'. The first edition was limited to only 300 copies. The preface to the work was written by a young student of alchemy, Eugene Canseliet, F.C.H. The preface began:

> 'For a disciple it is an ungrateful and difficult task to introduce a work written by his own Master. It is, therefore, not my intention to analyse here *Le Mystere des Cathedrales*, nor to underline its high tone and its profound teaching. I most humbly acknowledge my incapacity and prefer to give the reader the task of evaluating it and the Brothers of Heliopolis the pleasure of receiving this synthesis made so superbly by one of themselves. Time and truth will do the rest.'[1]

From this it would appear the author was Canseliet's alchemical Master and also there was a mysterious fraternity called the 'Brothers of Heliopolis'.

According to the French historian of alchemy, Jacques Sadoul, it would appear the work aroused little interest at its first appearance. It was followed two years later by a second volume, double the length of the first, entitled *Les Demeures philosophales et le symbolisme hermetique dans ses rapports avec l'Art sacre et l'esoterisme du Grand Oeuvre*. These two books have since acquired a reputation of comparing favourably with any of the classics of the alchemical literature and of being two of the best, clearest and most soundly based treatises on the subject. Be that as it may, Fulcanelli adhered strongly to the alchemical tradition the secrets of the art must never be disclosed to the profane.[2]

The first work, *Le Mystere des Cathedrals*, contained thirty-six illustrations, two in colour, by the artist Jean-Julien Champagne, although there are some forty-nine illustrations in the English edition, published by Neville Spearman in 1971. In this later edition, however, the black and white illustrations by Champagne have been replaced by photographs of the same subjects. In the English edition of *Les Demeures Philosophales*, however, there are forty-three illustrations and a frontispiece, most of which are drawn by Champagne.[3] Despite a spate of publications speculating upon the identity of the author whose pseudonym was 'Fulcanelli', who disappeared before the publication of his first book and whose identity was known only to Canseliet's limited circle of close friends, this has never been disclosed! All that was known about him was that 'the Master Fulcanelli' was elderly, distinguished, rich, immensely learned and, possibly, even of aristocratic or noble lineage.

There are in the literature, accounts of at least two transmutations carried out with small quantities of the Philosopher's Stone prepared by Fulcanelli. According to Canseliet, a small piece of lead was transmuted into gold by the addition of 'Fulcanelli's precipitate' and 'according to his instructions' in the little laboratory in which Canseliet worked at the Paris gasworks at Sarcelles. This transmutation was carried out by Canseliet himself, in the presence of Jean-Julien Champagne, the artist, and Gaston Sauvage, a chemist who was a contemporary and close friend of Canseliet.

An account of a second transmutation is given by Albert Riedel (*Frater Albertus*) in his book *The Alchemist of the Rocky Mountains* This was supposed to have taken place in the autumn of 1937 at the Castle de Lere, near Bourges, and was witnessed by the Castle's owner, Pierre de Lesseps, two physicists, a chemist and a geologist. According to Riedel, on this occasion

Fulcanelli was not only present in person but also transmuted a sample of lead into gold and, at the same time, some 100 grammes of silver into a like amount of uranium![4]

There is also a tradition that about this time Fulcanelli made contact with the research assistant of the French atomic physicist Andre Helbronner, warning him of the dangerous implications of manipulating nuclear energy. Stranger than this, Conseliet in a letter to Walter Lang stated that when he had worked with him as a young student of alchemy, Fulcanelli was already a very old man but carried his eighty years lightly. Thirty years later, in the 1950s, he was to see him again, briefly, and at a pre-arranged rendezvous, when he appeared to be no older than a man of fifty, no older than Canseliet himself at that time![5]

In a book entitled *Le Matin des Magiciens* by Louis Pauwels and Jacques Bergier, published by Editions Gallimard, Paris in 1960, what is claimed to be an exact account of a conversation between Bergier and an alchemist who he believed to have been Fulcanelli is given. This reads, in the English translation by Rollo Myers, published in 1963, as follows. According to Bergier, the alchemist believed to have been Fulcanelli, said:

'M. Andre Helbronnner, whose assistant I believe you are, is carrying out research on nuclear energy. M. Helbronner has been good enough to keep me informed as to the results of some of his experiments, notably the appearance of radio-activity corresponding to plutonium when a bismuth rod is volatilised by an electric discharge in deuterium at high pressure. You are on the brink of success, as indeed are several others of our scientists today. May I be allowed to warn you to be careful? The research in which you and your colleagues are engaged is fraught with terrible dangers, not only for yourselves, but for the whole human race. The liberation of atomic energy is easier than you think, and the radio-activity artificially produced can poison the atmosphere of our planet in the space of a few years. Moreover, atomic explosives can be produced from a few grammes of metal powerful enough to destroy whole cities. I am telling you this as a fact: the alchemists have known it for a very long time'. Bergier was about to interrupt when the alchemist interrupted him: 'I know what you are going to say, but it's of no interest. The alchemists were ignorant of the structure of the nucleus, knew nothing about electricity and had no means of detection. Therefore they have never been able to perform any transmutation, still less liberate nuclear energy. I shall not attempt to prove to you what I am now going to say, but I ask you to repeat it to M. Helbronner: certain geometrical arrangements of highly purified materials are enough to release atomic forces without having recourse to either electricity or vacuum techniques. I will merely read to you now a short extract (from *The Interpretation of Radium* by Frederick Soddy (1877-1956) and published in 1909) "I believe that there have been civilizations in the past that were familiar with atomic energy, and that by misusing it they were totally destroyed." He then continued "I would ask you to believe that certain techniques have partially survived. I would also ask you to remember that the alchemists' researches were coloured by moral and religious preoccupations, whereas modern physics was created in the 18th century for their own amusement by a few aristocrats and wealthy libertines. Science without a conscience... I have thought it my duty to warn a few research workers here and there, but have no hope of seeing this warning prove effective. For that matter, there is no reason why I should have any hope." On being asked the nature of his own researches the alchemist replied "You ask me to summarize for you in four minutes four thousand years of philosophy and the efforts of a life time. Furthermore, you ask me to translate into ordinary language concepts for which such language is not intended. All the same, I can tell you this much: you are aware that in the official science of today the role of the observer becomes more and more important. Relativity, and the principle of indeterminacy, show the extent to which the observer today intervenes in all these phenomena. The secret of alchemy is this: there is a way of manipulating matter and energy so as to produce what modern scientists call a 'field of force (force field). This field acts on the observer and puts him in a privileged position vis-à-vis the Universe. From this position he has access to the realities that are ordinarily hidden from us by time and space, matter and energy. This is what we call 'The Great Work'. The essential thing is not the transmutation of metals, but that of the experimenter himself. It's an ancient secret that a few men re-discover once in a century.'

In a book, *The Zelator: The Secret Journals of Mark Hedsel*, edited and with an introduction by David Ovason, there is the suggestion that Fulcanelli was 'quite recently' living in Florence.[6] But in the words of Walter Lang in his introduction to *Le Mystere des Cathedrales*, for all practical purposes Fulcanelli has vanished as though he never existed. Only his contributions to the literature of alchemy remain, together with a certain amount of speculation as to his identity.[7]

Le Mystere des Cathedrales

In the first of the three books written by Fulcanelli, the reader is introduced to the concept that in the architecture of the gothic cathedrals of France, and elsewhere, there resides a means of passing on esoteric knowledge, encoded in the structure and decoration of the buildings. The Gothic cathedral, according to Fulcanelli, should not be regarded as a work dedicated solely to the glory of God and Christianity but rather as a vast concretion of ideas and of tendencies in popular belief both sacred and profane.

The great cathedrals functioned during the medieval period not only for the worship of God but also as social centres for those dwelling in the city. The celebration of what may have passed for secular feasts such as 'The Feast of Fools' and 'The Feast of the Donkey', the former involving a grand procession and a hermetic fair and the latter with the donkey, the glorious Christ bearer, simulated the entry of Jesus into Jerusalem whose streets he once trod.

> 'This asinine power, which was worth to the Church the gold of Arabia, the incense and myrrh of the land of Saba.'

In medieval times the cathedrals were the refuge of all unfortunates, the poor, the sick and the hunted and were also available for the burial of the illustrious dead. The alchemists of the 14th century were, apparently, in the habit of meeting regularly either at the main Porch, at the Portal of St. Marcel or at the little Port-Rouge which was decorated with salamanders symbolic of fire. When Denis Zachaire in the 16th century set about cultivating contacts within the alchemical fraternity of Paris he confirmed they were in the habit of meeting regularly either at each others' lodgings or at the cathedral of Notre Dame.

Fulcanelli emphasised the term 'gothic' which related to the architecture of the cathedral, also had a different meaning from that of the ancient Germanic peoples, and should be considered as derived from *art gothique* or 'argotique', the argot or cant, a spoken cabala or 'green' language peculiar to those who wished to communicate their thoughts without being understood by outsiders. This green language was also used by philosophers such as Rabelais and Nostradamus. Argot is one of the forms derived from 'the language of the birds' and has a Sufi connotation, a language teaching the mystery of things and unveiling the most hidden truths. Fulcanelli claimed that Jesus revealed a knowledge of this language to his disciples through the Holy Spirit and it was thus through an 'innerstanding' of this hidden language that they became aware of the full meaning of his parables.

Returning to the symbolism of Gothic cathedrals, Fulcanelli pointed out the ground plan of a cathedral was cruciform and the word 'cross' was equivalent to the alchemical 'crucible'. This also gave an indication of the *materia prima*, which like Christ, suffers a Passion in which it dies, only to be revived, purified, spiritualised and finally reformed. *In hoc signum vinces*, (In this sign thou shalt conquer), for the Cross points the way by which the alchemist may obtain the FIRST STONE and bears, like Christ, the imprint of the nails. The symbol of the Cross together with the semi-circular apse of the cathedral, for instance, leads to the *crux ansata* or *ankh*, an emblem of Venus or copper related to the alchemical 'sulphur'. The first stone upon which Jesus built his Church, the rough ashlar, impure, gross and unpolished was not only the Apostle St. Peter, but also the Saviour himself!

Some cathedrals exhibited a further guide to alchemy in the pavement of their naves often ornamented by mosaic mazes or labyrinths symbolic of the two major mysteries of the Work, the finding of the path to reach the centre and then of the way to emerge, best accomplished with the aid of the thread of Ariadne. This guide or lodestone of *magnesia* pointed the way to the Philosopher's Stone, the object of the quest.

The orientation of the great cathedrals ensure the rose windows, which adorn the transepts, exhibit the colours of the work as the sun rotates from dawn through mid-day to sunset. These colours unfold in a circular progression from the shadows, the absence of light, or black, to the perfection of ruddy light passing through the colour white, '*considered as being the mean between black and red*' in the alchemical spectrum. Fulcanelli pointed out that in the Middle Ages the central rose window of the porches was called ROTA, the wheel, the alchemical hieroglyph of the time necessary for the coction of the philosophical matter and also the secret fire, the second agent which '*makes the wheel turn*' and produces the changes observed by the alchemist in the reaction vessel.

The Cathedral of Notre Dame in Paris

The patron saint of the great cathedral of Paris, built on the River Seine, is, like many of the great cathedrals of France, the Blessed Virgin Mary, the Mother of God. This, in itself, is of considerable significance if the power of the Philosopher's Stone is considered as a sacred gift from God. The bas-relief in the Great Porch of Notre Dame, occupying the place of honour among those representing the medieval sciences, is, however, thought also to represent 'Alchemy'. This is symbolised as a woman, her head touching the clouds, seated upon a throne and holding in her left hand a sceptre, the sign of royal power, while her right hand supports two books, one closed representing esotericism and the other open representing exotericism. Supported between her knees is a ladder of nine rungs, the *scala philosophorum*, indicating patience in the hermetic labour of nine stages.

Fulcanelli compares the symbolism unfolded by the carvings on the cathedral as a *Mutus Liber* not unlike that famous textless book of Jacob Sulat (or Saulat) published at La Rochelle in 1677. He implies that the Virgin Mother is none other than the personification of the primitive substance, used by the creator of all that is, for the furtherance of his designs. Many of the titles of the Virgin have deep alchemical connotations and if she represents form, the Incarnation, brought about by the influence of the Holy Spirit, the union of form and spirit, results in living matter '*and the Word was made flesh and dwelt among us*'. Mary, the mother of Jesus, was herself of the stem of Jesse and, according to Fulcanelli, the Hebrew word JES means FIRE.

The basement of the facade, which extends below the three porches, is largely dedicated to the science of alchemy. Fulcanelli was of the opinion that the carved images not only divulged the *subject of the Wise*, the processing of the secret solvent, but also a step-by-step guide to the making of the Elixir, from calcination to the ultimate coction or cooking. In *Le Mystere des Cathedrales* the order in which the figures succeed each other from the outside of the porch door is followed with a commentary upon each and a digression on the relevant aspect of alchemy or legend upon which the figure is based. It is almost a meditation on each Station of the Alchemical Cross. According to Canseliet, Fulcanelli's method of communication of the practical details of the work was somewhat different from that of his predecessors, for, while respecting the philosopher's law that imposes upon initiates the necessity for inviolable secrecy, he describes in detail all the operations after having divided them into several fragments. He thus takes each of the phases of the process, begins its explanation in one chapter, interrupts it to continue in another, and then completes it in a final passage, remote from the first.

Alchemy
Bas-relief on the Great Porch of Notre Dame, Paris

The work starts with the discovery of the mysterious fountain, the threefold fountain of the *Rosarium Philosophorum* which pours forth Virgin's Milk, Vinegar and *Aqua Vitae*, the beginning and first nature of all metals. From this fountain, springing from the foot of the Old Hollow Oak, comes the famous water which '*does not wet the hands*', perhaps the *Aqua Vitae* or 'brandy of the philosophers'! It is at this stage Fulcanelli make reference to two earlier alchemists – Basil Valentine and Nicholas Flamel. He drew attention to the latter's *Symbolical Figures*, the third of which depicts a hollow tree in a garden with many flower beds and with four blind men searching for the fountain without digging, one of whom weighs the water in his hand. Flamel's fourth figure relates to the massacre of the innocents, with a crowned king dressed in red regalia. Fulcanelli emphasises that the Crow is the official seal of the work in the same way that the Star is the signature of the initial subject, the *prima materia*. The Star not only refers to the Star of Bethlehem that guided the wise men on their journey to discover the birthplace of the King of the Jews but also to the Star Regulus of antimony that played such a significant part in the alchemical work of Isaac Newton. There is also a mystical connection with the Blessed Virgin Mary as the vessel in which the Incarnation took place – a connection that emphasised the special nature of Holy Alchemy.

Putrefaction

The first of the figures carved on pedestals and holding discs with emblems of the work is one holding a black CROW symbolising putrefaction, the coction of the philosophical REBIS. According to the alchemist Le Breton there are four putrefactions in the work involved, namely in the first separation, in the first conjunction, in the second conjunction and, finally, in the fixation of the SULPHUR. They are attended by an odour something like that of the grave, the *toxicum et venenum*, which is perceived not only by the sense of smell but by the understanding. It occurs, according to Bernard Trevisan, when 'our compost' is steeped in 'our permanent water' and a black cloud forms in the reaction vessel.

Philosophic Mercury

The second figure depicts a snake coiled on a staff which indicates the incisive and solvent nature of the philosophic mercury, avidly absorbing the 'metallic' sulphur. The staff or wand symbolised corporeal sulphur which, when combined with the mercury and placed in the reaction vessel and subjected to gradual cooking, gives rise to the series of stages necessary to achieve the Red Sulphur, followed by the Elixir and, finally, by the Universal Medicine.

Calcination

The third figure depicts a salamander, the incombustible and fixed central salt or metallic seed that can be extracted by percolation. The difference between ordinary calcinations and those of the alchemists is that while the former require high temperatures the latter require the more moderate heat of the SECRET FIRE - more like water than a flame. In this connection reference is made to the writings of Artephius (12th century) who wrote the *Liber Secretus* and to Pontanus (16th century) the author of the *Beschreibung des Secreten Philosophischen Feuers*.

The Universal Solvent

The fourth figure is that of Aries the Ram. This not only carries the alchemical connotation of the Spring season as appropriate for the commencement of the work but also of the nature of the solvent. This Fulcanelli describes as the 'Secret Fire enclosed in water'. Aries is also symbolic of the metallic principle and the adepts call this their steel or LODESTONE.

The Evolution of the Work

The fifth figure depicts an Oriflamme, or triple banner, displaying the colours of the work, black, white and red. These colours mark the gradual progression of the reaction, gradual because *Natura non facit saltus*, (Nature does not make jumps)! There are, however, many intermediate stages between the calcinations and the achievement of these colours which only serve as witness or an indication to the progress of the work. Black in the language of colours refers to Saturn (lead), earth, night, and death as a prelude to re-birth. White, as the light succeeds darkness, symbolises purity, simplicity and innocence, while red, the symbol of fire, indicates the predominance of spirit over matter. It is interesting that Paracelsus also ascribes colours to the four elements – black (or blue) to Earth, green to Water, yellow to Air and red to Fire. Fulcanelli also points out it is traditional to clothe the Blessed Virgin in blue, God in white and Jesus, the Son of God, in red.

Philosophy

The disc in the sixth figure bears the sign of the Cross which depicts the four elements and the two metallic principles. These are the Sun, the Moon and the philosophic Sulphur and Mercury. The seventh figure, although worn, is said to represent the Athanor (or furnace), and the Stone. The former also depicts the necessary support therein for the Philosopher's Egg, the reaction vessel.

The First Conjunction

The disc in the eighth carving shows a Griffin, a mythical monster with the head and chest of an eagle and the body of a lion. This is thought to be the hieroglyph of the First Conjunction. According to Eirenaeus Philalethes, the exact preparation of the flying eagles is the first degree of perfection, the knowledge of which requires an industrious and able spirit. The Philosophic Mercury is the Bird of Hermes and while the lion in the body of the Griffin represents terrestrial and fixed force, the eagle part represents the airy and volatile force. The eagle and the lion are both powerful but contrary natures and this gives some indication of the violence of this stage of the operation. The two champions attack and repulse each other and tear at each other fiercely until, the eagle having lost its wings and the lion its head, the antagonists form a single body, that of animated mercury, which is *midway between their two natures and homogeneous in substance.*

The Universal Solvent

The ninth subject depicts a woman showing, allegorically, the materials necessary for the construction of the Hermetic Vessel. She holds in her hand what appears to be the stave of a barrel, confirmed by the oak branch on her shield which also alludes to the mysterious spring, the fire of Nature, without which nothing can grow. This oak barrel also figures in the 12th Key of Basil Valentine and the oak is well known from the legend of Cadmus piercing the serpent or dragon, that guarded the fountain of Dirce, prior to sowing its teeth. Rounded inside like a half globe it forms a fitting vessel for the King and Queen to bathe in the water from the spring.

The Alchemical Lion

The next figure is that of a mail-clad knight with drawn sword and, on his disc, the king of the alchemical bestiary - the lion. At this stage in the preparation of the Stone, the lion is an extremely complex symbol. In the former figure the lion part of the Griffin represented the fixed basic part of the compound entering into a reaction to form animated mercury. The lion is, however, also a symbol of gold. The green lion may refer to the Alkahest or Secret Fire because it is immature, while the red lion is more mature having been its green predecessor brought by certain processes to that stage of perfection characteristic of hermetic Gold. In the words of Basil Valentine:

> *'Dissolve and nourish the red lion with the blood of the green lion, since the fixed blood of the red lion is made from the volatile blood of the green one, which makes them both of the same nature'.*

The lion has also been called the green vitriol, the universal emerald, the dew of May and even the vegetable stone. After this long time there is, however, no indication of the colour of the lion depicted on the disc in the cathedral.

The Sulphur that does not Burn

The next figure depicts a cock and a fox, symbols that derive from the Third Key of Basil Valentine of 'our sulphur' which does not burn, as does the natural product, although *'its brilliancy is seen far and near'*. To prepare an incombustible sulphur it is necessary to seek for 'our sulphur' in a substance that is itself incombustible which can *'only be after its body has been absorbed by the salt sea and again rejected by it'*. In this way the cock will swallow the fox and, having been drowned in the water and quickened by the fire, will in its turn be swallowed by the fox!

The next disc depicts a bull that refers to the zodiacal TAURUS beginning on 21 April. This is not only the male principle, or Sulphur, but also the second month of preparation so beautifully depicted in the *Mutus Liber* of Altus.

Dissolution

The first of the medallions, which ornament the lower row, depicts an unhorsed knight clinging to the mane of his horse. This allegory is concerned with the extraction of the fixed central and pure parts by the volatile or etheric parts in the philosophical dissolution. Fulcanelli describes it as the rectification of the spirit obtained and the action of this prepared spirit on the heavy matter. The horse represents the spiritual substance and the knight the weightiness of the gross metallic body. The absorption of the fixed by the volatile is carried out slowly and with difficulty, but it is only by this technique that the occult *salt* can be extracted from the red lion, helped by the spirit of green predecessor.

The First Matter

On the second medallion the Initiate holds up a mirror in one hand while in the other he holds up the horn of Amalthea – in Greek mythology a nymph, originally a goat, who was the nurse of Zeus – while beside him is the Tree of Life. The mirror symbolizes the beginning of the work, the Tree of Life its end, and the generous horn of plenty its result. The mirror of the art gives an indication of the nature of the *prima materia* being the attribute of truth, prudence and knowledge – a reflection of the alchemist himself! As Basil Valentine writes in his *Testamentum*:

> '*The whole body of the Vitriol must be recognised only as a Mirror of the philosophical Science*'

Weights and Proportions

The literature of alchemy says very little about the amounts of the various reactants used in the preparation of the Stone. Reference is sometimes made to the number of parts of lead or mercury transmuted by one part of the Stone but, in general, the alchemists are deliberately vague on the subject of weights and proportions with such statements as:

> '*Philosophic mercury results from the absorption of a certain part of sulphur by a determined quantity of mercury*'.

In the next medallion a pair of unequal scales is depicted, one scale outweighing the other in an apparent ratio of two to one.

The Mercury of the Philosophers

For some reason, at this stage, Fulcanelli draws attention to the possibility of substituting common gold for the metallic sulphur because the excess solvent can always be separated by distillation. Originally common gold was always excluded, despite the statement by Philalethes that '*as the end you look for is gold: so let gold be the subject on which you work, and none other.*' If the properties of gold are, however, modified by exaltation or transfusion it may be rendered fit for the work. Fulcanelli also mentions one of the cartouches, in which the Queen kicks down her servant Mercury, who comes with a cup in his hand to offer his services. This symbolises the solution of ordinary quicksilver in order to obtain from it the mercury of the philosophers, which is known as 'our mercury' to differentiate it from the liquid metal from which it comes. Fulcanelli goes on to observe:

> 'The *servus fugitivus* which we need is a solid, brittle mineral and metallic water, having the aspect of a stone and very easily liquefied. It is this water, coagulated in the form of a stony mass, which is the *Alkahest* and universal solvent.'

There follows a further medallion depicting the reign of Saturn which is in the form of an old man, stiff with cold, leaning on a block of stone, his left hand hidden in a form of muff. This represents the first phase of the second work, when the hermetic *Rebus* is enclosed in the centre of the Athenor, suffers the dislocation of its parts, finally to be mortified. It is the active and gentle beginning of the *fire of the wheel*, symbolised by cold and by winter, the embryonic season when the seeds, enclosed in the womb of the philosophic earth, are subject to the fermentative influence of humidity.

The sixth medallion, a repetition of the second, shows the adept with his hands joined as in prayer before an image of the head and shoulders of Nature in a mirror, the hieroglyph of the subject of the Wise.

The seventh medallion, on the right of the porch, depicts an old man ready to cross the threshold of a mysterious palace, having torn down the awning which had hidden the entrance from the eyes of the uninitiated. According to Fulcanelli:

> 'The first step in the practice has been achieved, the discovery of the agent capable of carrying out the reduction of the fixed body, of reincruding it, according to the accepted expression, in a form analogous to that of the first substance.'

The 'old man' is none other than 'our mercury', the secret agent, while the palace represents the living, philosophic or base gold. Reference is made to the allegory of the green and red lions, the solvent and the body to be dissolved, to Saturn, who it is said to have consumed his own children, and to the original colours of the components of the medallion – basically green with purple for the palace.

The encounter between the old man and the crowned king, the solvent and the body (volatile principle and fixed salt) is shown in the next medallion. This reaction may be either gentle with the gold being dissolved little by little and without violence, or vigourous, closely resembling chemical effervescence. A reaction of the Pontic water on the heavy matter is depicted in the literature by a fight to the death between dissimilar animals eagle and lion, previously depicted by Nicholas Flamel, cock and fox, by Basil Valentine and the remora and the salamander, by Cyrano de Bergerac.

Solution

Fulcanelli summarises the guiding principle of alchemy as one of depicting the various stages or ways in the practice of solution. The first way is by dissolving the alchemical gold by the solvent Alkahest and the second, solution of common gold in 'our mercury'.

The subject of the twelfth and last bas-relief is a warrior who drops his sword and stops speechless in front of a tree, at the foot of which a ram is rising up. The tree bears three enormous spherical fruits and a bird is seen coming out of its branches. The tree is the solar tree of Alexander Seton from which water must be extracted, while the warrior is the alchemist who has just accomplished the labour of Hercules, the 'Great Work'. The ram shows the adept has chosen the favourable season and the right substance, and the bird indicates the volatile nature of the compound. Nothing now remains but to imitate Saturn who in the words of the Cosmopolite:

> 'Drew ten parts of this water and immediately picked the fruit of the solar tree and put it in this water…. For this water is the Water of Life, which has the power to improve the fruits of this tree, so that from then on there will be no further need to plant or graft any, because by its scent alone it can make six other trees assume the same nature as itself.'

Fulcanelli points out that the loss of colour from the bas-reliefs over the centuries is to some extent compensated for by the foresight of Guillaume de Paris in having the motifs of the medallions reproduced on the panes of the central rose window. The glass, he says, complements the stone!

The Seven Planetary Metals

The Portal of the Virgin is, according to Fulcanelli, replete with still more alchemical imagery. On the side of a sarcophagus there are seven circles, symbolic of the seven planetary metals, arranged in a symmetrical pattern around the centre circle, apparently not a characteristic feature of the purely decorative motifs of Gothic art. Fulcanelli interprets these circles as symbolising the pattern of possible changes among the metallic planets. At the centre is the Sun (gold) and Saturn (lead) and the Moon (silver) at the ends. Then come Jupiter (tin) and Mercury (quicksilver) and, finally, on either side of the Sun Mars (iron) and Venus (copper).

The base of the Portal is divided into five niches with curious figurines between the outer curves of the arches. There are a dog and two doves – the Corascene bitch which both Artephius and Philalethes say one must know how to separate from the compost in the form of a black powder – and the doves of Diana hiding the spiritualization and sublimation of the philosophical mercury. A lamb symbolises the purification of the arsenical principle in the matter and a man turning round illustrates the alchemical maxim of *'solve et coagula'* – a saying which teaches how to set about and achieve the elementary conversion by volatilising the fixed and fixing the volatile!

A small square bass-relief, which synthesises and expresses the condensation of the universal Spirit, depicts the famous 'Bath of the Stars' in which the chemical sun and moon must both come to bathe in order to change their natures and become rejuvenated, also shows a child falling from a crucible. This is supported by an archangel with a halo and an outspread wing against a background of the night sky studded with stars. This Fulcanelli relates to the Massacre of the Innocents, an allegory also cited by Nicholas Flamel. He goes on to say:

> 'Without entering into details of the operative techniques – which no author has dared to do – I will, however, say that the Universal Spirit, embodied in minerals under the alchemical name of Sulphur constitutes the principle and the effective agent of all metallic tinctures. But one cannot obtain this spirit, the red blood of the children, except by decomposing what Nature had first assembled in them. It is, therefore, necessary that the body should perish, that it should be crucified and should die, if one wishes to extract the soul, the metallic life and the celestial dew imprisoned therein. And this quintessence, transfused into a pure, fixed and perfectly digested body, will give birth to a new creature, more splendid than any of those from which it proceeds. The bodies have no action on one another; the spirit alone is active.'

The Dew of May

The 'mineral blood', perhaps from the massacre of the innocents, needed to animate the fixed and inert body of gold is explained as being a condensation of the universal Spirit, the soul of all things which is capable of penetrating sub-lunary mixtures and making them grow. This could only be achieved at night under the protection of darkness, a clear sky and calm air, during springtime on earth. The philosophers gave it the name *'Dew of May'*. It was the life-giving humidity of the month of Mary, the Virgin Mother, which, according to Fulcanelli, *'can easily be extracted from a particular body, which is abject and despised.'* The nature of this dew came close to being the greatest secret of the work – the *Verbum dimissum* of Bernard Trevisan and the Lost Word of the medieval Freemasons. This section of the work concludes with a reference to the act of Creation when from the darkness of Chaos light was extracted and the divine Spirit moved on the waters of the Abyss – *Spiritus Domini ferebatur super aquas* and a verse:

> At *midnight* a *Virgin mother*
> Brings forth this *shining star*;
> At this miraculous moment,
> We call God our brother.

The Dry Way and the Moist Way

The single motif by the Porch of St. Anne is of particular significance. It depicts a Bishop, St. Marcellus, perched above an Athenor, where the philosophical mercury, chained in limbo, is being sublimated. It teaches the origin of the sacred fire. The Cathedral Chapter, by having the door in this porch closed all the year in accordance with a secular tradition, shows this is not the vulgar way but one reserved for the small number of philosophers – the elite of Wisdom.

There are two ways in alchemy – one short and easy called the 'dry way' and the other, longer and less rewarding, called the 'moist way'. In the dry way the celestial salt, which is the Philosopher's mercury, must be boiled for four days in a crucible over a naked fire, together with a terrestrial metallic body. Reference is also made to a further 'dry way' for which, according to Basil Valentine:

> '... neither great labour nor trouble is required and the expenses are small, the instruments of little worth. For this Art may be learnt in less than twelve hours and brought to perfection within the space of eight days, if it has its own principle within itself'.

There is also a shorter way called, according to Fulcanelli, 'the Regimen of Saturn' which has, perhaps, given rise to a basic maxim of alchemy *One Single Vessel, One Single Matter, One Single Furnace*'. The boiling of the work instead of necessitating the use of a sealed glass vessel, requires only a simple crucible. Fulcanelli also quotes the alchemist Cyliani, who, it is believed, may have been his 'father in alchemy':

> 'I would like to warn you here never to forget that only two matters of the same origin are needed, the one volatile, the other fixed; that there are two ways, the dry way and the moist way. I follow the latter one for preference as my duty although the former is very familiar to me: it is done with a single matter'.[8]

The Subject and Solvent of the Work

Another pillar with circular bezants depicts the matter, prepared and united in a single compound, which must be submitted to sublimation or the last purification by fire. This ensured that the earthly matters lost their cohesion and dispersed, while the pure incombustible principles emerged in a higher form – the Philosopher's Salt, called the 'King crowned with glory'. This, according to Hermes, is born in the fire and must rejoice in the marriage which follows so that hidden things may be made manifest. *Rex ab igne veniet, ac conjugio gaudebit et occulta patebunt.*

This pillar also gives an indication of the four elements with which the alchemist must work and the duality of the two natures whose union provides the Saturn of the Wise. Fulcanelli emphasised that the word Saturn(e) is an anagram of the word 'natures'.

The Extraction of Mercury and its Conjunction with Sulphur

In his final section on the symbolism of Notre Dame in Paris, Fulcanelli draws attention to a coat of arms, said to be that of St. Thomas Aquinas. It has the heraldic description:

> '... quartered French shield surmounted by a rounded segment. This supplementary piece shows a reversed matrass or, surrounded by a crown of thorns vert on a field sable. A cross or bears three roundels azure at base and arms dexter and sinister and has a heart gules at the

centre with a branch vert. Tears argent, falling from the matrass, are collected and fixed on this heart. The dexter chief quarter, halved or three stars purpure and azure seven rays or on a field tenne. In the sinister chief quarter a cloud violet on a field argent and three arrows of the same, feathered or, pointing towards the abyss. Dexter base three serpents argent on a field vert.'

This emblem is said to unveil the secrets relating to the extraction of mercury and its conjunction with sulphur.

The Cathedral of Notre Dame in Amiens

Fulcanelli also draws attention to the set of hermetic bas-reliefs in the central porch of the Cathedral at Amiens which are almost a faithful reproduction of those adorning the portal of Notre Dame in Paris – and even of the order in which they are given. While in Paris the figures hold discs, in Amiens they hold shields. In the porches of the Cathedral, dedicated to the Saviour, to St. Firminus and to the Virgin Mother respectively, there is also a series of quatrefoils illustrating alchemical themes.

The Fire of the Wheel

In the Porch of the Saviour is a quatrefoil depicting a philosopher, seated with his elbow resting on his right knee. In front of him are two cartwheels. He is keeping watch while the slow action of the fire of the wheel takes place close to him. This is interpreted as the sign of two revolutions, which must act in succession on the compound in order to ensure it attains the first degree of perfection. The first wheel corresponds to the humid phase of the operation (elixation) in which the compound remains molten until a light film is formed and gradually becomes thicker. The second period characterised by dryness (assation) is completed and perfected when the contents of the egg, which has been calcined, appears granulated or powdery, in the form of crystals, sand or ashes. The temperature at the beginning should not exceed that of the human body and is increased progressively up to around 300 deg C.

Philosophic Coction

This is depicted by the representation of a bare-headed figure, clean shaven and tonsured as a monk, wearing a calf-length tunic with a hood casting, aside his footwear and leaving what appears to be a little church with long, narrow windows, a cylindrical belfry and a false door. Fulcanelli interprets this carving of a church as an athenor, complete with its ash box. The tower, the dome and the belfry are symbolic of the secret furnace enclosing the philosophic egg. The figure also represents the coction itself as the alchemist is pointing with his right hand to a coal sack. The abandonment of his footwear indicates the extent to which prudence and silence must be carried in this secret task. Fulcanelli goes on to observe at the fourth degree of fire it becomes necessary to maintain a temperature as high as 1,200 deg C. (which is also indispensable in projection).

The Cock and the Fox

The allegory of the cock and the fox was, as mentioned, one dear to Basil Valentine. In his *Third Key* the fox running away with a hen and being attacked in return by the cock signifies Fixation of the Volatile and the Volatilization of the Fixed. The cock will swallow the fox, and, having been drowned in the water, and quickened by the fire, will in its turn be swallowed by the fox! A process symbolized by the alchemical axiom *Solve et Coagula*.

The First Matter

A second quatrefoil shows some dead trees, twisting and interlacing their knotted branches beneath a sky containing images of the sun, moon and a number of stars. This subject appears to refer to the first matter of the Great Art:

> 'The metallic planets, whose death, the Philosophers tell us, has been caused by the fire, and which have been rendered by fusion inert and without vegetative power, as trees are in winter'.

Fulcanelli then quotes from the testimony of *Le Psautier d'Hermophile* that metallic chaos, a product of the hands of nature, contains in itself all metals although it is not, itself, a metal. It contains gold, silver and mercury; but it is neither gold, nor silver, nor mercury.

The Philosopher's Dew

In the porch of the Virgin Mother there is a curious interpretation of the condensation of the universal spirit. An adept is regarding the stream of celestial dew falling upon a mass, thought by some to be a fleece. Fulcanelli considers it equally plausible that this mass is the mineral designated by the name *Magnesia* or philosophic lodestone. He then draws attention to the cryptogam plant, the *Nostoc*, which can be found in the early morning although it dries up rapidly in sunlight. The sun is the destroyer *par excellence* of all substances too young and feeble to resist its fiery power. To achieve success in alchemy daylight working is not to be recommended!

The Seven—Rayed Star

The hermetic study of the motifs of Amiens Cathedral is concluded in the porch of the Virgin Mother with an initiation scene. The master is indicating to three of his disciples a seven-rayed hermetic star. This Star of Bethlehem, the guide to the Magi in search of the Infant King, is also a guide to the Philosophers showing them the birth of the son of the sun. The riddle of Nicholas Rollin, Chancellor to Philip the Good, is cited:

> 'Only * indicated the science of its owner by this characteristic SIGN of the Work, the one and ONLY STAR.'

The City of Bourges

In the ancient city of Bourges attention is drawn by Fulcanelli to the great house of Jacques Coeur and the Lallement mansion. The motto of Jacques Coeur *'A vaillans cuer riens impossible'* (To the valiant heart nothing is impossible) when examined according to cabalistic rules discloses that it is a statement of the universal Spirit (ray of light), the common name for the basic matter. The three hearts, forming the centre of his coat of arms, emphasise the essential three repetitions for the complete perfection of the two Magisteries. In addition to the heart, the scallop shell was a favourite symbol of Jacques Coeur. This was a badge worn by pilgrims to the shrine of St. James at Compostella, in Northern Spain, a pilgrimage said to have been made by Nicholas Flamel. According to Fulcanelli, the shrine at Compostella was where all alchemists must begin:

> 'With the pilgrim's staff as a guide and the scallop shell as a sign, they must take this long and dangerous journey, half of which is on land and half on water. Pilgrims first, then pilots.'

Cbe Mineral Criad of the Great Work

In the Treasure Chamber of the house of Jacques Coeur is a sculptured group illustrating the meeting of Tristan (of Lyonnesse) and Isolde. In the centre of the motif there is a casket in the form of a cube at the foot of a bushy tree, the foliage of which conceals the crowned head of King Mark. This is the hieroglyph for the manufacture of the green lion which enables natural gold to be re-incruded (rendered raw), softened and restored to its original state in a saline, friable and very fusible form. It is to be noted that Isolde was the wife of both the old King and the young hero – the mineral triad of the Great Work and a reflection of the Oedipus legend. The Forest of the Dead King is likened to the Garden of the Hesperides.

In the Laillemant mansion, in Bourges, there is, apparently, every indication that alchemy was practiced, including a bracket depicting an alchemist with the vessel of the Great Work. This vessel is generally called the Philosopher's Egg or the green lion – the Egg because the covering or shell encloses the philosopher's *rebis*, composed of white and red in the same proportions as in a bird's egg. The second epithet, the green lion has not, however, been fully explained. Christoph Reibehand in his *Filum Ariadnes* ascribes the name 'green lion' to the vessel used in the coction and it would appear that the vase is considered in two ways, both in its matter and in its form.

Another bas-relief in painted stone represents St. Christopher and the child Jesus. According to Fulcanelli, the whole work of the art consists in processing the mercury of the philosophers until it receives the sign – called the Seal of Hermes, a device of criss-cross lines similar to those depicted in the belt of Offerus (St. Christopher). Fulcanelli emphasises that the laboratory, the place where the adept labours, and the Vase of the Work, the place where nature operates, are the two certainties which strike the initiate from the start of his visit and which make the Lallemant mansion one of the rarest and most intriguing of philosophical dwellings.

A further superb painted bas-relief, executed in the same style as that of St. Christopher, is of the Golden Fleece. The fable of the Golden Fleece is held to be a complete representation of the hermetic process that results in the Philosophic Stone. The Golden Fleece is both the matter prepared for the work and the final result, from which it differs only in purity, fixity and maturity. A distinction is drawn between the Philosopher's Stone and the Philosophic Stone where although similar in kind and origin the former is raw while the latter is perfectly cooked and digested.

After a rapid description of the images depicted in the ceiling of the chapel, Fulcanelli makes a brief reference to the postscript to a letter which he carried about with him for many years and which was reproduced in the second Preface of the 1957 edition:

'My old friend,
     This time you have really had the Gift of God; it is a great blessing and, for the first time, I understand how rare this favour is. Indeed I believe that, in its unfathomable depth of simplicity, the Arcanum cannot be found by the force of reason alone, however subtle and well trained it may be. At last you possess the Treasure of Treasures.   Let us give thanks to the Divine Light which made you a participant in it.   Moreover, you have richly deserved it on account of your unshakeable belief in Truth, the constancy of your effort, your perseverance in sacrifice and also, let us not forget... your good works.

     When my wife told me the good news, I was stunned with surprise and joy and was so happy that I could hardly contain myself: let us hope that we shall not have to pay for this hour of intoxication with some terrible aftermath.   But, although I was only briefly informed about the matter, I believed that I understood it, and what confirms me in my certainty is that the fire goes out only when the Work is accomplished and the whole tinctorial mass impregnates the glass, which, from decantation to decantation, remains absolutely saturated and becomes luminous like the sun.

You have extended generosity to the point of associating us with this high and occult knowledge, to which you have full right and which is entirely personal to you. We more than any, can appreciate its worth and we, more than any, are capable of being eternally grateful to you for it. You know that the finest phrases, the most eloquent protestations, are not as much as the moving simplicity of this single utterance: you are good, and it is for this great virtue that God has crowned you with the diadem of true royalty. He knows that you will make noble use of the sceptre and of the priceless endowment which it provides. We have for a long time known you as the blue mantle of your friends in trouble. This charitable cloak has suddenly grown larger and your noble shoulders are now covered by the whole azure of the sky and its great sun. May you long enjoy this great and rare fortune, to the joy and consolation of your friends, and even of your enemies, for misfortune cancels out everything. From henceforth you will have at your disposal the magic ring which works all miracles.

My wife with the inexplicable intuition of sensitives, had a really strange dream. She saw a man enveloped in all the colours of the rainbow and raised up to the sun. We did not have long to wait for the explanation. What a miracle! What a beautiful and triumphant reply to my letter, so crammed with arguments and – theoretically – so exact; but yet how far from the Truth, from Reality. Ah! One can almost say that he, who has greeted the morning star has for ever lost the use of his sight and his reason, because he is fascinated by this false light and cast into the abyss… Unless, as in your case, a great stroke of fate come to pull him unexpectedly from the edge of the precipice.

I am longing to see you, my old friend, to hear you tell me about the last hours of anguish and of triumph. But be assured that I shall never be able to express in words the great joy that we have felt and all the gratitude we have at the bottom of our hearts. Alleluia!

I send you my love and congratulations.

Your old…

He who knows how to do the Work by the one and only mercury has found the most perfect thing - that is to say he has received the light and accomplished the Magistery.'

The writer of this letter is unknown, although in view of the importance that it was to Fulcanelli it may have been from the most significant person in his life, perhaps even from his alchemical master and teacher, thought by some to have been the alchemist Cyliani. Little is known of Cyliani save that a new edition of '*his little work*' (Cyliani, *Hermes Devoilé*, Paris, F. Locquin, 1832) was re-published by Chacornac in 1915 and referred to by Canseliet in his Preface to the Second Edition of *Le Mystere des Cathedrales*, and by Fulcanelli himself in the same work when he quotes from this author concerning the two ways of preparation.

Les Demeures Philosophales

According to Fulcanelli, writing in Book One of the second part of his great alchemical trilogy, while chemistry is a science of facts, alchemy is one of causes. In this second work, Fulcanelli specifically addresses the commonly held belief that modern chemistry derives from alchemy. It is, he writes, derived not from chemistry but from Spagyrics, which, according to Dr. Johnson's *Dictionary of the English Language*, is a word coined by Paracelsus from the German *Spaher*, a searcher.

'Since Lavoisier, all the authors who have written on the history of chemistry agree to profess that our chemistry *comes by direct affiliation from old alchemy*. Consequently, the origin of the one is confused with that of the other, to such an extent that modern science would owe the positive facts on which it is built to the patient labour of the ancient alchemists'.

He goes on to contend that the real ancestor of our modern chemistry is ancient spagyrics and not the hermetic science itself. '*There is, he says, indeed a profound abyss between spagyrics and alchemy*'.

Early chemical science was a combination of the two exoteric arts spagyry and archemy that formed the basis of the laboratory practice of metallurgists, goldsmiths, painters, ceramic artists glassmakers, dyers, distillers and apothecaries. Archemy formed a special category among the ancient chemists who pursued some of the same aims as the alchemists, but the materials and the means they had at their disposal were uniquely chemical materials and means.

'To transmute metals into one another, to produce gold and silver from course minerals, or from saline metallic compounds, to force the gold potentially contained in silver and the silver potentially contained in tin to become real and extractable, was what the archemist had in mind. In the final analysis he was a spagyrist confined to the mineral realm and who voluntarily neglected animal quintessences and vegetable alkaloids.'

Apparently medieval law forbade the private possession of furnaces and chemical apparatus without preliminary permission so that much experimentation was driven underground which formed side activities held to be unworthy of the true philosopher but yet contributing a wide variety of experience and knowledge to the developing chemical science. Fulcanelli gives as an example of an archemic process which involves the effect of violent reactions such as the solution of metallic silver in concentrated nitric acid. This reaction, when carried out in the manner described, gives rise to a small quantity of *a fine black sand* at the bottom of the reaction vessel. This deposit has the appearance and the superficial characteristics of the purest gold, but not the correct density! It might be thought that the larger the scale of this operation the greater the yield of this black deposit. This is, however, not the case, although precipitation of the silver from the silver nitrate solution and repeating the process will give a further yield of the deposit. It is said that the newly produced metal remains capable of taking and keeping, by contraction, the increased density the adult metal possesses (i.e. 19.3).

This is summed up by Fulcanelli:

'If one wants to have some idea of the secret science, let him bring his thoughts back to the work of the farmer and that of the microbiologist, since ours is placed under the dependence of analogous conditions. For, as Nature gives the farmer the earth and the grain, and the microbiologist the agar-agar and the spore, similarly she gives the alchemist the proper metallic terrain and the appropriate seed. If all the circumstances favourable to the regular process of this special culture are rigorously observed, the harvest cannot but be abundant...'

In summary, alchemical science, of an extreme simplicity in its materials and its formula, nevertheless remains the most unrewarding, the most obscure of all, by reason of the exact knowledge of the required conditions and the required influences. There is its mysterious side, and it is towards the solution of this most difficult problem that the efforts of all the sons of Hermes converge.'

Attempts to Identify Fulcanelli

Numerous attempts have been made to elucidate the mystery of Fulcanelli in terms of the known protagonists in the history of the alchemist. Those closest to him in the persons of Eugene Canseliet, Jean-Julien Champagne, Gaston Sauvage, Jules Boucher and the Parisian bookseller Pierre Dujols were, at first, considered as possible identities for the elusive Fulcanelli. Even F. Jolivet Castelot, the President of the Alchemists' Society of France around 1914, was considered as a candidate for the mysterious *nom de plume*!

Robert Amberlain, a student of the occult, having read both of Fulcanelli's books and seeking to publish his own contribution, *Dans l'ombre des Cathedrales*, contacted Jean Schemit, Fulcanelli's publisher, and extracted from him a statement to the effect that the negotiations over his books had been carried out on Fulcanelli's behalf by Jean-Julien Champagne, the talented illustrator, as well as by Eugene Canseliet. Canseliet, however, has denied that Champagne was involved in these negotiations. There are, also, discrepancies in Amberlain's case which have been pointed out by Kenneth Rayner Johnson and also by Canseliet while he was yet alive. Canseliet certainly knew the true identity of his master which, despite much speculation, still remains a mystery.

The 'Finis Gloria Mundi'

It is mentioned in the preface to the Second Edition of *Les Demeures Philosophales* that this particular was not, in fact, Fulcanelli's last book. A third part, entitled *Finis Gloria Mundi* (The End of the World's Glory) had, apparently, been written, and then retracted by the author. Had this been published it would have raised the previous two volumes to the status of a most extraordinary alchemical trilogy. It is held that Fulcanelli, who apparently succeeded in achieving the Philosopher's Stone in 1923, prepared three books to be published in the course of time by his sole disciple Eugene Canseliet F.C.H., a master of alchemy himself. The *Mystere des Cathedrales* was published in 1926, and *Les Demeures Philosphales* in 1930. The *Finis Gloria Mundi* remained unpublished. The manuscript which had been entrusted to Canseliet for safe keeping was claimed by him to have been stolen from his house some time between 1960 and 1970.

Archibald Cockren

In England a physiotherapist named Archibald Cockren achieved some interest by carrying out alchemical experiments and producing 'curative oils' from most of the planetary metals. In 1940 he published an account of his work, combined with an excellent introduction to alchemy, in a book entitled *Alchemy Rediscovered and Restored*.[9]

Very little is known of Cockren's antecedents, although the basic details of his career are given in the Foreword to Cockren's book by Sir Dudley Myers who knew him during the 1930s. Other, sometimes conflicting, information derives from the writers C.A. Burland, Ithell Colquhoun and C.R. Cammell. An account of his work as an alchemist has been recently given by Patricia Tahil.[10]

It is thought that Cockren may have been an Australian immigrant born in the 1880s who came to England at the turn of the century. The first that was heard of him is that he qualified in 1904 as a physiotherapist at the National Hospital for Paralysis and Epilepsy, in London, for 'all purposes of massage, remedial exercises and electrical treatment'. After a spell at the Great National Central Hospital he entered private practice, setting up his consulting rooms in the West End of London in about 1908. He was, apparently, one of the leading pioneers of electro-massage. His friends are of the opinion that he was an intelligent straightforward man who had the courage to experiment in an obscure and practically discredited field. According to C.A. Burland, he was incapable of intentional falsehood. His work was clear and simple.

The First World War resulted in his spending two years in the Russian Hospital for British Officers in South Audley Street, until in 1917 he began working in the Prisoners of War Hospital where he gave distinguished service there and in the Millbank Military Hospital. In 1918 he was transferred to the Australian army, and served on the Peace Conference staff of the Australian Prime Minister in 1919. Following this, he returned to private practice. In the 1930s he was living in Boundary Road, London, NW8 where he was visited by Edward Garstin, a member of the Quest Society run by G.R.S. Mead and Gerard Heym (one of the founders of the Society for the Study of Alchemy and Early Chemistry). He is said to have had a well-appointed laboratory in the Holborn area, furnished by the generosity of Mrs. Meyer Sassoon, to whom he dedicated his book. This laboratory was destroyed by enemy action during the London Blitz in 1940. Cockren's book *Alchemy Rediscovered and Restored* was published in 1940 and has since been reprinted. Despite reports that he was killed by a direct hit on his Holborn laboratory, it would appear he survived, eventually dying somewhere near Eastbourne during the 1960s.

The Alchemy of Archibald Cockren

Cockren was firmly of the opinion that alchemy, as demonstrated by two of its most important exponents Basil Valentine and Paracelsus, was concerned not only with the attainment of the Philosopher's Stone, but also with the preparation of medicines, or elixirs, by a process which involved, as in the Hermetic doctrine, the separation of the ethereal from the gross – the true secret of the spagyric art.

In the years following the First World War, he could, apparently, distinguish only two systems of medicine – allopathy and homeopathy. The former is defined in the Oxford English Dictionary as 'the curing of a diseased action by the inducing of another of a different kind, yet not permanently diseased'. It is a term applied by homeopaths to the ordinary or traditional medical practice and to a certain extent is in common use to distinguish it from homeopathy.

Homeopathy is based on the principle of 'like curing like'. Although the principles of homeopathy were known to Hippocrates and Paracelsus, it was Samuel Hahnemann in the late 18th century who, appalled by the existing medical practices which so often did more harm than good, sought for a method that would be safe, gentle and effective. Based on experiments on themselves with small doses of reputedly poisonous or medicinal substances, called provings, Hahnmann and his followers carefully noted the symptoms they produced and then treated patients showing similar symptoms with these substances in very small quantities, often with very encouraging results. To avoid side-effects, Hahnmann experimented to establish the smallest effective dose when, to his surprise, he found the more the remedy was diluted the more active it became. An important principle of homeopathy and other forms of what is now known as 'alternative medicine' is that patients are treated rather than the disease. Once a multidimensional picture of each patient has been built up, infinitesimal doses of the appropriate remedy will often help in self-healing.

Alchemists have always maintained the first essential of a really effective healing agent is that it should contain the vital principle or quintessence of the herb or metal used. However, the homeopath does not provide this element in his preparations while the allopath, the conventional medical practitioner, administers his remedies in too crude a form. Cockren believed that in the administration of a metal, for instance, the body of the metal is worthless as a medicine for it cannot heal - it is the essence or quintessence alone that is curative. It was to the preparation of these quintessences of various metals that Cockren devoted much of his work.

Medicine from Metals

Cockren observed in his book:

'Only too often the body is poisonous, and until that gross part of the metal be broken up, its administration is definitely harmful. Probably one of the most common forms of metallic poisoning is that of mercury, but remove the harmful parts of the metal and the healing essence is free to do its work thoroughly. Nitrate of silver is a caustic poison, but remove the gross part of the metal and the essence of the silver is a cure for diseases of the brain. Lead salts are poisonous, it is true, and in many cases their administration has resulted in death from lead poisoning, but remove the poisonous matter and the remaining essence, which is clear, sweet-smelling, and aromatic in taste, forms a cure for all diseases of the spleen. Copper, when the gross body of the metal is removed and its essence unlocked, is invaluable for the nervous system and the kidneys; likewise tin for the liver, iron for all inflammatory diseases and the bile, and gold for the heart and general circulation. As with the other metals, gold is only suitable for a medicine when the salts of gold are reduced into the oil of gold and distilled into a golden liquid. Then and only then is gold tolerated and utilized by the human body. The salts of gold used at the present day can never be assimilated, for by their ordinary method of preparation they can never be properly distilled and purified.'

Cockren drew attention to work carried out at the Rockfeller Institute on the use of colloidal metals in medicine. Colloidal iron is more readily absorbed by the body, while copper in this form is a powerful agent in the reduction of neuralgic and nervous conditions.

Cockren in the Laboratory

Cockren maintained the practice of alchemy in the laboratory was a far from easy task and it was only by continuous experimentation and constant comparison with the writings of the alchemists that any success might be achieved. He wrote:

'Looking back on the years of persistence in the face of countless difficulties and failures which ever confront the would-be alchemist, one can well question the wisdom of pursuing such a course. At last, however, it does seem that these labours may not have been entirely in vain, for from these experiments has gradually emerged the vision of the benefit this art could be to man who, in his present state of imperfection, with its accompanying suffering of mind and body, would seem to require some assistance on his way through life.

Come with me, therefore, to my little laboratory with its array of alembics, crucibles and sandbaths, and hear something of the struggles of the would-be alchemist and of the mysteries he seeks to unravel.

After a careful study of Basil Valentine's *Triumphal Chariot of Antimony*, I decided to make my first experiments with antimony. I soon found, however, that on arriving at a crucial point, the key had almost invariably been deliberately withheld, and a dissertation on theology inserted in its place. Gradually, however, I came to realise that the theological discourse was not without object, but actually the means of veiling a valuable clue of some kind. After much labour, a fragrant golden liquid was finally obtained from the antimony, although this was merely a beginning. The *alkahest* of the alchemist, the First Matter, still remained a mystery.

Then followed processes with iron and copper. After purification of the salts or vitriols of these metals, of calcinations, and the obtaining of a salt from the calcined metal by a special process, followed by careful distillation and re-distillation in rectified spirits of wine, the oil of these meals was obtained, a few drops of which used singly, or in conjunction, proved very efficacious in cases of anaemia and debility which the ordinary iron medicine failed to touch.

The conjunction of iron and copper proved to be an elixir of very stimulating and regenerating character, the action being such as to clear the body from toxins, and I well remember on taking a few drops one evening that the prospect of a spell of fairly strenuous mental work, even after a really laborious day, seemed to hold no terrors for me!'

But still the alkahest remained an enigma, and so further experiments were made with silver and mercury. For those with silver, fine silver was reduced with nitric acid to the salt of the metal, carefully washed in distilled water, sublimated by a special process, finally yielding up a white oil which had a very soothing effect on highly nervous cases.

In the case of mercury, the metal on being reduced to its oil, produced a clear crystalline liquid with great curative properties, but unlike common mercury, no poisonous qualities.

After this I decided to work upon fine gold – gold that is without any alloy. This was dissolved in *Aqua Regia* and reduced to the salts of gold; these were washed in distilled water, which in its turn was evaporated in order to remove its very caustic properties. It was at this point that a very real difficulty arose, for when these salts of gold lose their acidity, they slowly but surely tend to return to their metallic form again. Nevertheless, an elixir was finally produced from them by distillation, although even then a residue of fine metallic gold remained behind in the retort.

Having got so far I realised that without the *alkahest* of the philosophers the real oil of gold could not be obtained, and so I went back and forth in the alchemist's writings to obtain the clue. The experiments which I had already made considerably lightened my task, and one day, while sitting quietly in deep concentration the solution to the problem was revealed to me in a flash, and at the same time many of the enigmatical utterances of the alchemists were made clear.

Here then, I entered upon a new course of experiment, with a metal for experimental purposes with which I had had no previous experience. This metal, after being reduced to its salts and undergoing special preparation and distillation, delivered up the Mercury of the Philosophers, the *Aqua Benedicta*, the *Aqua Celestis*, the Water of Paradise. The first intimation I had of this triumph was a violent hissing, jets of vapour pouring from the retort into the receiver like sharp bursts from a machine-gun, and then a violent explosion, whilst a very potent and subtle odour filled the laboratory and its surroundings. A friend has described this odour as resembling the dewy earth on a June morning, with the hint of growing flowers in the air, the breath of the wind over heather and hill, and the sweet smell of the rain on the parched earth.'

This gas Cockren was able to condense into a clear inflammable, golden-coloured liquid – the *Athoether* described by the Comte St. Germain – essential in the preparation of the oil of gold, a deep amber liquid of oily consistency. He considered the *Athoether* and the red tincture to be the Mercury and Sulphur described by the alchemists. These male and female principles conjoin to form a deep amber liquid – Philosophic Gold – a far more potent elixir than the oil of gold. He found that this amber liquid '*literally shines and reflects and intensifies rays of light to an extraordinary degree*'.

'And now to the final goal, the Philosopher's Stone. Having found my two principles, the Mercury and the Sulphur, my next step was to purify the dead body of the metal, that is, the black dregs of the metal left after the extraction of the golden water. This was calcined and carefully separated and treated until it became a white salt. The three principles were then conjoined in certain exact quantities in a hermetically sealed flask in a fixed heat neither too hot or too cold, care as to the exact degree of heat being essential, as any carelessness in its regulation would completely spoil the mixture.'

On conjunction the mixture takes on the appearance of a leaden mud, which rises slowly like dough until it throws up a crystalline formation rather like a coral plant in growth. The 'flowers' of this plant are composed of petals of crystal which are continually changing in colour. As the heat is raised, this formation melts into an amber-coloured liquid which gradually becomes thicker until it sinks into a black earth at the bottom of the glass. At this point (the sign of the Crow in alchemical literature) more of the ferment of mercury is added. In this process, which is one of continual sublimation, a long-necked, hermetically sealed flask is used, and one can watch the vapour rising up the neck of the flask and condensing down the sides. This process continues until the state of 'dry blackness' is attained. When more of the mercury is added, the black powder is dissolved and from this conjunction it seems that a new substance is born, or, as the early alchemists would have expressed it, a Son is born. As the black colour abates, colour after colour comes and goes until the mixture becomes white and shining; the

White Elixir. The heat is gradually raised yet more, and from white the colour changes to citrine and finally to red – the Elixir Vitae, the Philosopher's Stone, the medicine of men and metals. From their writings it appears that many alchemists found it unnecessary to take the Elixir to this very last stage, the citrine coloured solution being adequate for their purpose.'

It is of interest to note that an entirely different manifestation comes into being after the separation of the three elements and their re-conjunction under the sealed vase of Hermes. By the deliberate separation and unification of the Mercury, Sulphur and Salt, the three elements appear as a more perfect manifestation than in the first place.

Charles Richard Cammell, a friend of Archibald Cockren, watched the growth of this crystal of gold from the black mass in the hermetically sealed vessel over a period of six months and relates the phenomenon to that described by Paracelsus:

'It is also possible for gold to be so acted upon by the industry and art of the skilled alchemist that it will grow in a curcurbite with many wonderful branches and leaves, which experiment is very pleasant to behold, and full of marvels. The process is as follows: let gold be calcined by means of *Aqua Regia* so that it becomes a chalky lime; which place in a curcurbite, pouring in good and fresh *Aqua Regia* and water of gradation so that it exceeds four fingers across. Extract it again with the third degree of fire until nothing more ascends. Again pour over it distilled water, and once more extract by distillation as before. Do this until you see the Sol rise in the form of a tree with many branches and leaves. Thus there is produced from Sol a wonderful and beautiful shrub which alchemists call the Golden Herb, or the Philosopher's Tree. This process is the same with other metals, save that the calcinations may be different, and some other aqua fortis may have to be used. This I leave to your experience. If you are practiced in Alchemy you will do what is right in these details.'

Paracelsus, **Concerning the Nature of Things**, Book II, para. 4)

Sources Quoted by Cockren

Cockren's book is written in three parts, the first historical, the second theoretical and the third practical. There are extensive extracts or quotations from *Aureus or the Golden Tractate* by Hermes Trismegistus, *The Book of the Revelation of Hermes* interpreted by Theophrastus Paracelsus. The work is introduced by the *Smaragdine Table* of Hermes Trismegistus. In the historical section mention is made of a number of eminent alchemists throughout history: Zozimos the Panopolite, Morienus and King Kalid, Geber, Rhasis, Alfarabi, Avicenna, Artephius, Arnold of Villanova, Albertus Magnus, Thomas Aquinas, Raymund Lully, John Cremer, Isaac Hollandus, Nicholas Flamel, Paracelsus, Jean Baptista van Helmont, Denys Zachire, John Frederick Helvetius, Roger Bacon, Sir George Ripley, Thomas Charnock, Edward Kelly and Dr. John Dee, Thomas Vaughan, Eirenius Philalethes, Sir Isaac Newton, Alexander Seton, Sendivogius and the Comte de St. Germain. From his experimental work Cockren is strongly influenced by the work of Basil Valentine as special reference is made to his works, particularly *The Triumphal Chariot of Antimony* and the *Twelve Keys*.

Discussion

Archibald Cockren apparently worked for about forty years before he may have achieved the final goal of the alchemists - the Philosopher's Stone. He does not refer to any experiments to establish the potency of the Stone or of its use other than for the cure of human ailments. There is no reference in his writing to the transmutation of base metals into silver or gold, even to establish the success of his work.

True to the alchemical tradition Cockren makes no mention of his *prima materia* save that it was a metal with which he had no previous experience. In the theoretical section of his book he is guided by the Emerald Tablet of Hermes Trismegistus and the three principles of mercury, sulphur and salt. These represent the spirit (Aether), soul (Blood) and body (Ashes) and the King and Queen conjunction of which Sol is the father and Luna the mother. He relates the seven basic metals to their planetary forces, and holds that all metals are in a constant state of progression. Gold, the perfect metal, stands at the summit of perfection whilst all other metals are on the way towards eventually becoming gold; thus the alchemist merely accomplishes by art what Nature does slowly over many years. Metals are living, breathing substances containing mercury, sulphur and salt in varying proportions to account for their differences in properties.

It must always be remembered that the language of alchemy is not that of modern chemistry for it expresses its truths in depth by signs and symbols and as mental concepts based on archetypes common to all mankind throughout the ages. Alchemy is an art although it employs many of the techniques of a science and not all who aim at its goal 'hitte the marke'. According to Cockren:

> 'Alchemy brings us the vision of the heights to which man may attain; it teaches us that he is Triune, that is Spiritual, Mental and Physical; that his future is greater than at present can be envisaged; that Life is Law and Wisdom.'

Bibliograpy

Amberlain, Robert, **Jean-Julien Champagne, alias Fulcanelli:** Dossier Fulcanelli, Les Cahiers de la Tour Saint-Jacques, No. 9, Paris, 1962

Black, Robert M., **The Secret Art of Alchemy**, S.R.I.A., 1991

Canseliet, Eugene, **ALCHIMIE** Etudes diverses de Symbolisme hermetique et de pratique Philosophale, Jean-Jacques Pauvert, 1964

Canseliet, Eugene, **L'Alchimie et son Livre Muet**, Reimpression integrale de l'edition originale de La Rochelle 1677. Introduction et commentaries par Eugene Canseliet F.C.H. disciple de Fulcanelli, Jean-Jacques Pauvert et Francoise Harmel, 1986

Fulcanelli, **Le Mystere des Cathedrales**, Neville Spearman, London, 1971

Fulcanelli, **The Dwellings of the Philosophers**, Archive Press and Communications, Boulder, Colorado, 1999

Ovason, David, **The Zelator**: The Secret Journals of Mark Hedsel, Arrow Books, London, 1998

Ovason, David, **The Secrets of Nostradamus**: The medieval Code of the Master Revealed in the Age of Computer Science, Century Books, London, 1997

Pauwels, Louis and Bergier, Jacques, **The Dawn of Magic**, Anthony Gibbs & Phillips, London, 1963

Rayner Johnson, Kenneth, **The Fulcanelli Phenomenon**, Neville Spearman, 1980

Riedel, Albert (Frater Albertus Spagyricus), **The Alchemist of the Rocky Mountains**, Paracelsus Research Society Salt Lake City, Utah, 1976

Roob, Alexander, **The Hermetic Museum: Alchemy and Mysticism**, Benedikt Taschen, Cologne, 1996

Sadoul, Jacques, **Alchemists & Gold**, Neville Spearman, London, 1972

Chapter 15

Summary and Conclusions

1.   According to the laws of natural science, transmutation of one element into another is possible but has only been achieved on a very small scale. For example, bombardment of a target of platinum with fast neutrons results in the formation of minute quantities of an isotope of gold. Chemical reactions, carried out in conventional apparatus, such as used in a modern chemical laboratory, are concerned only with the valency electrons and are unable to affect the nucleus which determines the elemental properties.

2.   A question which has not so far been answered is whether any gold was ever actually produced by the alchemists by treating base metals with that enigmatic catalyst, or reagent, the Philosopher's Stone. Specimens of 'alchemical gold' are held in various museums throughout the world, for instance in the British Museum (London) and the Kunsthistorisches Museum (Vienna). Whether these samples of gold have been subjected to analysis, using modern analytical techniques, is uncertain, but in view of the claims made in the past for their extreme purity, such analyses would be of interest. The abundance of the various isotopes which might be ascertainable by mass spectrometric analysis would also yield valuable information. Such results as have been reported, particularly those involving measurement of the density of the material, have been disappointing.

3.   There is no doubt that the alchemists themselves believed that transmutation was possible. There are a number of accounts of transmutations carried out by reliable although, perhaps, gullible scientific observers such as John Frederick Helvetius and Johann Joachim Becher. Both Isaac Newton and Robert Boyle were firm believers in the possibility of transmutation, the former devoting much of his time to alchemical studies and the latter being recorded as having participated in a demonstration of the transmutation of lead into gold. Even as recently as 1922 Eugene Canseliet, using a minute quantity of the Stone, given to him by Fulcanelli, was reported as being able to perform an alchemical transmutation of a small piece of lead which resulted in 100 g. of gold.

4.   It would appear from the literature that once the Philosopher's Stone had been prepared transmutation was, thereafter, possible in a repeatable manner. It was the preparation of, and the philosophy behind, the Philosopher's Stone that appears to have been subject to rather different laws from those governing the behaviour of chemical change as are currently understood. Transmutation could be carried out by anyone, while the preparation of the Stone was limited to only a very few individuals who had received some form of spiritual enlightenment.

5.   This enlightenment may have been brought about by certain mental exercises connected with the symbolism of the work and by being directed onto the right lines by a 'father in alchemy'. It may well have been a gift in the same way that some people claim the ability to scry in a crystal or possess certain psychic powers, such as mediumship, which others do not have.

6.   Alchemy was of the nature of a spiritual exercise carried out on a series of planes and governed by cosmic forces which the alchemist had to take into account. It is for this reason that appreciation and understanding of alchemy is almost beyond reason. Its nature transcends the purely intellectual. Any attempt to enclose the concept in a neat abstract thought pattern is doomed to failure as it is not possible to build a coherent picture of the alchemical process. It may

well be that as Carl Jung has pointed out, some of the basic patterns of alchemical thought are buried in the individual sub-conscious, somewhere among the complex structure of body, mind and spirit comprising human personality and on the four Kabbalistic planes of *Assiah*, *Yetzirah*, *Briah* and *Atziluth*. The concept of a common unconscious, limited access to which may explain certain of the phenomena of extra-sensory perception such as telepathy and clairvoyance, may also have been a factor in the spread of alchemical thought and motifs around the world over the centuries.

7.   Alchemy was certainly not a science as commonly defined. Nor was it all an illusion based on the ignorance of the nature of matter and chemical change. It was very much an activity based on the interaction of psychic forces, generated in the mind of the alchemist, with actual matter. As this became more noble so did the mind of the alchemist, but it would seem that in so doing he lost the material side of his nature, becoming more spiritual and transcendent and, in fact, less of a being in the material world. This might account for certain phenomena such as the mysterious longevity and nebulous nature of such characters as Nicolas Flamel, the Comte de Saint-Germain and Fulcanelli.

8.   Although to the present day there is a continuing interest in alchemy, the scepticism of the scientific community has done much to ensure its art remains secret and hidden. Pronounced disbelief is known to effect even such phenomena as water divining. Psychic phenomena of several different kinds often refuse to manifest themselves in the presence of sceptics and under so-called 'strict laboratory conditions'. Indeed, the whole basis of modern science and technology would be undermined were the tenets of alchemy to be established beyond all reasonable doubt. The concept that chemistry is only a minor branch of alchemy would not be acceptable to the majority of chemists. The laws which govern the material world do not necessarily control the other spiritual worlds, and causation upon which material science depends appears to play a less significant role in alchemy.

9.   A common alchemical tenet is: *'When the pupil is ready, the Master will appear'*. This may account for the prevalence of alchemical texts by both 'true' alchemists and would-be puffers. May they not have gone into print in the hope that a Master would contact them and confirm and enlighten their path on the road to perfection?

10.   From what it has been said it would appear that the production of precious metals such as gold was not the sole purpose of the true alchemist. There is very little in the literature concerning the other forms of the Stone and in Western alchemy the Elixir of Life has only recently received serious attention with the work of Fulcanelli, Archibald Cockren and Armand Barbault. The cure of disease and the healing of the sick was, however, always paramount in Rosicrucian circles. Paracelsus and the iatrochemists were as practicing physicians primarily concerned with these aspects. It was, however, the regeneration of the alchemist himself that became the objective of those workers seeking the secrets of Holy Alchemy.

11.   On the physical plane the most important factor was the starting material, the *prima materia*, generally held to be an ore related to gold (not necessarily chemically related) and the identity of this substance is nowhere mentioned specifically. On the spiritual plane, the starting material appears to have been the alchemist himself, although linked in some mystical way to the *prima materia*.

12.   This survey has endeavoured to give some account of alchemy and the alchemists, the principles of Holy Alchemy and the motivation of some of its principal practitioners in their own words. It has been found impossible to express the subject in conventional analytical terms

because alchemy is not like that! The impression that alchemy gives is one of profound respect and adoration of the creature for the Creator and an identification on His behalf with the works of nature, animal, mineral and vegetable. In the language of Epigram XLII of *Atalanta Fugiens* of Michael Maier:

> Let nature be your guide, and with your art
> > Follow her closely, Without her you will err.
> Let reason be your staff; experience lend
> > Power to your sight that you may see afar.
> Let reading be your lamp, dispelling dark,
> > That you may guard 'gainst throngs of things and words.

APPENDIX A

PHILALETHE'S RULES FOR THE SECRET ART

Rule I.

Whatever any Sophister may suggest unto you, or you may read in any Sophistical Author; yet let none take you from this ground, (viz.) That as the end you look for is Gold: so let Gold be the subject on which you work, and none other.

Rule II.

Let none deceive you with telling you, that our *Gold* is not common, but Philosophical; for *common Gold* that was dead, which is true: But as we order it, there is made a quickening of it, as a grain of Corn in the Earth is quickened.

So then in our work, after six Weeks, Gold that was dead, becomes quick, living, and spermatical: and in our composition, it may be called *Our Gold*, because it is joyn'd with an Agent that will certainly quicken it: So a Condemned Man, is called a Dead Man, though at present living.

Rule III.

Besides Gold, which is the Body or Male, you must have another Sperm, which is the Spirit and Soul; or Female, and this is *Mercury*, in Flux and Form like to common *Argent Vive*, yet more clean and pure.

There are many, who instead of *Mercury*, will have strange Waters or Liquors, which they stile by the name of *Philosophical Mercury*; Be not deceived by them, for what a Man sows, that he must look to reap: If thou shalt sow thy Body in any Earth, but that which is Metalline and Homogeneal to it; thou shalt instead of a *Metalline Elixir*, reap an unprofitable *Calx*, which will be of no value.

Rule IV.

Our *Mercury* is in substance one with common Argent Vive, but far different in Form; For it hath a Form Coelestial, Fiery, and of excellent Virtue: and this is the Nature which it receives by our Artificial Preparation.

Rule V.

The whole Secret of our Preparation, is that thou take that Mineral which is next of kin to Gold, and to *Mercury*, Impregnate this with Volatile Gold which is found in the reins of *Mars*, with this purify your *Mercury* until seaven times are past, then it is fitted for the King's Bath.

Rule VI.

Yet know, that from seaven times to ten, the M*ercury* is made better and better, and is more active, being by each Preparation actuated by our true *Sulphur*; which if it exceed in number of Preparations, becomes too fiery, which instead of dissolving the Body, will Coagulate itself.

Rule VII.

This *Mercury* thus actuated, is after to be distilled in a Glass retort twice or thrice; and for this reason, because some Atoms of the Body may be in it, which were insensibly left in the Preparation of the *Mercury*, afterwards it is to be cleansed well with Vinegar and *Sal-armoniack*, then is it fit for the work.

Rule VIII.

Chuse your Gold for this work pure and clean from any mixture: if it be not so when you buy it, make it so by Purgation; then let it be made fine, either by Filing, Malleating, Calcining with Corrosives, or any other way, by which it ay be made most subtle.

Rule IX.

Now come to your mixture, in which take of the aforesaid Body so chosen and prepared, one Ounce of *Mercury*, as above taught animated, two Ounces or three at the most, mix them in a Marble which may be warmed so hot as water will heat it; grind both together till they be well incorporated, then wash the mixture with Vinegar and Salt till it be very pure; and lastly, Dulcifie it with warm water, and dry it carefully.

Rule X.

Know now, that whatever we say out of Envy, our way is none other, and we protest, and will protest, that neither We, nor any of the Ancients knew any other way; for it is impossible that our secret can be wrought by other Principles, or any other disposition than this. Our Sophism lies only in the two kinds of Fire in our work: the Internal secret Fire, which is God's instrument, hath no qualities perceptible to man, of that Fire we speak often, and seem yet to speak of the External heat; and hence arise among the unwary many Errours. This is our Fire which is graduated, for the External heat, is almost linear all the work, to the White work, it is one without alteration, save that in the seven first days we keep the heat a little slack for certainty and security sake, which an experienced Philosopher need not do.

But the Internal governing heat is insensibly graduated hourly, and by how much is daily vigorated by the continuance of Decoction, the Colours are altered, and the Compound maturated: I have unfolded a main knot unto you, take heed of being insnared here again.

Rule XI.

Then you must provide a Glass Tun, in which you may perfect your work, without which you could never do any thing: Let it be either Oval or Spherical, so big in reference to your Compound, that it may hold about twelve times the quantity of it within its Sphere, let your Glass be thick and strong, clear and free of flaws, with a neck about a Span of Foot long; In this Egg put your matter, sealing the neck carefully, without flaw, or crack, or hole, for the least vent will let out the subtle Spirit, and destroy the work.

You may know the exact Sealing of your Glass thus, when it is cold, put the neck where it is sealed, into your mouth, and suck strongly; if there be the least vent, you will draw out the Air, that is in the Vial, into your mouth, which when you take the Glass from your mouth, is again suckt into the Glass with a hissing, so that your ear may perceive the noise; this is an undoubted tryal.

Rule XII.

You must then provide yourself with a Furnace, by wise men called an *Athanor*, in which you may accomplish your work; nor will any one serve in your first work; But such a one in which you may give a heat obscurely red at your pleasure, or lesser, and that in its highest degree of heat, it may endure twelve hours at the least.

This if you would obtain; Observe, First, that your nest be no bigger then to contain your dish with about an Inch vacancy at the side where the Vent-hole of your *Athenor*, is for the Fire to play.

Secondly, Let your Dish be no bigger then to hold one Glass with about an inch thickness of Ashes between the Glass and side, remembering the word of the Philosopher, One Glass, One Thing, One Furnace; for such a Dish standing with the bottom level to the vent-hole, which in such a Furnace ought to be but one, about three Inches Diameter, sloping upwards, will with the stream of Flame, which is always playing to the top of the Vessel, and round about the bottom, be kept always in a glowing heat.

Thirdly, If your Dish be bigger, your Furnace vent must be within a third part, or a fourth as big as your Platter is Diameter, else it cannot be exactly, nor continually heated.

Fourthly, If your Tower be above six inches square at the Fire-place, you are out of proportion, and can never do rightly as to the point of heat; For if you cause it (if above that proportion) to stream with flame, the heat will be too big: And if it stream not, it will not be big enough, or very hardly.

Fifthly, Let the top of your Furnace be closed to an hole which may but just serve for casting in of Coals about three Inches Diameter or Square, which will keep down the heat powerfully.

Rule XIII.

These things thus ordered, set in your Glass with your matter, and give Fire as Nature requires, easie, not too violent; beginning there where Nature left. Now know, that nature hath left your Materials in the Mineral Kingdom; therefore though we take comparison from Vegetables and Animals; Yet you must understand a Parallel in the Kingdom, in which the Subject you would handle is placed: As for Instance, if I should Analogize, between the Generation of a Man, and the Vegetation of a Vegetable, you must not understand, as though the heat for one, were to be measured by the other; for we know, that in the ground Vegetables will grow, which is not without heat, which they in the Earth feel, even in the beginning of the Spring; yet would not an Egg be hatched in that heat, nor could a man feel any warmth, but rather to him a numbing cold.

Since then you know that your work appertains all to the Mineral Kingdom; you must know what heat is fit for Mineral Bodies; and may be called a gentle heat, and what violent; First, now consider, where Nature leaves you, not only in the Mineral Kingdom, but ion it to work on Gold and *Mercury*, which are both incombustible: Yet *Mercury* being tender, will break all Vessels, if the Fire be over extreme; Therefore though it be incombustible, and so no Fire can hurt it, yet also it must be kept with the Male Sperm in one Glass, which if the Fire be too big, cannot be, and by consequence the work cannot be accomplished. So then from the degree of heat that will keep Lead or Tin constantly molten, and higher, so high as the Glass will endure without danger of breaking, is a temperate heat; and so you begin your degrees of heat according to the Kingdom in which nature hath left you.

As then the highest degree of heat which the root of a Tree feels in the bowels of the Earth; is not by far comparable to the lowest degree of heat an Animal hath; So the highest degree of heat a Vegetable will endure without burning, is too low for the first degree of Mineral heat as to our Work.

Rule XIV.

Know, that all your progress in this Work is to ascend in *Bus & Nubi*, from the Moon up to the Sun; that is in *Nubibus*, or in *Clouds*: Therefore I charge thee to sublime in a continual vapour, that the Stone may take Air, and live.

Rule XV.

Nor is this enough, but for to attain our permanent Tincture, the water of our lake must be boyled with the Ashes of *Hermes Tree*; I charge thee then to boyl night and day without ceasing, that in the troubles of the stormy Sea, the Heavenly Nature may ascend, and the Earthly descend.

For verily, if we did not Boyl, we would never name our work Decoction, but Digestion; For where the Spirits only Circulate silently, and the Compound below moves not by an Ebullition, that is only properly named Digestion.

Rule XVI.

Be not over hasty, expecting Harvest too soon, or the end soon after the beginning: For if thou be patiently supported, in the space of fifty days at the farthest, thou shalt see the Crows Bill.

Many (saith the Philosophers) do imagine our Solution to be an easie work; But how hard it is, they can only tell, who have tried and made Experience: Seest thou not a Grain of Corn, sow it, and after three days thou shalt only see it swell'd; which being dry'd is the Corn as it were before: Yet thou canst not say it was not cast into its due Matrix; for the Earth iss its true place, but only it wanted its due time to Vegetate.

But things of an harder Kernel lie in the ground a far longer time, as Nuts and Plumb-stones, for each thing hath its season; And this is a true sign of a natural Operation, that it stays its season, and is not Precipitate: Dost think then, that Gold the most solid Body in the world? will change its Form in a short time; Nay, thou must wait and wait until about the 40th day utter blackness begins to appear; when thou seest that, then conclude thy Body is destroy'd, that is, made a living Soul, and thy Spirit is dead, that is Coagulated with the Body; but till this sign of Blackness, both the Gold and the *Mercury* retain their Forms and Natures.

Rule XVII.

Beware that thy Fire go not out, no not for a moment, so as to let thy Matter be cold, for so Ruine of the Work will certainly follow.

By what has been said, thou mayest gather, that all our work is nothing else but an uncessant boyling of thy Compound in the first degree of liquefying heat, which is found in the Metalline Kingdom, in which the Internal Vapours shall go round about thy matter, in which fume it shall both die, and be revived.

Rule XVIII.

Know, that when the White appears, which will be about the end of Five Months, that then the accomplishment of the White Stone approacheth; Rejoice then, for now the King hath overcome death, and is rising in the East with great Glory.

Rule XIX.

Then continue your Fire until the Colours appear again, than at last you shall see the fair Vermillion, the Red Poppy: Glorifie God then, and be thankful.

Rule XX.

Lastly, you must boyl this Stone in the same water, in the same proportion, with the same Regimen, (only your Fire shall then be a little slacker) and so you shall increase Quantity and Goodness at your pleasure.

Now the only God the Father of light, bring you to see this Regeneration of the light, and make us to rejoice with him for ever hereafter in light. Amen.

'CLAVIS': Keynes MS 18 English Translation
Kings's College, Cambridge

First of all know antimony to be a crude and immature mineral having in itself materially what is uniquely metallic, even though otherwise it is a crude and undigested mineral. Moreover, it is truly digested by that sulphur that is found in iron and never elsewhere.

Two parts of antimony with iron give a regulus which in its fourth fusion exhibits a star; by this sign you may know that the soul of the iron has been made totally volatile by virtue of the antimony. If this stellate regulus is melted with gold or silver by an ash heat in an earthen pot, the whole regulus is evaporated, which is a mystery. Also, if this regulus is amalgamated with common mercury and is digested in a sealed vessel on a slow fire for a short time – two or three hours – and then ground for 1/8 hour in a mortar without moisture while being warmed moderately, until it spits out its blackness, until the water, then it may be washed to deposit the greatest part of its blackness, until the water, which in the beginning becomes quite black, is scarcely more tinged by the blackness. This can be done by flushing it with water many times. Let the amalgam be dried, again placed near the fire, and kept in the above-mentioned heat for three hours. Afterwards let it be ground again as before in a dry and warm mortar. It pushes out new blackness, which must be washed away again; this must be repeated continually until the whole amalgam becomes like shining and cupellated silver, whereas at first it had a dark leaden colour.

Then distill this mercury which has been so washed and amalgamate over gain seven or nine times, and in each amalgamation see to the heating, grinding, and washing as many times as before. Distill the whole as before. On the seventh time you will have a mercury dissolving all metals, particularly gold. I know whereof I write, for I have in the fire manifold glasses with gold and this mercury. They grow in these glasses in the form of a tree, and by a continued circulation the trees are dissolved again with the work into new mercury. I have such a vessel in the fire with gold thus dissolved, where the gold was visibly not dissolved by a corrosive into atoms, but extrinsically into a mercury as living and mobile as any mercury found in the world. For it makes gold begin to swell, to be swollen, and to putrefy, and to spring forth into sprouts and branches, changing colours daily, the appearances of which fascinate me every day. I reckon this is a gerewat secret in Alchemy, and, I judge, it is not rightly to be sought from artists who have too much wisdom to decide that common mercury ought to be attacked through reiterated cohobation by the regulus of Leo [that is of iron and antimony]. That unique body, that regulus, however, is familial with mercury seeing that it is closest to that mercury you have known and recognized in the whole mineral kingdom, and hence most closely related to gold. And this is the philosophical method of meliorating nature in nature, cosanguinity in cosanguinity.

With regard to this operation, look at the Letter responding to Thomas of Bologna, and you will find this question fully solved.

Another secret is that you need the mediation of the virgin Diana [quintessence, most pure silver]; otherwise the mercury and the regulus are not united.

The regulus is made from antimony four ounces, nine parts, iron two ounces, four parts, this is a good proportion. Do not neglect to have a mass of antimony greater than that of iron, for if an error is made here you will be disappointed. Make the regulus by casting in nitre bit by bit; cast in between three and four ounces of nitre so that the matter may flow.

It is not a good idea to prepare in one crucible a greater quantity than the above measure of antimony. The antimony is ground, then cupelled together with iron, whatever others may say or write.

Little nails may be used and especially the ends of those broken from horseshoes. Let the fire be strong so that the matter may flow [like water], which is easily done. When it flows, cast in a spoonful of nitre; and when that nitre has been destroyed by the fire, cast in another. Continue that process until you have cast in three or four ounces. Then pile up the charcoals about the crucible, taking care that they do not fall into it. Increase the fire as much as the fusion of common silver requires, and keep it at that state for 1/8 hour. [The matter ought to be like a subtle water if you have laboured correctly.] Then pour the matter out into a cone. The regulus will subside. Separate the ashy scoria from it. Keep the cooled material in a dry vessel.

It is a sign of a good fusion if the iron is completely fused and if the scoriae break up by themselves into powder.

Beat the regulus and add it to two, or at the most 2 ½, ounces of nitre. Grind the regulus and the nitre together completely and again melt. Throw away the arsenical and useless scoriae.

Grind the regulus a third and fourth time with at most one ounce of nitre and melt in a new crucible, and on the fourth time you will have scoriae tinged with a golden colour and a stellate regulus.

N.B. In the last three time the scoriae must be thrown away because they are arsenical however, they are useful; in surgery.

N.B. In the last three fusions the regulus must be beaten, and ground and mixed with nitre. Some cast the nitre into a crucible, but this is not recommended, for, firstly the fusion is as a result prolonged and the regulus is not without some loss of itself by exhalation. Secondly, nitre thrown in in this way stays on the surface and in time it cools the regulus. And since nitre flows easily, it may flow at first and encrust so that it will not flow again without a large fire. If that happens, the best part of the regulus perishes in the conflagration, whence it is that sometimes a star perishes because it is falsely ascribed to a constellation. You will see that the regulus mixed with nitre in this way flows easily with it; and you will not see it become hard in any manner, except for the difference in the depuration, which is far greater if it is mixed than if the nitre is just tossed in.

Take of this regulus one part, of silver two parts, and melt them together until they are like fused metal. Pour out, and you will have a friable mass of the colour of lead.

N.B. If the regulus is joined with silver, they flow more easily than either one separately and they remain fused as long as lead even though there are thus two parts of silver, which is then changed into the nature of antimony, friable and leaden.

Beat this friable mass, this lead, and cast it together with the mercury of the vulgar into a marble mortar. The mercury should be washed (say ten times) with nitre and distilled vinegar and likewise dried (twice), and the mortar should be constantly heated just so much as you are able to bear the heat of with your fingers. Grind the mercury ¼ of an hour with an iron pestle and thus join the mercury, the doves of Diana mediating, with its brother, philosophical gold, from which it will receive spiritual semen. The spiritual semen is a fire which will purge all the superficial impurities of the mercury, the fermental virtue intervening. Then take a little beaten sal ammoniac and grind with the mercury. When it is fully amalgamated, add just enough humidity to moisten it, and this one philosophical sign will appear to you: that in the very making of the mercury there is a great stink. Finally, wash your mercury by pouring on water, grinding, decanting, and again pouring on fresh water, until few faeces appear.

(Dobbs, B.J.T., *The Foundations of Newton's Alchemy*, pp. 251-255)

Notes and References

Chapter 1: An Outline of Alchemy

1. Bynum, W.F., Browne, E.J. and Porter, Roy, *Dictionary of The History of Science*, London, The Macmillan Press, 1981, pp. 9-10

2. Terms used by Joseph Needham when discussing Chinese alchemy. 'Aurifiction' describing the simulation of metals to appear as gold by colouring or surface treatment, while 'aurifaction' relates to the transmutation of base metals into a metal with many of the properties of gold.

3. Read, John, *Prelude to Chemistry*, London, G. Bell and Sons Ltd., 1939, p. 40

4. Sherwood Taylor, F., *The Alchemists*, London, William Heinemann, 1952, (reprinted 1976 by Granada Publishing Limited) refers to the **Gold-making of Cleopatra** and the page of symbolic drawings one of which contains the symbols of gold, silver and mercury enclosed in two concentric circles, within which appear the words *One is the serpent which has its poison according to two compositions* and *One is All and through it is All and by it is All and if you have not All, All is Nothing* (p. 55).

5. An excellent source for alchemy in ancient and medieval India is the *History of Hindu Chemistry* by Acharya Prafulla Chandra Ray (Indian Chemical Society, Calcutta, 1956). The work discusses chemistry in pre-historic India, during the Vedic and Ayurvedic period and, following a transitional period that during the Tantric era and the iatrochemical period.

6. Atwood, M.A., *The Suggestive Inquiry Into the Hermetic Mystery*, London, Trelawney Saunders, 1850 - originally published anonymously by Mary Anne South, the daughter of Thomas South of Bury House, Gosport, Hampshire. After a number of copies of the work had been sold or issued to public libraries the remaining issue of the book was stopped, recalled (at a cost of £250) and burnt on the lawn of Bury House together with an unpublished alchemical poem composed by Thomas South himself. Some copies of the book survived and were republished in a revised edition and with an introduction by Walter Leslie Wilmhurst in 1918. A modern edition was issued by the Julian Press of New York in 1960. A Latin version of the *Tabula Smaragdina Hermetis* is given on p. 7 with its English translation as quoted.

7. *Splendor Solis*: Alchemical Treatises of Solomon Trismosin, Adept and Teacher of Paracelsus, including 22 allegorical Pictures Reproduced from the Original Paintings in the Unique Manuscript on Vellum, dated 1582, in the British Museum. With Introduction, Elucidation of the Paintings, aiding the Interpretation of their Occult meaning, Trismosin's Autobiographical Account of his Travels in Search of the Philosopher's Stone, A Summary of his Alchemical Process called 'The Red Lion', and Explanatory Notes by J.K. (Julius Kohn). London, Kegan Paul, Trench, Trubner & Co., Ltd. (*c.*1920). This was a publication of the Harley Manuscript 3469. A new translation, by Joscelyn Godwin was issued in1981 by Adam McLean in a limited edition as part of the *Magnus Opus Hermetic Sourceworks*, Edinburgh. Julius Kohn, an Austrian émigré, was a pupil of the Revd. W.A. Ayton one of the early members of the Hermetic Order of the Golden Dawn.

8. Discussed in *A History of Magic and Experimental Science* by Lynn Thorndike, New York, Columbia University Press 1923 onward. (Vol. II, pp. 14-49).

Chapter 2: The language of Alchemy

1. Fulcanelli, *Le Mystere des Cathedrales*, London, Neville Spearman, 1971, p. 42 *et seq*. The language of the Birds is also discussed by David Ovason in *The Secrets of Nostradamus*, London, Century Books, 1997, ch. 4.

2. According to Ferguson the commentary on the Emerald Tablet has been printed fairly frequently. The translation into English was made from a 1541 text of *De Alchemia* one time thought to have been the earliest appearance of the Tablet. In 1923, however, E.J. Holmyard discovered an Arabic text of the Tablet in Jabir's *Second Book of the Element of the Foundation* dating from the 8th century A.D.

3. The translations are taken from Mrs. Atwood's *Suggestive Enquiry* (p. 20) in which they are reprinted from their publication in *The Critic*, (new series), 1845, no. 13, p. 352.

4. Michael Maier, quoted in his work *Symbola Auria Mensae*.

5. Taken from N. Barnaud '*Commentariolum in Enigmaticum quiddam Epitaphium Bononiae studiorum, ante multa secula marmoreo lapidi insculptum* in the *Theatrum Chemicum* vol. 5.

6. A further dichotomy exists between alchemy which set out to produce the Philosopher's Stone and the transmutation of base metals into gold and a pseudo-alchemy which had the objective of making base metals appear as silver or gold. The former being described as 'aurifaction' and the latter as 'aurifiction'.

7. For details of Newton's alchemical studies see Betty Jo Teeter Dobbs: *The Foundations of Newton's Alchemy or 'The Hunting of the Greene Lyon'* Cambridge, Cambridge University Press, 1975 and *The Janus Faces of Genius*: *The Role of Alchemy in Newton's Thought*. Cambridge University Press, 1991. The alchemical interests of the Hon. Robert

Boyle are discussed in *The Aspiring Adept: Robert Boyle and his Alchemical Quest* by Lawrence M. Principe, Princeton, New Jersey, Princeton University Press, 1998.

8.   Waite, Arthur E., *The Secret Tradition in Alchemy*, London, Stuart and Watkins, 1969 gives a masterful account of the subject in his inimitable manner.

9.   MacGregorMathers, S.L., *The Kabbalah Unveiled*, London, Routledge and Kegan Paul, 11th impression, 1970, p. 9.

10.   Principe, Lawrence, 'Chemical Translation' and the Role of Imputities in Alchemy: Examples from Basil Valentine's Triumph-Wagen', *Ambix*, 1987, vol. 34, pp. 21-30.

Chapter 3: The Philosopher's Stone

1.   Description taken from *The Lure and Romance of Alchemy* by C.J.S. Thompson, (London, George G. Harrap, 1932, pp. 70-71).

2.   *The New Chemical Light drawn from the Fountain of Nature and of Manual Experience* by Michael Sendivogius is reprinted in *The Hermetic Museum*, volume II, pp. 79-158 from which this passage is taken.

3.   Elias Ashmole, *Theatrum Chemicum Britannicum*, London, Nath; Brooke, 1652.

4.   Jean-Baptiste van Helmont the physician and chemist was born in Brussels in 1577. In 1618, while working in his laboratory at Vilvorde was visited by an unknown alchemist who appears to have been visiting several established men of science with the object of promoting the 'truth' of the Philosopher's Stone. Van Helmont was held to be the greatest chemist of his day and one who was difficult to deceive. According to Louis Figuier 'he was incapable of imposture himself and had nothing to gain by lying about the episode.' As the experiment was carried out by van Helmont himself in the absence of the alchemist, it is hard to suspect any kind of fraud. It is possible that the subsequent published account may have been fictitious.

Chapter 4: The Theory of Transmutation

1.   Arnold or Arnald of Villanova (1240-1313) studied in Aix-en-Provence, became a medical student at Montpellier and concluded his studies at the Sorbonne in Paris. He was a contemporary of Albertus Magnus and Roger Bacon. The statement on the Philosopher's Stone is taken from his *Speculum* and is quoted by M.A Atwood in her *Suggestive Enquiry* p. 72.

2.   Eirenius Philalethes' *An Open Entrance to the Closed Palace of the King* (in *The Hermetic Museum* volume II, p. 165 for the two-fold nature of 'our gold'.

3.   Jung, C.J., *Mysterium Conjunctionis* (Collected Works Vol. 14), London, Routledge and Kegan Paul, 1963.

4.   Burkhardt, T., *Alchemy*, London, Stuart and Watkins, 1967.

5.   The first part of *Ripley Reviv'd* by Eirenaeus Philalethes (Printed for William Cooper, at the Pelican in Little Britain, 1677) is *An Exposition upon Sir George Ripley's Epistle to King Edward IV* in the opening passages of which Philalethes draws a series of conclusions from Staves IX, X, XI, and XII. Philalethes goes on to comment: 'These Conclusions are but few in number but of great weight or concernment; the Amplification, Illustration and Elucidation therefore of them, will make a son of Art truly glad. '

6.   *The Hermetic Museum* volume II, p. 87.

Chapter 5: Alchemical Operations and Apparatus

1.   Ashmole, Elias, *Theatrum Chemicum Briannicum*, New York, Johnson Reprint Corporation.1967 (Reprint of the 1652 London Edition) pp. 107-193.

2.   Thompson, C.J.S., *The Lure and Romance of Alchemy*, London, G.G. Harrap, 1932, p. 108

3.   Samuel Johnson, *A Dictionary of the English Language*, London, Longman, Hitch and Hawes, 1755 (Facsimile reprint by Longmans Group U.K. Limited, 1990)

4.   Rowse, A.L., *Simon Forman: Sex and Society in Shakespeare's Age*, London, Weidenfeld and Nicolson, 1974.

5.   Canseliet, Eugene, *L'Alchimie et son Livre Muet (Mutus Liber)*, Paris, Jean-Jacques Pauvert and Francoise Harmel, 1986.

6.   Cockren, A., *Alchemy Rediscovered and Restored*, London, Rider and Company, 1956.

7.   Sherwood Taylor, F., *The Alchemists*, Frogmore, St. Albans, Granada Publishing, 1976, pp. 95-96.

Chapter 6: An Interpretation of Alchemy

1. Atwood, M.A., *Hermetic Philosophy and Alchemy*, (Revised Edition with an Introduction by Walter Leslie Wilmhurst), New York, The Julian Press, 1960.

2. Waite, A.E., *The Secret Tradition in Alchemy: its Development and Records*, London, Stuart and Watkins, 1969, pp. 395-397.

3. Lord Rutherford, *The Newer Alchemy*, Cambridge, Cambridge University Press, 1937, p. 62 writes 'The amount of transformation produced is usually on a minute scale and only rarely is the quantity of matter produced either visible or weighable'.

4. Jacobi, Jolan, *The Psychology of C.G. Jung*, London, Kegan Paul, 1942.

5. Quotation from '*Aqua Vitae: Non Vitis*: or The Radical Humiditie of Nature: Mechanically, and Magically dissected, by the Conduct of Fire, and Ferment' (British Museum MS Sloane 1741) reprinted in *The Works of Thomas Vaughan* edited by Alan Rudrum, Oxford, Clarendon Press, 1984, p. 588.

6. King, Francis, *Ritual Magic in England*, London, Neville Spearman, 1970, contains a chapter describing the alchemical rebirth of the Golden Dawn and gives a description of the ritual process noted in the Golden Dawn manuscript known as Z2.

7. Regardie, Israel, *The Golden Dawn*, Wisconsin, Hazel Hills Corporation, 1970, volume III, pp. 184-192.

Chapter 7: Alchemy - Hoax or Hyperchemistry?

1. Lord Rutherford, *The Newer Alchemy*, Cambridge, Cambridge University Press, 1937.

2. Holmyard, E.J., *Alchemy*, Harmondsworth, Penguin Books, 1957, pp. 93-95.

3. Waite, A.E., *The Hermetic Museum Restored and Enlarged*, London, John M. Watkins, 1953, Vol. II, p. 197.

4. Westcott, William Wynn, (Ed.), *Nicholas Flamel His Exposition of the Hieroglyphical Figures which he caused to be Painted upon an Arch in St. Innocents Church yard in Paris: Concerning both the Theory and Practice of the Philosophers Stone.*

5. Helvetius, John Frederick, *Golden Calf which the World worships and adores*, Westcott *op. cit.* pp. 271-300.

6. Holmyard, E.J., *op. cit.* pp. 253-261.

7. Sherwood Taylor, F., *The Alchemists*, St. Albans, Granada Publishing, 1976.

8. Manget, J.J., *Bibliotheca Chemica Curiosa*, 1702, Vol. 1, pp. 196-210.

9. Van Helmont, J.B., *Oriatrike or physick refined*, London, 1662, pp. 751-752.

10. Partington, J.R., *A History of Chemistry*, London, Macmillan and Co. Ltd., 1961, Vol. Two, p. 217.

11. Sherwood Taylor, F., *op. cit.* pp. 133-137.

12. Principe, Lawrence M., *The Aspiring Adept; Robert Boyle and his Alchemical Quest*, Princeton, Princeton University Press, 1998.

13. Holmyard, E.J., *The Works of Geber*, London, J.M. Dent, 1928.

14. Holmyard, E.J. and Mandevill, D.C., *Avicennae de Congelatione et Conglutinatione Lapidum: being sections of the Kitab al-Shifa*, Paris, Paul Geuthner, 1927.

15. Newman, William R., *The Summa perfectionis of the Pseudo-Geber: a critical Edition, Translation and Study*, Leiden, E.J. Brill, 1991.

16. Principe, Lawrence M., 'Chemical Translation and the role of impurities in alchemy: examples from Basil Valentine's Triumph-Wagen', *Ambix*, 1987, Vol. 34, Pt.1, pp. 21-30.

17. Lord Rutherford, *op. cit.* (2)

18. Fulcanelli, *Les Demeures Philosophales*, Boulder, Colorado, Archive Press, 1999.

19. Eliade, Mircea, *The Forge and the Crucible*, Chicago, University of Chicago Press, 1978.

20. Barbault, Armand, *Gold of a Thousand Mornings*, London, Neville Spearman, 1975.

21. Karpenko, V., 'Coins and Medals made of Alchemical metal', *Ambix*, 1988, Vol. 35, Part 2, pp. 65-76.

Chapter 8: Aspects of Greek Alchemy

1. Orpheus is believed by some to have been an historical personality while others held him to have been a god or imaginary hero. Legend has it that he was married to Euridyce and on her death descended into Hades in an attempt to induce Persephone to let his wife return to him. Such were his musical skills that Persephone agreed provided that he did not look at Euridyce as she followed him back. He could not resist the temptation, however, and Euridyce remained in Hades. (Oxford Reference Encyclopedia, 1998)

2. Partington, J.R., *A History of Chemistry*, 1970, Vol. 1, p. 4

3. Little is known of the life of Pythagoras who lived (*c*.550 - *c*.500 B.C.) It would seem that he emigrated from his native Samos to southern Italy, where he founded a sect characterised by common beliefs and dietary restrictions in pursuit of esoteric knowledge.

4.   Jowett, B., *The Dialogues of Plato*, Oxford. The Clarendon Press, 1871

5.   Aristotle remained with Plato until the latter's death in 348 / 7.5Berthelot, Marcellin, I*ntroduction a l'etude de la chemie des anciens et du moyen age*, Paris, 1889 pp. 37-38.

6.   Sherwood Taylor, F., 'The Alchemical Works of Stephanos of Alexandria, Part I', *Ambix*, 1937, Vol. 1, pp. 116-139.

7.   Sherwood Taylor. F., 'The Alchemical Works of Stephanos of Alexandria. Part II', *Ambix*, 1938, Vol. 2, pp. 39-49.

8.   Holmyard, E. J., *Alchemy*, Harmondsworth, Penguin Books, 1957, p. 29

9.   Read, John, *Prelude to Chemistry*, London, G. Bell and Sons, 1939,

10.   Sherwood Taylor, F. *The Alchemists*, Frogmore, St. Albans, Granada Publishing, 1976, opposite p. 95, 'The Gold making of Cleopatra', (Courtesy of Journal of Hellenic Studies)

11.   Sherwood Taylor, F., 'The Origins of Greek Alchemy', *Ambix*, 1937.Vol. 1, pp. 30-47.

12.   Read John, *Prelude to Chemistry*, London, G. Bell and Sons, 1957, p. 40.

13.   Sherwood Taylor, F., *op. cit.*, pp. 57-60.

14.   Sherwood Taylor, F., 'Translation of "The Visions of Zosimos", *Ambix*, 1937, Vol. 1., pp. 88-92.

15.   Jung, C.J., *Alchemical Studies,* (The Collected Works, Vol. 13), pp. 57-105

Chapter 9: Aspects of Chinese Alchemy

1.   A general bibliography includes:

Burkhardt, T., *Alchemy*, London, Stuart and Watkins, 1967

Feifel, E., *Pao Phu Tzu* (*Nein Phein*), Monumenta Serica, 1941, Vol. 6, p. 113, 1944, Vol. 9, p.1, 1946, Vol.11, p. 1

Holmyard, E.J., *Alchemy*, Harmondsworth, Penguin Books, 1957

Hommel, R.P., *China at Work*, New York, Day, 1937

Ho Ping-Yu and Needham, J., 'The Laboratory Equipment of the Early Medieval Chinese Alchemists', *Ambix*, 1959, Vol. VII, pp. 57-115

Karlgren, B., *The Book of Documents* (*Shu Ching*)', Bull. Museum of Far Eastern Antiquities, 1950, Vol. 22, p. 1

Needham, J., *Science and Civilisation in China*, Cambridge, Cambridge University Press, Vol. 3 and Vol. 5, Pts. 2 & 3,

Read, J., *Prelude to Chemistry*, London, G. Bell and Sons, 1939

Sherwood Taylor, F., *The Alchemists*, London, Heinemann, 1952

Ware, J.R., *Alchemy, Medicine and Religion in the China of 320 A.D.; the Nei Phien of Ko Hung*, Cambridge, Mass., M.I.T. Press, 1966

Wilhelm, R., *I Ching or Book of Changes* London, Routledge & Kegan Paul, 1951

Wu Lu-Chhiang and Davis, T.L., *An Ancient Chinese Treatise on Alchemy entitled Tsan Thung Chhi written by Wei Po-Yang about 142 A.D.*, ISIS, 1932, Vol. 18, p. 210

idem., *An Ancient Chinese Alchemical Classic: Ko Hung on the Gold Medicine and on the Yellow and on the White*, Proc. British Academy, 1935, Vol. 70, p. 221

2.   Wilhelm, R., *op. cit.*

3.   Read, J., *op. cit.* pp. 122-123

4.   Wu and Davis, *op. cit.*

5.   Ware, J.R., *op. cit.*

6.   *ibid, op. cit.*

7.   Needham, J., *op. cit.*, p. 123

8.   Needham, J., *op. cit.*

9.   Wilhelm, R., *op. cit.*

10.   Wilhelm, R. and Jung, C.J., p. 19

11.   Needham, J., *op. cit.* Vol. II, p. 149

12.   Wilhelm, R. and Jung, C.J., *op. cit.* p. 54

Chapter 10: Aspects of Indian Alchemy

1.   Cresswell, C.W., *The Way of Tantra*, Trans. Metropolitan College SRIA, 1982, pp. 13-19

2.   Ashmole, Elias, *Theatrum Chemicam Briannicum*, 1652

3.   Needham, Joseph, *Science and Civilisation in China,* Cambridge, Cambridge University Press, 1974, 1976, Pts. II & III

4.   Allegro, John M., *The Sacred Mushroom and the Cross*, London, 1970

5.   Todd, R.G. (Ed.), *Extra Pharmacopoeia*, Martindale, 25th Edition, London, The Pharmaceutical Press, 1967

6.   Ray, P., *History of Chemistry in Ancient and Medieval India* incorporating the *History of Hindu Chemistry* by Acharya Prafulla Ray, Calcutta, Indian Chemical Society, 1956).

7.   Wujastyk, Dominic, 'An Alchemical Ghost: The Rasaratnakara by Nagarjuna', *Ambix*, 1984, Vol. 31, pp. 70-83

8.   Cresswell, C.W., *op. cit.*

9.   Mookerjee, Ajit and Khanna, Madhu, *The Tantric Way: Art, Science and Ritual*, London, 1977

Further references include:

Avalon, Arthur (Sir John Woodroffe), *The Serpent Power*, Madras, 1924

British Herbal Medicine Association, *British Herbal Pharmacopoeia*, Vols. 1 & 2, Cowling, 1976-9,

Eliade, Mircea, *The Forge and the Crucible*, Chicago, University of Chicago Press, 1978

Garstin, E.J. Langford, *The Secret Fire: An Alchemical Study*, London, 1932

Gerson, Scott, *Ayurveda: The Ancient Indian Healing Art,* Shaftesbury, 1993

Rawson, Philip, *Tantra: The Indian Cult of Ecstacy*, London, 1973

Shepherd, H.J., 'Alchemy: Origin or Origins', *Ambix*, 1970, Vol. 17, p. 69

Taton, Rene, *Ancient and Medieval Science*, London, 1957

Chapter 11: Aspects of Islamic Alchemy

A general bibliography includes:

Atwood, M.A., *Hermetic Alchemy and Philosophy*, New York, 1960, Julian Press.

Burkhardt, Titus, *Introduction to Sufism*, London, 1976, The Aquarian Press.

Fuck, J.W., 'The Arabic Literature on Alchemy according to An-Nadim', *Ambix*, 1951, Vol. 4, pp. 81-144

Hamarneh, Sami K., 'Arabic-Islamic Alchemy', *Ambix*, 1982, Vol. 29, pp. 74-87.

Holmyard, E.J., *Alchemy*, 1957, Penguin Books, Harmondsworth.

Holmyard, E.J., *The Works of Geber Englished by Richard Russell, 1678: A New Edition*, 1928, London, J.M. Dent and Sons.

Holmyard, E.J. and Mandeville, D.C., *Avicennae de Congelatione et Conglutinatione Lapidum being sections of the Kitab Al-Shifa*, 1927, Paris, Paul Geuthner.

Plessner, Martin, 'The History of Arabic Literature', *Ambix*, 1972, Vol. 19, pp. 209-213.

Redgrove, H. Stanley, *Alchemy: Ancient and Modern*, 1922, London, William Rider and Son.

Shah, Idries, *The Sufis*, 1964, London, The Octagon Press.

Shah, Idries, *The Way of the Sufi*, 1974, Harmondsworth, Penguin Books.

Sherwood Taylor, F., *The Alchemists*, 1976, St. Albans, Granada Publishing.

Stapleton, H.E., *An Alchemical Compilation of the Thirteenth Century, A.D.*, Memoirs of the Asiatic Society of Bengal, 1910, Vol. III, No. 2, pp. 57- 94.

Stapleton, H.E., 'The Antiquity of Alchemy', *Ambix*, 1953, Vol. 5, pp. 1-43.

Stavenhagen, Lee, *A Testament of Alchemy: being the Revelations of Morienus to Khalid ibn Yazid,* 1974, Hanover, University Press of New England.

Wilson, C. Anne, 'Jabirian Numbers, Pythagorean Numbers and Plato's Timaeus', *Ambix*, 1988, Vol. 35, pp. 1-13.

Chapter 12: Aspects of European Alchemy

1.   Information on the life and works of Thomas Aquinas relevant to his interest in alchemy is contained in the following references:-

Holmyard, E.J., *Alchemy*, 1957, Harmondsworth, Penguin Books.

Honderich, T. (Ed.), *The Oxford Companion to Philosophy*, 1995, Oxford University Press.

Sherwood Taylor, F., **The Alchemists**, 1952, London, William Heinman.

Thorndyke, Lynn, *A History of magic and Experimental Science*, 1923, New York, Columbia University Press.

Thurston, H.J. and Attwater, D., **Butler's Lives of the Saints**, 1981, London, Burns and Oates.

Von Franz, Marie-Louise (Ed.), *Aurora Consurgens – A Document attributed to Thomas Aquinas on the Problem of Opposites in Alchemy*, 1966, London, Routledge and Kegan Paul.

2.   References to the work of Basil Valentine may be found in the following:

Atwood, M.A., *A Suggestive Enquiry into the Hermetic Mystery*, New York, 1960

Brehm, E., 'Roger Bacon's Place in the History of Alchemy', *Ambix*, 1976, Vol. 23, pp. 53-58.

Bridges, J.H., *The Life and Work of Roger Bacon: An Introduction to the Opus Majus*, London, 1914, Williams and Norgate.

Bridges, J.H. (Ed.), *The Opus majus of Roger Bacon,* Vols. 1 and 2, Oxford, 1897, Vol. 3, 1900.

Brewer, J.S., (Ed.), *Fr. Rogeri Bacon, Opera quaedam hactenus inedita*, London, 1859.

Little, A.G., *Roger Bacon Essays: Contributed by various writers on the occasion of the Commemoration of the seventh century of his birth*, Oxford, 1914.

Sadler, L.V., 'Alchemy and Greene's Friar Bacon and Friar Bungay', *Ambix*, 1975, Vol. 22, pp. 111-124.

Thorndyke, L., *History of Magic and Experimental Science*, New York, 1923, Columbia University Press, Vol. II, pp. 616- 91.

Woodruff, F. Winthrop, *Roger Bacon: A Biography*, London, James Clarke, n.d.

3.   References to Nicholas Flamel may be found in the following:

Burland, C.A., *The Arts of the Alchemists*, London, 1967, Weidenfeld and Nicholson.

Holmyard, E.J., *Alchemy*, Harmondsworth, 1957, Penguin Books.

Mangeti, Jo. Jacobi, *Bibliotheca Chemica Curiosa*, (1792 edition reprinted), 1976, Arnoldo Forni Editore.

Read, John, *Prelude to Chemistry*, London, 1939, G. Bell and Sons.

Sadoul, Jacques, *Alchemists and Gold*, London, 1972, Neville Spearman.

Westcott, William Wynn, *Nicholas Flammel, His Exposition of the Hieroglyphical figures which he caused to be Painted upon an Arch in St. Innocents Church Yard in Paris: Concerning both the Theory and Practice of the Philosopher's Stone*, (1624 edition reprinted 1889).

Chapter 13: The English Alchemists

1.   Holmyard, E.J., *Alchemy*, Marmonsworthy, Penguin Books, 1957, pp. 114 - 115.

2.   Brewer, J.S., (Ed.), *Fr. Rogeri Bacon, Opera quaedam hactenus inedita*. Vol. I, London, Longman, Green, Longman and Roberts, 1859

3.   Thorndyke, Lynn, *History of Magic and Experimental Science*, New York, Columbia University Press, 1923-1958, Vol. 2, p. 678.

4.   Sadler, Lynn Veach, 'Alchemy and Greene's Friar Bacon and Friar Bungay', *Ambix*, 1975, Vol. 22, pp. 111-124

5.   Bacon, Roger, *Opus maius*, (Ed. Bridges) ii, p. 167

6.   Holmyard, Eric John, *Alchemy*, Harmondsworth, Penguin Books, 1957, p. 117

7.   Little, A.G., (Ed.), *Roger Bacon: Essays contributed by Various Writers on the Occasion of the Commemoration of the Seventh Centenary of his Birth*, Oxford, Clarendon Press, 1914, p. 300

8.   Bacon, Roger, *De Arte Chymiae*, pp. 389-395

9.   Little, A.G., *op. cit.*, pp. 285-320

10.   Holmyard, E.J., *op. cit.*, p. 183

11.   McLean, Adam, *The Hermetic Journal*, 1980, (8), p. 35

12.   *ibid, idem*, 1984, (24), pp. 17-24

13.   Anon, *Collectanea Chemica: being certain select treatises on Alchemy and Hermetic Medicine*, London, Vincent Stuart, 1963

14.   Nierenstein, M. and Chapman, P.F., *Enquiry into the Authorship of the Ordinall of Alchimy*, Isis, 1932, Vol. 18, pp. 290-321

15.   Reidy, J., 'Thomas Norton and the Ordinall of Alchimy', *Ambix*, 1957, Vol. 6, pp. 59-85

16.   Manget, Jean Jacques, *Bibliotheca Chemica Curiosa*, 1702, Vol. II, p. 285 *Thomas Norton Tractatus Crede Mihi seu Ordinale dictus*.

17.   Sherwood-Taylor, J., 'Thomas Charnock', *Ambix*, 1946, Vol. II, pp. 148-176

18.   Pritchard, Allan, 'Thomas Charnock's Book Dedicated to Queen Elizabeth', *Ambix*, 1979, Vol. 26, pp.56-73

19.   Burland, C. A., *The Arts of the Alchemists*, London, Wiedenfeld and Nicholson, 1967

Chapter 14: Contemporary Alchemists

1.   Fulcanelli, *Le Mystere des Cathedrales*, London, Neville Spearman, 1971, p. 5

2.   Sadoul, Jacques, *Alchemists and Gold*, London, Neville Spearman, 1972

3.   Fulcanelli, *The Dwellings of the Philosophers*, Boulder, Colorado, Archive Press and Communications, 1999

4.   Johnson, K. Rayner, *The Fulcanelli Phenomenon*, London, Neville Spearman, 1980, p. 163

5.   Pauwels, L. and Bergier, Jacques, *The Dawn of Magic*, London, Anthony Gibbs and Phillips, 1963

6.   Ovason, David, *The Zelator: The Secret Journals of Mark Hedsel*, London, Arrow Books, 1999

7.   Fulcanelli, *op. cit.* (ref.1), pp. 19-31

8.   Cyliani, *Hermes devoilé*, Paris, F. Loquin, 1832

9.   Cockren, A., *Alchemy Rediscovered and Restored*, London, Rider and Company, 1956

10.   Tahil, Patricia, *The Hermetic Journal*, 1981, vol. 13, pp. 35-39.

Index

Compiled by Peter Hamilton Currie – Member of the Society of Indexers

Page numbers in bold type denote the more important references; page numbers in italics denote illustrations or their captions; passim (e.g. 29-35 *passim*) conveys that the subject is referred to not continuously but in scattered passages throughout the pages; **q** stands for 'quoted' and **n** for 'notes'. In connection with dates: **b** – born; **d** – died; *fl* (*floruit*) – flourished; *c* (*circa*) – about (this date). Where other than English names are represented, they are done so in the manner commonly represented where English is spoken, with cross referencing to the native equivalents or variants thereof.

www.ingramcontent.com/pod-product-compliance
Lightning Source LLC
Chambersburg PA
CBHW080019240326

41598CB00075B/325